AIR AND SPACEBORNE RADAR SYSTEMS: AN INTRODUCTION

AIR AND SPACEBORNE
RADAR SYSTEMS:
AN INTRODUCTION

Air and Spaceborne Radar Systems: An Introduction

Philippe Lacomme
Jean-Philippe Hardange
Jean-Claude Marchais
Eric Normant

Translated from the French
by
Marie-Louise Freysz and Rodger Hickman

William Andrew
publishing

Published in the United States of America by William Andrew Publishing, LLC
13 Eaton Avenue
Norwich, NY 13815
(800) 932-7045
www.williamandrew.com

President and CEO: William Woishnis
Vice President and Publisher: Dudley R. Kay
Production Manager: Kathy Breed

Production services, page composition and graphics: *TIPS* Technical Publishing
Printed in the United States.

10 9 8 7 6 5 4 3 2 1

SciTech is an imprint of William Andrew for high-quality radar and aerospace books.

Library of Congress Catalog Card Number: 2001087624
Photos used in part opening pages are courtesy of THALES Airborne Systems.

This book may be purchased in quantity discounts for educational, business, or
sales promotional use by contacting the Publisher.

This book is co-published and distributed in the UK and Europe by:

The Institution of Electrical Engineers
Michael Faraday House
Six Hills Way, Stevenage, SGI 2AY, UK
Phone: +44 (0) 1438 313311
Fax: +44 (0) 1438 313465
Email: books@iee.org.uk
www.iee.org.uk/publish
IEE ISBN: 0-85296-981-3

TABLE OF CONTENTS

FOREWORD

The history of airborne radar is almost as old as that of radar itself. The improvement in detection range provided by an airborne platform was realised early during the Second World War, and the development of the cavity magnetron at almost the same time allowed higher radar frequencies and, hence, directive antennas to be used. Nowadays, radars on aircraft have a great variety of functions: from navigation and meteorological purposes, to more specialised purposes on military aircraft associated with surveillance and weapon delivery. Development of processing techniques such as coherent Moving Target Indication and Synthetic Aperture Radar have been matched by huge advances in technology, such as digital processing and solid-state phased arrays. More recent decades have seen the development of satellite-borne radars for geophysical environmental monitoring and surveillance applications.

A book that brings together a detailed theoretical treatment and a systems-level engineering understanding of the subject is both unusual and of great potential value to the radar community. The structure of the book combines a coverage of the principles of radar with a discussion of different applications and missions, showing how the design of the radar is adapted to each. The final chapters are devoted to a view of future technological developments and the ways that airborne and spaceborne radars may be expected to develop in response to new types of targets and missions. The French radar industry has played a significant role in the development of many of the innovations in airborne and spaceborne radar. The authors of this book are acknowledged as experts in the field and they provide a uniquely European perspective on the subject.

For all of these reasons, this book will be of value to a wide audience, both as a reference to radar engineers and those responsible for the specification and procurement of airborne and spaceborne radar systems, and as a textbook in graduate-level courses on radar.

HUGH GRIFFITHS
PROFESSOR, UNIVERSITY COLLEGE LONDON
IEE PGEL5 COMMITTEE, IEEE RADAR SYSTEMS PANEL

For over half a century, radar has been a permanent feature of surveillance activities. Practically unaffected by meteorological conditions, it operates independently of sunlight, while its detection ranges and the angular domain it covers make it an essential tool for continuous surveillance of a very wide area. Over the last fifty years, radar operational capability and performance have continued to improve, and one can safely assume that this will hold true for the coming decades.

This book, devoted to airborne and spaceborne radar, avoids a purely theoretical approach and is certainly not intended for an "elite" group of specialists. Rather, it is a practical tool that we hope will be of major help to technicians, student engineers, and engineers working in radar research and development. The many users of radar, as well as systems engineers and designers, should also find it of interest.

Airborne and spaceborne radar systems, themselves highly complex systems, are fitted to mobile and often rapidly changing platforms that contain many other items of equipment. Radar can therefore not be considered as a separate entity. Its design must ensure its "compatibility" with the systems of which it forms a part, and with the dense electromagnetic environment to which it is often exposed. Naturally, and most importantly, it must also satisfy operating requirements.

Radar technology evolves at a rapid pace and can quickly appear obsolete. For this reason it is only briefly developed in this work. However, we have taken the major trends into account when describing the next generation of radars, as their feasibility is largely dependent on these new developments.

The book is divided into five parts:

- General Principles
- Target Detection and Tracking
- Ground Mapping and Imagery
- Principal Applications
- Radars of the Future

Following a historical overview and a reminder of the main principles behind radar, the functions, modes, properties, and specific nature of modern airborne radar systems are studied in detail. Next, the book examines radar's role within the mission system when carrying out missions assigned to the aircraft or the satellite. The fourth section covers the possibilities of radar as well as its limitations and constraints. Finally, given changing operational requirements and the potential opened up by technological development, the final section describes how radar may evolve in the future.

REMARK

As airborne and spaceborne radars are often used in military applications, and in order to comply with security regulations, in this book we refrain from quoting existing systems or equipment that are either under development or in use. Explanations and examples are therefore based on the laws of physics (i.e., information that is in the public domain) and on hypothetical "equipment."

PART I
GENERAL PRINCIPLES

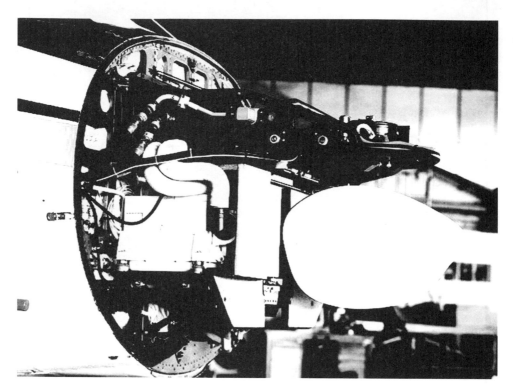

Maritime Patrol Radar (Ocean Master)

1

THE HISTORY AND BASIC
PRINCIPLES OF RADAR

1.1 HISTORY

In 1887 the German physicist Heinrich Hertz discovered electromagnetic waves and demonstrated that they share the same properties as light waves. These electromagnetic waves are often known as "Hertzian waves."

In the very early 1900s, Telsa in the US and Hülsmeyer in Germany proposed detection of targets by the use of radio waves.

The principle behind RADAR (Radio Detection And Ranging), based on the propagation of electromagnetic waves or, more precisely, that of radio-frequency (RF) waves, was described by the American Hugo Gernsback in 1911. In 1934 the French scientist Pierre David successfully used radar for the first time to detect aircraft. In 1935 Maurice Ponte and Henri Gutton, during trials carried out onboard the *Orégon*, part of the Compagnie Générale Transatlantique fleet, detected icebergs using waves with a 16 cm wavelength (λ). In 1936 Professor Kunhold (Germany) detected aircraft.

Radar came into its own during the Second World War as the ideal technique for detecting the enemy, both day and night. As early as 1940 the British RAF, led by Watson Watt, developed a dense network of ground-based radars. This clinched their victory in the Battle of Britain, as it provided sufficient warning to deploy fighter planes under optimum conditions. The German army also set up its own ground-based radar network, which, from 1942 onward, they used to transmit the position of detected targets to the fighter control center. In order to intercept and shoot down the waves of allied bombers deployed at night, German fighter pilots used either daytime fighters to attack allied planes tracked by light from ground projectors, or night fighters equipped with radar.

The first ever operational warplane equipped with an airborne radar was the Messerschmitt Me 110 G-4 in 1941. Its Telefunken radar, the FUG 212, used a bulky antenna comprising a number of dipoles located outside the aircraft, on the nose. By June 1944 the German fighter unit possessed over 400 aircraft of this type with a radar range of approximately 5 km, this range being limited by the altitude at which the carrier was flying.

By 1944 the American Naval Air Service was equipped with a Corsair with a radar pod on the right wing, while the American Air Force had a Northrop P-61A Black Widow fitted with a Western Electric radar system.

During the night of July 24–25, 1943, 800 RAF bombers carried out a raid on Hamburg. During this raid the bombers carried out the first ever operational chaff launch (metal strips whose dimensions vary depending on the wavelength of the radar they attempt to confuse). This operation rendered German ground-based and airborne radars totally non-operational, blinded by an excess of objects to detect. It marked the beginning of electronic warfare.

Radar operators noted that the British Mosquito fighter planes and the Japanese Zero fighter planes, both wooden constructions, were particularly difficult to detect; they were the original stealth aircraft.

In 1943 Allied surface ships fitted with radar were used to detect German submarine snorkels, causing the German navy to suffer heavy losses.

Later the main steps in radar technological evolutions were

- pulse compression (in the early '60s)
- pulse Doppler radar (late '60s)
- digital radars ('70s)
- medium PRF radar (late '70s, early '80s)
- multimode programmable radar (mid-'80s)
- airborne electronically scanned antenna radar ('90s)

The first radar images of the Earth were obtained in 1978 using Synthetic Aperture Radar (SAR), operating in the L-band ($\lambda \approx 30$ cm) and mounted on the American satellite Seasat. Resolution of the images obtained, both day and night, was close to 25 m.

1.2 BASIC PRINCIPLES

Radar is a system that transmits an electromagnetic wave in a given direction and then detects this same wave reflected back by an obstacle in its path.

1.2.1 BASIC CONFIGURATION

Figure 1.1 illustrates the first basic radar design. The various components of radar include: for transmission, a transmitter sending a continuous sinusoidal wave to a transmitting antenna and, for reception, an antenna plus a high-gain receiver and a detector whose output signal is displayed using a radar display such as a CRT.

The role of the transmitting antenna is to concentrate the energy transmitted in a chosen direction in space (beam center). The transmitting antenna gain, G_t, is maximum along the axis and varies depending on the direction (see Chapter 3).

The receiving antenna collects the transmitted energy backscattered by the target in the same chosen direction. This receiving antenna has a gain G_r. Supposing the two antennas are identical, $G_t = G_r$.

The wave transmitted (in this case continuously) is propagated to and from the target at the speed of light, c. In a non-magnetic medium, the following is true:

$$c = \frac{299.7925.10^{6}}{(K_e)^{1/2}} \text{ m/s}$$

In a vacuum, the dielectric constant K_e is equal to one. In air, its value varies slightly depending on temperature, composition, and pressure. At sea level it equals 1.000 536. In practice, the speed of light for radars is taken to be 300 000 km/s.

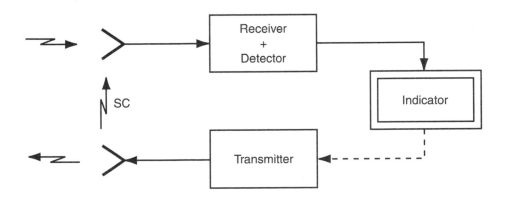

To ensure that the receiving channel only detects the signal backscattered by the obstacle or target, it must be decoupled from the transmission channel. An antenna, whatever the technology it uses, has a radiation pattern composed of a main lobe and side and far lobes (see Figure 1.2).

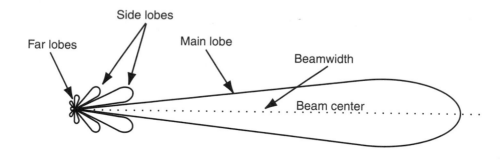

FIGURE 1.2 ANTENNA DIAGRAM

For the radar shown in Figure 1.1, despite the fact that both antennas are operating in the same direction, they have a leakage, in this case due to the far lobes. For example, if the far lobes of both antennas are 40 decibels below the maximum level of the main lobe (along the beam center), the isolation of the two channels is equal to 80 dB. Under such conditions, if the signal backscattered by the obstacle and received by the receiving channel is stronger than that caused by spurious coupling, the obstacle will be detected. In practice, numerous other factors come into play. These will be dealt with in turn, and in particular in Chapter 3.

The radar shown in Figure 1.1 is a bistatic system. Although transmission and reception are adjacent, they do not physically overlap. This frequently used concept (e.g., for launching semi-active missiles) will be examined in a later section.

1.2.1.1 RANGE MEASUREMENT

If the radar transmission is a pure continuous wave with frequency f_0, the backscattered wave will have the same frequency (if the relative velocity between radar and target is equal to zero), whatever the range. However, the greater the target range, and the lower the Radar Cross Section (RCS) of the target, the weaker the received signal. The RCS characterizes the backscattering coefficient of the target.

The target range can be obtained using one of several methods:

- by calculating the time between the detected target echo and the transmitted wave

- by calculating the difference in frequency between the received echo and the transmitted wave in the case of linear frequency modulation
- by calculating the differential phase of the double detection of an echo obtained using two transmissions of different frequencies (Chapter 8.6)

The following sections give a rapid overview of the first two methods.

TIME MEASUREMENT

In order to obtain the target range by calculating the time between the transmitted wave and the detected echo, the radar signal should be emitted in short pulses as shown in Figure 1.3.

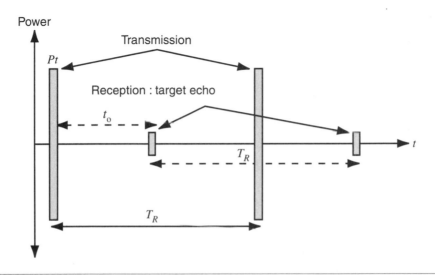

FIGURE 1.3 PULSE MODULATION

A radar using this type of transmission is known as a "pulse radar." It periodically transmits microwaves with peak power P_t. The interval between two pulses is known as the interpulse period, T_R. Under such conditions, measurement of time t_o, equal to the wave propagation time on the two-way path between radar and target, gives the range R between the radar and the target.

Note that the frequency of the wave transmitted has no influence on this measurement:

$$R = \frac{c \cdot t_0}{2}$$

FREQUENCY SHIFT MEASUREMENT

In order to obtain the radar-target range by calculating the difference in frequency between the transmitted wave and the detected echo, transmission must be linearly frequency-modulated (Figure 1.4).

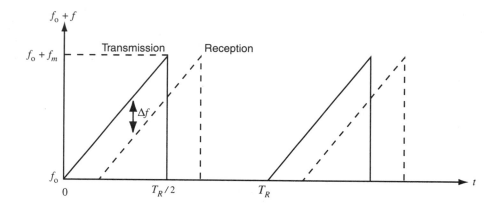

FIGURE 1.4 LINEARLY FREQUENCY-MODULATED TRANSMISSION

Ignoring the Doppler effect, the range between the radar and the target is given by

$$R = c \cdot \frac{T_R}{2} \cdot \frac{\Delta f}{f_m}$$

in which

f_m =   maximum modulation frequency

Δf =   difference between transmission and reception frequency

$c \cdot T_R/2$ =   range domain without range ambiguity

For example, T_R = 100 µs, Δf = 0.2 f_m, and R = 3 km.

1.2.1.2 THE INVENTION OF MONOSTATIC RADAR

By pulsing the radar on and off, it is possible to use the same antenna for both transmission and reception. During transmission, the highly sensitive input to the reception channel must be protected from the powerful transmission level. This avoids saturation, or worse, destruction of the receiver circuits. Figure 1.5 shows the basic block diagram of this solution. The circulator, a specifically adapted microwave circuit, transmits energy from (1) to (2) but not to (3). It also transmits energy from (2) to (3) but not to (1). To function correctly, the load impedance at each pair of terminals must equal its characteristic impedance. In (1) the incident signal is produced by the transmitter and transmitted to the antenna. Only a fraction of this signal arrives at (3) at the receiver input. This is due to imperfections in the circulator and to mismatching, such as partial reflection onto the antenna (Standing Wave Ratio: SWR). This also holds true for (2) to (3) and (3) to (1). In practice, protection can attain 30 dB.

Under such conditions, if the transmitter supplies 100 kW of peak power, the transmission levels at the receiver input are excessive (100 W peak), resulting in saturation or destruction. Additional protection is therefore required. This protection, placed at the input to the receiver, must act synchronously with the transmitter. It can comprise pre-ionized gas tubes and/or semi-conductor diodes. As well as magnetron transmitters, this circulator/protection system often uses passive microwave components known as Duplexers (combining a magic-T and a circulator)—Transmit-Receive (TR) and Anti-Transmit-Receive (ATR) (Bentéjac 1992).

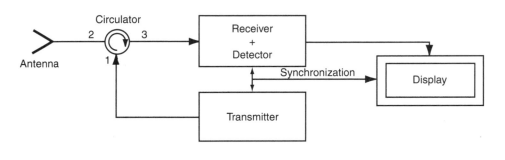

FIGURE 1.5 BASIC RADAR CIRCUIT 2

WAVEFORM

The first generations of airborne radars almost exclusively used magnetron transmitters. Magnetrons are microwave "oscillator" tubes that deliver high peak power (of the order of 100 kW), with mean power approximately 1,000 times weaker (100 W). For a magnetron to oscillate at its own frequency, it must be triggered by a modulator supplying it with a high-power "rectangular" pulse. This pulse is generally 1 µs. Given the magnetron "form factor," it can only be reproduced every 1 ms. A radar fitted with this type of transmitter is known as Low Pulse Repetition Frequency (LPRF) and will be unambiguous in range if used exclusively within a 150 km range domain (i.e., with a PRF of 1 000 Hz).

As a first approximation, the waveform is determined by the transmitter. Receiver protection and CRT display sweeping (as well as certain receiver and processing circuits) should function synchronously.

THE TRANSMITTER

Radar transmission can be obtained using either resonating microwave tubes such as magnetrons or amplifier tubes such as certain klystrons or Traveling Wave Tubes (TWT). Solid-state transmitters have recently been used; these deliver low peak power and can be used, for example, for missile homing heads and active radar antennas.

Returning to the magnetron transmitter, its main characteristics are as follows (all other considerations being equal):

- low cost, bulk, and weight
- high peak power/mean power ratio
- good efficiency levels
- low magnetron duty cycle (50 to 200 Hz)
- fixed frequency oscillations, linked to mechanical aspects but variable in temperature and with non-negligible phase and amplitude noise levels. This type of magnetron is known as a tunable magnetron. So called "coaxial" magnetrons have more stable frequencies. Some special types of magnetron can be frequency modulated by a few percent using a small motor

THE ANTENNA

The first airborne antennas were composed of a set of dipoles, giving a certain amount of directivity. They were rapidly replaced by antennas fitted with parabolic reflectors with a feed at the center of the reflector (Figure 1.6).

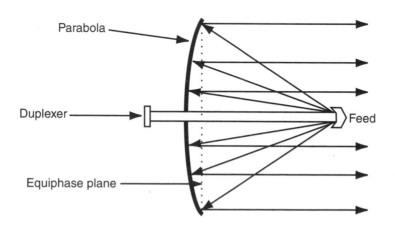

FIGURE 1.6 PARABOLIC DISH ANTENNA

Energy transmitted from the duplexer via the feed, which can be a mini-horn, illuminates the entire parabola, which, via reflection, forms a beam with parallel rays. This ensures optimal directivity. The feed, which causes slight blockage, illuminates just the parabola. On reception, the waves backscattered by the target follow the same trajectory but in the opposite direction. Antenna gain and directivity are thus doubly influential.

The gain and directivity of the antenna main lobe depend on the dimensions of the antenna in relation to the wavelength used (λ) and the efficiency (η). Gain (G) is defined as the ratio between the energy radiated along the radioelectric axis and that radiated by an omnidirectional antenna (isotropic). Where S is equal to the antenna surface area, the gain is as follows:

$$G = 4\pi \frac{S}{\lambda^2}\eta$$

For a circular parabolic dish antenna with a 60 cm diameter, $\lambda = 3$ cm, and $\eta = 70\%$,

$$G = 2{,}800 = 34.5 \text{ dB}.$$

Antenna directivity is characterized by the aperture of the main beam. The narrower the beam, the greater the directivity. Directivity plays a vital role in determining the direction of the target "seen" by the radar.

Two factors should be taken into consideration:

- total beamwidth measured between the two beam zeros (θ_{nn})
- beamwidth at 3 dB (θ_{3dB}). This beamwidth is by far the most frequently used (Figure 1.7)

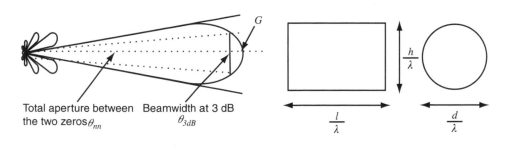

Total aperture between   Beamwidth at 3 dB
the two zerosθ_{nn} θ_{3dB}

FIGURE 1.7 BEAMWIDTH

Beamwidth depends on the size of the antenna, the wavelength used, and the illumination function. For a circular antenna, beamwidth is circularly symmetric. Uniform illumination gives the following approximate equations:

$$\theta_{nn} = 2\frac{\lambda}{l} \qquad \theta_{3\,dB} = 0.88\frac{\lambda}{l} \text{ in one plane, with in } \theta \text{ radians}$$

$$\theta_{nn} = 2\frac{\lambda}{h} \qquad \theta_{3\,dB} = 0.88\frac{\lambda}{h} \text{ in the other plane}$$

A circular antenna with uniform illumination gives:

$$\theta_{nn} = 2.3\frac{\lambda}{d} \qquad \theta_{3\,dB} = 1.02\frac{\lambda}{d}$$

If illumination is optimized in order to reduce side and far lobes in relation to the main lobe (Gaussian), the beamwidth of a circular antenna at 3 dB is as follows:

$$\theta_{3\,dB} = 1.25\frac{\lambda}{d}$$

To illustrate this point, for a 60 cm diameter parabola with a 3 cm wavelength, the beamwidth at 3 dB is 3.57°.

To close the subject of antennas, for this chapter at least, we simply point out that

- scanning the search volume requires antenna pointing using an antenna controller
- increasing the accuracy of angular measurement involves using methods such as plot center or monopulse beam sharpening

THE RECEIVER AND THE DETECTOR

Figure 1.8 shows the block diagram for a receiver associated with a detector. This figure also includes a frequency diagram.

The pulse emitted at frequency f_0 is backscattered by the target(s), passes through the antenna and the duplexer, and is then applied to the microwave mixer. This mixer, which is essentially a diode (crystal detector) with nonlinear characteristics, also receives the continuous wave f_{OL} from an oscillator. Whenever f_0 exists (either transmission pulse leakage or echoes), an intermediate-frequency f_i signal appears on output from the mixer. This intermediate frequency represents the difference between f_0 and f_{OL}. The f_i harmonic component is filtered by the intermediate frequency amplifier (matched filter) whose bandwidth B is matched to the bandwidth of the transmission signal.

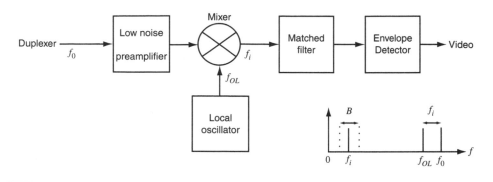

FIGURE 1.8 RECEIVER

The envelope detector supplies the display with a monopolar video composed of input thermal noise increased by the preamplifier and mixer noise as well as by the target echoes (Figure 1.9).

To illustrate this point, some possible values are shown below:

$$f_0 = 10\ 000 \text{ MHz}$$

$$f_{OL} = 9\ 900 \text{ MHz}$$

$$f_i = 100 \text{ MHz}$$

$$B = 2 \text{ MHz}$$

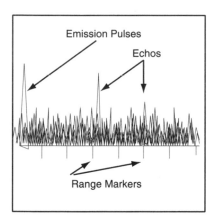

Type A : Amplitude — Range

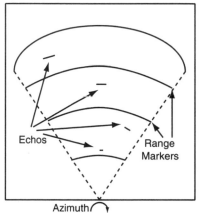

PPI sectored : Range — Azimuth
Luminous echoes
(here, without noise)

FIGURE 1.9 INFORMATION SHOWN ON THE DISPLAY

THE DISPLAY

From the outset, the use of radar created the need for the operator to be able to view the information supplied by the radars. This led to the introduction of Cathode Ray Tube (CRT) displays. These displays have changed greatly over time, together with the characteristics of the tubes and electronic circuits they use. How information is displayed is influenced more by operational needs and possibilities than by the radar itself.

Returning to the first radar systems, the displays were A-scope or Plan Position Indicators (PPI), over 360°, sectored, etc. (Figure 1.9). The A-scope presentation shown here is said to be "raw," as it has not been submitted to any particular selection or processing. Whenever the display uses luminosity, the operator "processes" the signal, sometimes assisted by a threshold, by afterglow, or by the CRT memory; it "recognizes" echoes superimposed on the noise.

1.2.2 CHOICE OF A WAVELENGTH, f_{OL}

Radars operate over an extremely wide range of frequencies, from 40 MHz to 100 GHz. This range of frequencies thus covers the HF-, VHF-, UHF-, L-, S-, C-, X-, Ku-, K-, Ka-, V-, and W-bands. Choosing the wavelength for a specific radar involves a trade-off between a number of factors such as

- the properties of electromagnetic waves (see Chapter 4)
- operational aims and applications
- available volume and technology, etc.

As will be shown in a later section, the wavelength used for most airborne radars is situated in X-band, that is, in the 8-12.5 GHz frequency band (λ: 2.4-3.75 cm).

2

INITIAL STATEMENTS OF
OPERATIONAL REQUIREMENTS

2.1 INTRODUCTION

Nowadays airborne and space-based radars have some civilian applications, but most of them are defense oriented. Airborne civilian applications concern mainly weather radars of liners and Exclusive Economic Zone surveillance (EEZ). Space-based civilian radars are used for global earth resource management. In this section we focus on defense missions and military radar systems that gather the main technical issues.

2.2 MISSIONS

We can divide operational missions into four main missions:

- surveillance
- reconnaissance
- targeting
- weapon delivery

Radars are involved in all four missions.

2.2.1 SURVEILLANCE

Surveillance aims to give to decision makers, at a strategic level, the information they need to answer these questions: Is there a threat? What is the threat? What target do we attack in what conditions?

These decisions are the result of the fusion of many information sources, the radar being one of them. This information has to be disseminated to all levels of decision makers and is the input of the Communication Command Control and Information system (C^3I).

In Air Defense the surveillance is carried out by Airborne Early Warning systems (AEW), which are in charge of detecting any airborne threat with sufficient notice to be able to react on time. They need not only to detect the target but also to track it, to identify it as an enemy, and to localize it with enough accuracy to designate it to interceptors (fighters in general). They require a very long range (several hundred nautical miles) and have to deal with thousands of tracks.

Ground Surveillance aims to acquire the general battlefield situation. It relies on the Moving Target Indicator System (MTI)—which enables you to detect and track ground-moving vehicles (tanks and trucks) and helicopters—and on Synthetic Aperture Radar (SAR) imaging radars—which give a high-resolution picture of fixed echoes (steady vehicles, buildings, bridges, airfields, etc.).

These ground surveillance missions are performed

- by satellite SARs, which give very large accessibility and a very low update rate (hours to days)
- or by airborne standoff systems from large jet aircrafts, which give real-time access at the cost of a limited terrain coverage (specially in hilly regions) due to quite low grazing angles
- or by High or Medium Altitude Long Endurance (HALE or MALE) Unmanned Vehicles (UAV) at a shorter range

2.2.2 RECONNAISSANCE

Once an action is decided, reconnaissance aims to give to field players, at a tactical level, the information they need: Where are (and where will be) the targets, what are their defenses, what military means to use, what kind of weapon is the best suited, when to conduct the attack, etc.

Due to short reaction time in Air Defense, this task is generally carried out directly by the AEW system. In fact in some cases the AEW platform is a C^3I system itself that controls the fighters in real time.

For ground targets, due to slower evolution of the situation and to a more complex environment (collateral damage avoidance, mask), a specific mission is needed.

This mission uses SAR/MTI systems either fitted in POD carried by a manned aircraft (business jet or fighter), or carried by an unmanned vehicle (UAV).

The main objective of these missions is to re-acquire the targets (in case they are moved as Ballistic Missile Launchers), and to identify and locate them accurately. Once again fusion of different sensors (ESM, optical, etc.) is generally needed to assess the global tactical situation, including Air Defense threat assessment.

All this information is gathered at the C^3I center where the interceptions or the strike missions are preplanned.

2.2.3 FIRE CONTROL AND TARGETING

Air Defense or Air Superiority is carried out with interceptors or fighter aircrafts that take off from airbase a few minutes after the AEW alert. They use their nose-mounted radar first to detect the target in the search domain designated by the Air Defense system at a distance ranging from 30 to 100 NM.

Then they have to track these targets to extract the cinematic parameters (position, velocity vector) in order to compute if the targets are in the missile firing domain and to display this information to the pilot (see Figure 2.1).

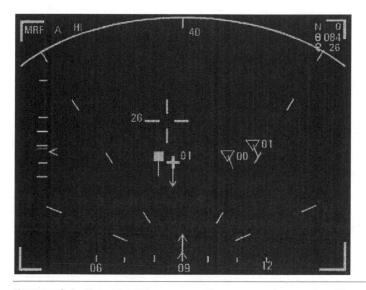

FIGURE 2.1 FIGHTER HEAD-DOWN DISPLAY IN AIR-TO-AIR MODE

Identification is needed before firing to avoid collateral or fratricide kill. This identification relies on cooperative means such as IFF transponders (Identification of Friend or Foe) or on Non-Cooperative Target Recognition (NTCR) given directly by the radar signature of the target. All this information from the radar and other sensors (on-board or received from other platforms through a tactical data link) are merged to give the pilot a global tactical situation picture (Tactical Situation Awareness).

Before missile launch, the fighter radar gives the missile the information on the target parameters (cinematic parameters, predicted position at interception time, etc.) and continues to track the target in order to detect any maneuver and to update the designation via the fighter-missile data link. Then the missile seeker comes into action to steer the missile to the target.

Electromagnetic seekers are radars in charge of acquiring the right target and tracking it in order to steer the missile to it.

Fighter radars and seekers have to counter the threat of Electromagnetic Counter Measures (ECM) or Jammers, which aim either to prevent the detection of targets or to deceive the tracking.

Air-to-Ground strikes are carried out with fighters to destroy fixed or mobile ground objectives. The weapons range from conventional to LASER-guided bombs or long-range Air-to-Ground missiles. These fighters used to rely on optical fire control systems (visible or infrared), but as these are subject to severe adverse weather limitations and are quite short-range, the radars are more and more the preferred solutions for all weather standoff weapon delivery.

Fixed targets like bridges or buildings can be designated in geographical coordinates (WGS 84) to the weapon at the Mission Preparation level with data supplied by Surveillance or Reconnaissance missions (if available with sufficient accuracy). In these cases, a weapon with GPS guidance can reach an acceptable accuracy (about 10 m). More and more, however, target recognition prior to weapon delivery is needed because targets of interest can be moved between the reconnaissance mission and the strike (Ballistic Missile Launchers or Sol-Air Defense for example). In these cases the fighter relies on high-resolution radar (SAR modes—1 m resolution or less) to achieve this task at long range (up to 100 km) in all weather conditions. These high-resolution modes are also required for damage assessment (DA) to evaluate the results of the strike.

In addition to high-resolution imaging modes needed for target classification, the radar has to supply an accurate localization of the targets. This requires a very good knowledge of the platform velocity, which is given either by a high- performance inertial unit (coupled to GPS) or by specific radar modes that give a high-accuracy measurement of the platform velocity from ground Doppler velocity estimation. After the weapon is released, it is necessary to evaluate the result of the strike. This damage assessment can be performed with a high-resolution SAR imaging mode.

2.3 CARRIERS AND WEAPONS

2.3.1 CARRIERS

This book is concerned only with platforms that can be equipped with radar. In a military context, these platforms include the following:

- satellites
- aircraft
- helicopters
- active homing head missiles (seekers)
- Unmanned Air Vehicles (UAV)
- smart munitions

With regard to civilian platforms, only the first three are relevant to this study.

2.3.2 WEAPONS

A wide and ever-increasing range of aeronautical weapons has been developed:

- guns
- rockets
- conventional bombs (smooth, braked, anti-runway)
- laser-guided missiles or bombs
- Air-to-Air missiles
- Air-to-Ground missiles
- Air-to-Sea missiles
- cruise missiles
- lethal UAVs
- etc.

Smart munitions need targeting with the help of one or more passive, semi-active, or active sensors. Radars belong to the last category.

2.4 SYSTEM FUNCTIONS

Each type of carrier requires its own basic system functions. The choice of carrier depends on the type of mission to be accomplished, as well as performance requirements, which call for specific functions provided by equipment, sensors, and weapons, either existing or yet to be developed. All these devices form the weapon system. Each integrated component must communicate and operate coherently with the other components in the system. The weapons system must therefore have a centralized command center and a highly efficient communications network. Moreover, the carrier weapons system itself forms part of a wider operational system and therefore needs to communicate with this larger system and "act" coherently with it. Despite its complexity and importance within the weapons system, radar is only a part of the system. It can therefore only be designed in relation to the missions, carriers, weapons, specific functions, and performance requirements.

An example of this system dependency is a combat aircraft that must be able to carry out the following missions in all weather conditions:

- sky policing
- air superiority
- interception and combat
- penetration and Suppression of Enemy Air Defense (SEAD)

Due to the variety of the missions, the fighter requires a multi-function weapons system of a general architecture as shown in Figure 2.2.

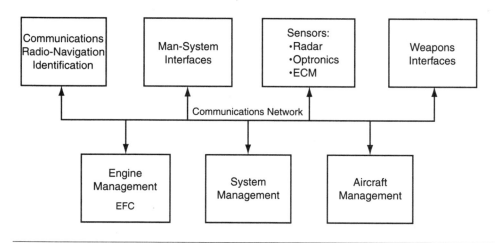

FIGURE 2.2 GENERAL ARCHITECTURE

The aim of this book, which is devoted entirely to radar, is not to analyze every possible mission and function of radar systems. Indeed, this vast subject is under continual development, and definitions vary according to the air force and country in question.

However, the missions assigned to airborne radars can be summarized as follows:

- Air-to-Air (A-A)
- Close Combat (CC) within a maximum range of 10 NM (one nautical mile = 1,852 m)
- Air-to-Ground (A-G)
- Air-to-Sea (A-S) (or Air-to-Surface)

Often the pilot will decide that two functions need to be carried out simultaneously, one to accomplish the main mission and the other to ensure safety.

2.5 DEFINITIONS OF FLIGHT CONDITIONS

evasive: target movement intended to reduce enemy firing domain prior to missile release (e.g., 2g)

side stepping: target movement designed to avoid the incoming missile (e.g., 9g)

Very Low-Altitude flight (VLA): < 500 feet

Low-Altitude flight (LA): 500 to 2 000 feet

Mid-Altitude flight (MA): 2 000 to 30 000 feet

High-Altitude flight (HA): 30 000 to 70 000 feet

Very High-Altitude flight (VHA): > 70 000 feet

3

The RADAR Equation

3.1 Introduction

One of the radar engineer's main concerns is to determine the radar power budget, which consists of three parts:

- transmission of energy to the target
- backscatter of part of that energy back to the radar
- reception of this backscattered signal by the radar

The power budget enables calculation of the parameters needed to ensure the required range performance. One major aspect is energy backscattering from the target. This is an almost unpredictable phenomenon and must be dealt with statistically.

3.2 Signal Transmission and Reception

3.2.1 The Role of the Antenna on Transmission

The role of the antenna on transmission is to concentrate the energy transmitted along a chosen direction in space.

$P_1(\vec{k})$ is the power transmitted in direction $\vec{k}$ by a directive antenna, and $P_2(\vec{k})$ is the power transmitted by an omnidirectional antenna in that same direction. The transmission source, P_t, is the same for both antennas (see Figure 3.1).

By definition, the gain, $G_t(\vec{k})$, of the antenna in direction $\vec{k}$ is given by the ratio

$$G_t(\vec{k}) = \frac{P_1(\vec{k})}{P_2(\vec{k})}.$$

P_t is the power transmission source.

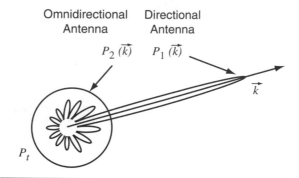

FIGURE 3.1 GAIN ON TRANSMISSION

The power transmitted within solid angle $d\Omega$ in the direction $\vec{k}$ by the omnidirectional antenna is

$$dP_2 = \frac{P_t}{4\pi} d\Omega .$$

The power transmitted inside $d\Omega$ by the directive antenna (Figure 3.2) is

$$dP_1 = G_t(\vec{k})dP_2 = G_t(\vec{k})\frac{P_t}{4\pi} d\Omega .$$

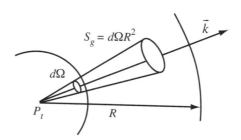

FIGURE 3.2 WAVE PROPAGATION IN FREE SPACE

In free space, this energy is retained in angle $d\Omega$. At range R, the area intercepted by $d\Omega$ on a sphere with a radius R is

$$dS = d\Omega R^2.$$

The power density, W, per area unit is therefore

$$W = \frac{dP_1}{dS} = G_t(\vec{k})\frac{P}{4\pi R^2} .$$

3.2.2 ROLE OF THE ANTENNA ON RECEPTION

If an energy sensor, with geometric area S_g normal to $\vec{k}$, is placed at range R in solid angle $d\Omega$, the power crossing S_g is as follows:

$$P_r = W \cdot S_g$$

In reality, an antenna, with area S_g, only captures part of P_r (due to losses, weighting function, etc.). By definition, the effective area S_{ef} is an area such that

$$P_r = W \cdot S_{ef}.$$

S_{ef} is the ideal geometric area of an antenna capturing P_r with a power density W.

For the same antenna, either transmitting or receiving, this gives the ratio

$$G_t(\vec{k}) = \frac{4\pi S_{ef}}{\lambda^2}, \tag{3.1}$$

where λ is the wavelength.

3.2.3 REFLECTION FROM THE TARGET

The target receives part of the transmitted energy. The incident EM field excites currents on the target, which then reradiates the energy in directions determined by its shape and material construction, and in a manner that depends (often very strongly) on the geometry and polarization of the incident field. In short the target acts very much like an "inefficient antenna" and usually does not reradiate most of the energy in the backward direction (toward the radar). This is called "target scattering" and will be studied macroscopically with the aid of a *model*.

On reception the target acts as an antenna with an area $S_{ef} = \sigma$ aimed at the transmitter. The power captured by this antenna is radiated omnidirectionally without loss.

The value of σ, known as the Radar Cross Section (RCS), is such that the power captured by the radar receiver is the same as when the model is used in place of the real target.

This example is an ideal illustration of backscattering for this particular configuration. However, as we shall see later, the value of σ represents the target for this configuration only. The slightest alteration of this configuration can cause major modifications to σ.

3.3 RADAR EQUATION IN FREE SPACE

Now let us reconsider the power budget of the link (in free space): a radar transmitting power P_t in the direction of a target located at distance R with an antenna gain of G_t.

The power density at the target is as follows:

$$W = G_t \frac{P_e}{4\pi R^2}$$

The power received by the target is $P_c = W \cdot \sigma$, where σ is the effective area of the target considered as a receiving antenna. The target diffuses this power isotropically in accordance with the model.

The power received by the radar receiver antenna, which has an effective area S_{ef}, is

$$P_r = \frac{P_c}{4\pi R^2} S_{ef},$$

giving the power budget

$$P_r = P_t \frac{G_t S_{ef} \sigma}{(4\pi)^2 R^4}.$$

Replacing S_{ef} with $\dfrac{G_r \lambda^2}{4\pi}$ yields the power budget

$$P_r = P_t \frac{G_t G_r \lambda^2 \sigma}{(4\pi)^3 R^4}.$$

REMARKS

- For a monostatic radar (one that uses the same antenna for transmission and reception),

$$G_t = G_r = G.$$

- P_e and P_r designate either peak power and mean power.
- For ease of measurement, transmitted power is generally measured directly at the transmitter output, and received power is measured directly at the receiver input. Microwave elements between the transmitter and the antenna on the one hand, and the antenna and the receiver on the other, create losses l (with $l > 1$) that must be taken into account (see Figure 3.3).

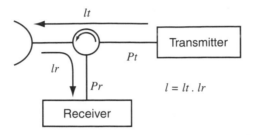

FIGURE 3.3 MICROWAVE LOSSES

The radar equation is therefore generally written as

$$P_r = P_t \frac{G^2 \lambda^2 \sigma}{(4\pi)^3 R^4 l}.$$ (3.2)

3.4 THE RADAR CROSS SECTION OF A TARGET

It is quite difficult to accurately estimate the value of the target cross section σ given its extreme sensitivity to the various parameters to be taken into account (shape, frequency, presentation, polarization, type of material, etc.). This value is usually obtained by measurement. In order to illustrate this phenomenon, we shall use an example that permits this type of calculation.

3.4.1 EXAMPLE OF THE DOUBLE SPHERES

The sphere is one of the few objects that allows direct and exact calculation.

In this case, $\sigma = \pi a^2$, where a is the radius of the sphere.

For our purposes, we shall take an object comprising two spheres with an RCS σ_1 and σ_2 at a distance d from each other (Figure 3.4).

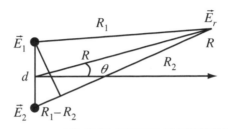

FIGURE 3.4 DOUBLE SPHERES

The received field is the sum of the fields $\vec{E}_1$ and $\vec{E}_2$ from each of the spheres; that is, $\vec{E} = \vec{E}_1 A_1 e^{j2\pi\varphi_1} + \vec{E}_2 A_2 e^{j2\pi\varphi_2}$, where A_1 and A_2 are the signal amplitudes given by the radar equation

$$A_1 = \sqrt{\frac{P_e G^2 \lambda^2 \sigma_1}{(4\pi)^3 R_1^4 l}} \ldots A_2 = \sqrt{\frac{P_e G^2 \lambda^2 \sigma_2}{(4\pi)^3 R_2^4 l}},$$

and where $\varphi_1 = \dfrac{4\pi R_1}{\lambda}$ and $\varphi_2 = \dfrac{4\pi R_2}{\lambda}$ are phase shifts due to propagation.

Given that $R_1 \approx R_2, A_1 \approx A_2 = A$, we can say that

$$\vec{E} = \vec{E}_1 A e^{j\varphi_1}\left(1 + \frac{\sqrt{\sigma_2} e^{j(\varphi_2 - \varphi_1)}}{\sqrt{\sigma_1}}\right). \tag{3.3}$$

The total RCS equals

$$\sigma = \sigma_1 \left|1 + \frac{\sqrt{\sigma_2}}{\sqrt{\sigma_1}} e^{j(\varphi_2 - \varphi_1)}\right|^2.$$

We shall now examine variations with $\varphi_2 - \varphi_1$:

$$\varphi_2 - \varphi_1 = 2\pi\frac{2(R_2 - R_1)}{\lambda} = 2\pi\frac{2d\sin\theta}{\lambda}, \tag{3.4}$$

where d equals the distance between the reflectors, and θ is the angle at which the double spheres are seen. Figure 3.5 shows variations in RCS, σ, as a function of θ. RCS, σ, fluctuates rapidly between

$$\sigma = \left|\sqrt{\sigma_1} + \sqrt{\sigma_2}\right|^2 \text{ for } \frac{2d}{\lambda}\sin\theta = 2\pi k$$

and

$$\sigma = \left|\sqrt{\sigma_1} - \sqrt{\sigma_2}\right|^2 \text{ for } \frac{2d}{\lambda}\sin\theta = 2\pi\left(k + \frac{1}{2}\right).$$

For example, for $d = 3m^2$ and $\lambda = 0.03m^2$, the difference between maximum RCS and minimum RCS is $\Delta\theta = 2.5$ mrd, less than 0.15 degrees!

Moreover if $\sigma_1 \approx \sigma_2$, modulation depth is significant (from $\sigma = 4\sigma_1$ to $\sigma \approx 0$). We should therefore expect the RCS to be extremely sensitive to each of the

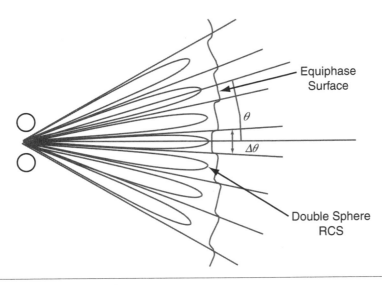

FIGURE 3.5 RCS OF TWO SPHERES

characteristic parameters of the configuration, such as wavelength, aspect angle, target movement, etc.

Note that even a slight change in wavelength can considerably modify the value of RCS in a given direction. This characteristic can be exploited to avoid detection losses due to unfavorable combinations of phase if the radar uses frequency agility.

3.4.2 GENERAL EXAMPLE

In the case of a real target, many reflectors combine to form the same number of double spheres or multispheres. This gives a highly complex backscattering pattern with almost uncontrollable parameters (see Figure 3.6).

Given its extreme sensitivity to this configuration, RCS is presumed to be an unpredictable variable macroscopically characterized by various parameters such as the following:

- mean value
- standard deviation
- distribution function
- autocorrelation

For practical reasons (contractual commitments on range capability), radar experts have established four models of typical targets:

SWERLING MODEL I

Target fluctuation is of Rayleigh type. Its RCS probability density function (PDF) is

$$p(\sigma) = \frac{1}{\overline{\sigma}} e^{-\frac{\sigma}{\overline{\sigma}}}.$$

The level is constant throughout the entire dwell time (10 ms to 100 ms). Random variations occur between one antenna scan and the next (1 s to 10 s).

This is the case of a complex target (with many equivalent scatterers) illuminated at a fixed frequency.

SWERLING MODEL II

The target follows the same fluctuations as in Model I, but the levels are decorrelated from one pulse to the next.

This is the case for a complex target illuminated with frequency agility.

SWERLING MODEL III

The target fluctuates in accordance with the function

$$p(\sigma) = \frac{4\sigma}{\overline{\sigma}^2} e^{-2\frac{\sigma}{\overline{\sigma}}}$$

with decorrelation from one scan to the next (the case of a target with a single dominant scatterer).

SWERLING MODEL IV

Target fluctuation is the same as in Model III but with decorrelation from pulse to pulse.

EXAMPLES OF RCS

combat aircraft (head on) σ = 1 to 5 m^2

transport aircraft (tail on) σ = 10 to 100 m^2

aircraft carrier σ = 100 000 m^2

Figure 3.6 shows real aircraft RCS measurements in S band.

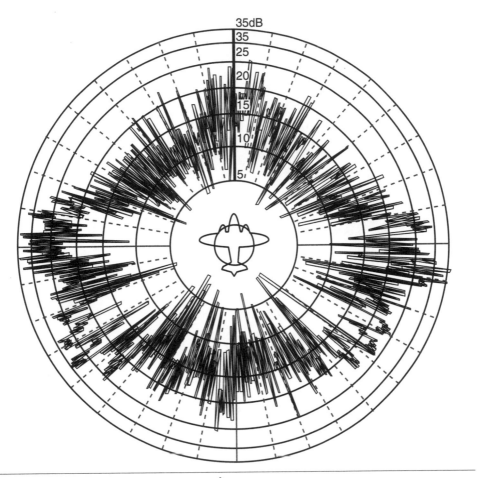

FIGURE 3.6 AIRCRAFT RCS IN S-BAND, λ =10 CM (RIDENOUR 1947)

3.5 MATHEMATICAL MODELING OF THE RECEIVED SIGNAL

The power of the received signal is not sufficient to define the optimal processing required for its detection, and we must determine the mathematical expression of this signal.

The transmitted signal $u_e(t)$ is composed of a modulation $u(t)$ (which can be complex) modulating a carrier wave with a frequency f_0.

The mathematical expression of $u_e(t)$ is $u_e(t) = \Re e[u(t)e^{j2\pi f_0 t}]$, where $\Re e$ represents the real part of this expression.

The signal received at time t comes from the signal transmitted at time $t - \tau(t)$, where $\tau(t)$ is the delay caused by the outward and return distance to and from the target (Figure 3.7).

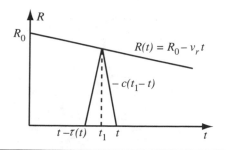

FIGURE 3.7 CALCULATION OF S(T)

If the target is a point object (constant RCS) moving at radial velocity v_r starting at distance R_0 (for $t = 0$), then $R = R_0 - v_r t$ (Figure 3.7). If t_1 is the solution of $R_0 - v_r t_1 = -c(t_1 - t)$, we can write

$$\tau(t) = 2(t - t_1)$$

and

$$t - \tau(t) = \frac{c + v_r}{c - v_r} t - \frac{2R_0}{c} .$$

The real received signal is written as

$$s_r(t) = \Re e[s(t) e^{j2\pi f_0 t}], \tag{3.5}$$

in which

$$s(t) = A u\left(\frac{c + v_r}{c - v_r} t - \frac{2R_0}{c}\right) e^{j2\pi \frac{2v_r}{\lambda\left(1 - \frac{v_r}{c}\right)} t} e^{j\varphi} = I(t) + jQ(t) ,$$

where A is attenuation due to propagation

$$A = \sqrt{\frac{G^2 \lambda^2 \sigma}{(4\pi)^3 R^4 l}}$$

and $\varphi = -2\pi \dfrac{2R_0}{\lambda}$ is a constant phase term.

The (real) received signal can be expressed as

$$s_r(t) = I(t)\cos(2\pi f_0 t) + Q(t)\sin(2\pi f_0 t) .$$

If we use phase amplitude demodulators (PAD) to obtain the product of $s_r(t)$ by a wave at f_0 on the one hand, and the same wave phase shifted by $\pi/2$ on the other—that is, the product of $s_r(t)$ and $\cos(2\pi f_0 t)$, and the product of $s_r(t)$ and $\sin(2\pi f_0 t)$—after low-pass filtering (elimination of components at $2f_0$), we obtain the in-phase component $I(t)$ and the quadrature phase component $Q(t)$ of the signal.

REMARKS

- $s(t)$ carries all the information concerning the received signal. From now on, we shall only take into account modulations and presume that the transmitted signal is $u(t)$ and the received signal is $s(t)$.
- For real targets $v_r/c \ll 1$, $s(t)$ is expressed as follows:

$$s(t) = Au\left(t - 2\frac{R_0 - v_r t}{c}\right)e^{j2\pi\frac{2v_r}{\lambda}t}e^{j\varphi}$$

- In a normal situation of narrow bandwidth operation (a few percents) and over short processing periods (a few milliseconds), the equation

$$t_0 = 2\frac{R_0 - v_r t}{c},$$

representing the delay due to the distance traveled to and from the target, is presumed constant. We can say that

$$s(t) = Au\left(t - t_0\right)e^{j2\pi f_D t}e^{j\varphi}, \tag{3.6}$$

- where f_D is the Doppler frequency of the target.
- $s(t)$ is a complex signal whose components $I(t)$ and $Q(t)$— such that $s(t) = I(t) + jQ(t)$—are obtained from the real signal $s_r(t)$ via synchronous demodulation using a quadrature mixer (I/Q mixer).

The received signal, $s(t)$, is therefore a replica of the transmitted signal $u(t)$ after the following transformations:

- attenuation due to propagation,

$$A = \sqrt{\frac{G^2\lambda^2\sigma}{(4\pi)^3 R^4 l}}$$

- delay due to propagation,

$$t_0 = \frac{2R}{c}$$

- frequency shift due to the Doppler effect,

$$f_D = \frac{2v_r}{\lambda}$$

- phase rotation

$$\varphi = -2\pi \frac{2R_0}{\lambda}$$

dependent on R_0 and λ. This constant equation only concerns signal processing in specific applications such as *FSK* range measurement (see Chapter 7) or improving range resolution using step-frequency techniques (Chapter 14).

3.6 DIRECTION OF ARRIVAL AND MONOPULSE MEASUREMENT

The signal reflected from the target is characterized not only by its power and variations over time as previously described, but also by the direction of arrival of the reflected wave. For a *point target* this wave is spherical (Figure 3.11), and at the antenna input it can be considered a plane wave front.

An approximate idea of target direction is given by the position of the antenna at the moment the target is detected. However, this inaccurate measurement (linked to antenna aperture) does not give the sign and amplitude of the pointing error (and therefore cannot be used to close the angular tracking loop).

The theory used to measure the direction of arrival can be found in numerous papers, including that of F. Le Chevalier (1989), which can be used as a reference. This measurement is based on monopulse angular difference.

At the elementary pulse level, this device supplies the signals $\vec{\Sigma}$ and $\vec{\Delta}$ (known as sum and difference). These are linear combinations of the signals received by two antenna elements, with a small offset (Figure 3.8) such that

$$\vec{\Sigma} = \vec{S_A} + \vec{S_B} \text{ and} \vec{\Delta} = \vec{S_A} - \vec{S_B} .$$

The angular difference measurement is given by the following equation:

$$\Delta = \frac{\vec{\Sigma} \cdot \vec{\Delta}}{|\Sigma|^2} = \frac{(G_A(\alpha)\vec{S} - G_B(\alpha)\vec{S})(G_A(\alpha)\vec{S} + G_B(\alpha)\vec{S})}{\left|G_A(\alpha)\vec{S} + G_B(\alpha)\vec{S}\right|^2} = \frac{G_A(\alpha) - G_B(\alpha)}{G_A(\alpha) + G_B(\alpha)}$$

This equation is based exclusively on the direction of arrival, α (angular difference with respect to the axis), and the patterns $G_A(\alpha)$ and $G_B(\alpha)$. It is known as amplitude monopulse.

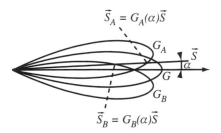

FIGURE 3.8 ANGULAR DIFFERENCE MONOPULSE

In modern phased-array radars, the sum, elevation difference, and circular difference channels are obtained using four quadrants of the antenna as shown in Figure 3.9; this is phase monopulse.

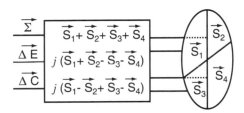

FIGURE 3.9 PHASE MONOPULSE (PHASED-ARRAY ANTENNA)

The angular difference signal is approximated by

$$\Delta = \frac{\vec{\Sigma} \cdot \vec{\Delta}}{|\Sigma|^2} \approx \tan(q\alpha)$$

where q is a coefficient determined by the antenna (see Figure 3.10).

> **Note:** *If the received signal is a jammer signal transmitted by the target, the angular measurement made on the jamming signal is the same as for the useful target. This is one of the major advantages of monopulse angular difference, which enables, in any case, the direction of the jammed target to be known if the jammer is carried by the target itself.*

3.6.1 ANGULAR FLUCTUATION (GLINT)

Figures 3.5 and 3.11 show the equiphase area, that is, the position of the points in space where the signal phase given by Equation 3.4 is constant.

The antenna monopulse angular difference measures the perpendicular to the equiphase surface. In the case of a point target (a single reflector), the equiphase surface is a sphere centered on this point.

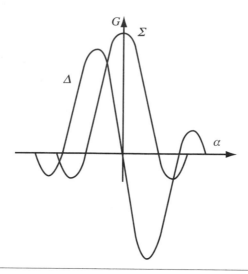

FIGURE 3.10 SUM AND DIFFERENCE SIGNALS

The direction indicated is correct. However, in the case of a complex target made up of several reflectors, for the points in the direction where Equation 3.3 is zero (and RCS is zero), the equiphase area is deformed; the phase rotates by 2π when θ varies by $\Delta\theta$. The perpendicular to the equiphase area does not point at the target, in particular when RCS is at minimum level (Figure 3.11). The difference signal is no longer in phase with the sum signal.

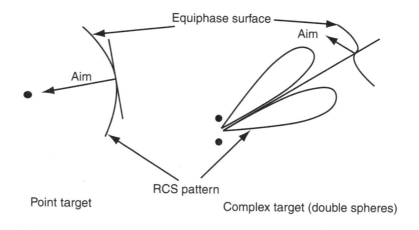

FIGURE 3.11 ANGULAR GLINT

The fluctuation in direction of a complex target caused by target noise is known as glint. Glint greatly reduces angular tracking accuracy of this type of target.

4

PROPAGATION

4.1 INTRODUCTION

Immediately after the introduction of radar, it became clear that the power of the received signal did not always obey the radar equation as described in Chapter 3. In some cases the signal was far weaker than predicted by the equation, and in others it was much stronger, producing spectacular detection ranges. Engineers working in telecommunications were already well aware of this phenomenon of abnormal propagation, caused by the atmosphere but also by the proximity to terrestrial objects. (Note that the radar equation was calculated for propagation in free space only.)

This chapter deals with the influence of the atmosphere and the ground on the propagation of radar signals.

4.2 ROLE OF THE GROUND

4.2.1 THE REFLECTION PHENOMENON

The ground is not electrically neutral. It acts as a refractive, reflecting medium for radio waves, producing phenomena of reflection, diffraction, and shadowing.

Reflection properties are well known in optical science. A ray that is reflected from a flat surface P makes an angle, r, with the normal to the plane equal to the incidence angle, i (see Figure 4.1).

Seen from the receiver R, the reflected wave appears to come from a fictitious point known as the image, symmetrical to the actual source S in relation to the plane P.

The ratio between the reflected field $\vec{E_r}$ and the incident field $\vec{E_i}$ gives the reflection coefficient ρ of the plane.

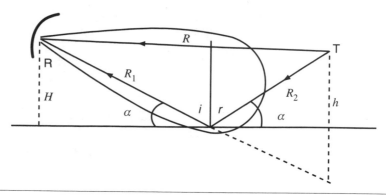

FIGURE 4.1 GROUND REFLECTION

The fact that radar waves share the same reflection properties explains the double sphere phenomenon (see Chapter 3) between the target, T, and its image, I. This phenomenon is all the more noticeable when target, T or receiver R are close to ground level.

If E_i is the field received via the direct path, and E_r is the field received via the indirect path (after reflection), the total of the two fields can be calculated as in Section 3.4.1:

$$\vec{E} = \vec{E_i} A e^{j\varphi_i}(1 + \rho e^{j(\varphi_2 - \varphi_1)}),$$

where $\vec{E_r} = \rho \vec{E_I}$ with ρ being the complex reflection coefficient.

In this case,

$$\varphi_2 - \varphi_1 = 2\pi \frac{R_r - R}{\lambda} = 2\pi \frac{\Delta}{\lambda},$$

where R is the length of the direct path, R_r is the length of the reflected path, and $R_r = R_2 + R_1$, with Δ being the difference between these two paths.

Using the simple hypothesis of flat ground

$$\sin\alpha = \frac{h}{R_2} = \frac{H}{R_1}$$

(where α is the depression angle and a complement of i and r), we can state that

$$R^2 = R_1^2 + R_2^2 + 2R_1 R_2 \cos 2\alpha,$$

with $R_1 + R_2 = R + \Delta$ and Δ small compared to R:

$$\Delta = \frac{R_1 R_2}{R}(1 - \cos 2\alpha) = \frac{2R_1 R_2}{R}\sin^2\alpha$$

Hence,

$$\Delta = \frac{2hH}{R}.$$

The resulting power is then

$$P = P_d\left|1 + \rho e^{2\pi j\frac{2hH}{R\lambda}}\right|^2. \qquad (4.1)$$

This phenomenon will therefore produce either a strengthening or a weakening of the received signal, depending on whether the direct or reflected waves are combined in phase or in phase opposition at receiver R, independently of the characteristics of target T.

The result will be a coverage diagram with peaks and zeros (Figure 4.2), with the apparent antenna gain being modulated by this phenomenon.

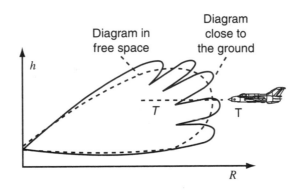

FIGURE 4.2 INFLUENCE OF THE GROUND ON COVERAGE

A target T moving along a trajectory T will pass through the different lobes, and the received signal will fluctuate slowly, modulating its own fluctuations.

As shown in Figure 4.3, there are four possible paths for the waves:

1. the radar-target-radar direct path
2. the radar-target-ground-target path

3. the radar-ground-target-radar path

4. the radar-ground-target-ground-radar reflection path

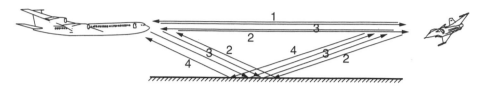

FIGURE 4.3 MULTIPATH EFFECT OF THE REFLECTION

This means that if the reflection coefficient is close to one, which is the case over a steady sea at low grazing angle (see Figure 4.5), the electromagnetic field can be four times higher than that of a direct path alone, depending on the relative phase of the four signals.

In fact you can consider that there are two transmitters (the real one and its image through the reflection) and two targets (the real one and its image), as shown in Figure 4.4.

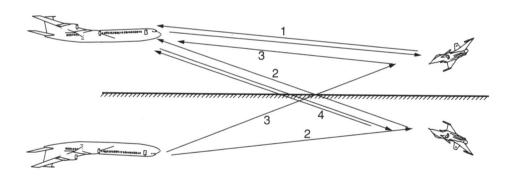

FIGURE 4.4 MULTIPATH EFFECT INTERPRETATION

In the best case (all paths in phase), the received power could be 16 times higher over a steady sea than in free space. In the general case, the four signals don't add exactly in phase, and the reflection coefficient is less than one (rough sea); nevertheless, over the sea we can expect a range increase of 30% to 40% due to reflection effect.

REMARK

Reflection is a particular disturbance for radars operating close to the ground (or the sea), such as naval fire control radars, because the point targeted oscillates between the target and its image and can even *be located outside this segment* due to the glint phenomenon (see Chapter 3).

It is therefore essential to take into account the reflection coefficient ρ in order to anticipate these phenomena.

The coefficient ρ depends on a great number of factors, including the following:

- the nature of the terrain
- the wavelength
- the polarization of wave and incidence angle

THE NATURE OF THE TERRAIN

This is a vital component, particularly when wavelengths λ are small in relation to the roughness of the terrain, which is generally the case for radar waves. Wave reflection is far better from a smooth sea or a lake than it is from a field, a forest, or a mountainous region. Similarly, weather conditions bring changes for the same terrain (snow, crops waiting to be harvested, wet ground, etc.).

THE WAVELENGTH

This has already been discussed: reflection properties depend on the relationship between wavelength and obstacle height (e.g., the height of sea waves).

The greater the ratio, the greater the reflection.

THE POLARIZATION OF WAVE AND INCIDENCE ANGLE

Reflection phenomena are greater for waves whose incidence forms a low grazing angle (a phenomenon that can easily be observed optically over a stretch of water). Reflection phenomena also depend on polarization. Figure 4.5 shows one example of variations in ρ with these parameters.

Note that, for these measurements, the coefficient ρ is close to -1 (magnitude 1, phase 180°) at low grazing angles, whatever the polarization.

Relation (4.1) is therefore

$$P \cong P_d \left(2\pi \frac{2hH}{R\lambda} \right)^2 .$$

Under these conditions, strong attenuation of the received signals is recorded for targets or radars with low altitudes. This is all the more noticeable at short wavelengths (λ).

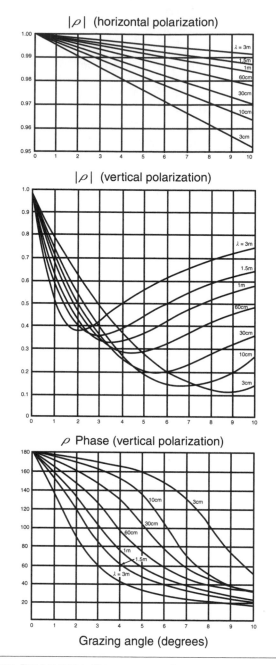

FIGURE 4.5 GROUND REFLECTION COEFFICIENT ρ

The first maximum of the diagram appears for

$$2\pi \frac{2hH}{R_T\lambda} = \pi$$

at range

$$R_T = \frac{4hH}{\lambda}.$$

R_T is known as the transition range.

Beyond R_T the received power is further attenuated in comparison with normal transmission by the factor:

$$\left(\frac{2hH}{R\lambda}\right)^2.$$

The decrease in received power with R thus obeys a function R^{-4} for a single trajectory. For a two-way radar path (transmit and receive), the decrease in received power is therefore proportional to R^{-8} (instead of R^{-4} for a normal radar path) beyond the transition range $R_T = (4hH)/\lambda$.

Figure 4.6 shows the variation in received power with range.

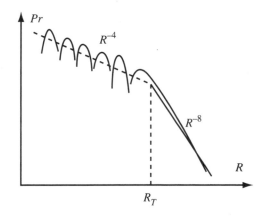

FIGURE 4.6 POWER VARIATION WITH RANGE

4.2.2 THE PRESENCE OF OBSTACLES—DIFFRACTION

The phenomenon of diffraction is well-known in optics: when an electromagnetic wave encounters an obstacle, energy is retransmitted in all directions, in particular behind the obstacle.

The effect of diffraction remains limited at short wavelengths. Terrestrial obstacles mask targets located behind them, as these obstacles lie in areas of shadow.

These obstacles also act as reflectors, often extremely powerful ones. Chapter 5 will study their properties.

4.3 THE ROLE OF THE TROPOSPHERE

The troposphere is the lower, non-ionized part of the atmosphere in which aircraft travel. It acts as a refractive medium.

4.3.1 NORMAL PROPAGATION

The atmosphere is characterized by a refractive index, n, that is close to one but varies with air density, and thus with temperature and altitude.

Under normal conditions, a standard atmosphere is defined whose refraction index, n, is a decreasing monotonic function of altitude h with a gradient of

$$\frac{dn}{dh} = \frac{-0.25}{R_T},$$

where R_T is the earth's radius.

The principle of refraction is well known in optics. The angle of incidence i and the angle of refraction r are linked by the equation $n_1 \cdot \sin i = n_2 \cdot \sin r$, where n_1 and n_2 are indexes of the mediums 1 and 2.

By dividing the atmosphere into successive "slices" (see Figure 4.7) whose index decreases with h, the wave path will gradually be deviated and directed towards the Earth.

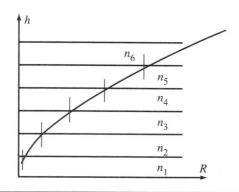

FIGURE 4.7 REFRACTION IN STANDARD ATMOSPHERE

Consequently, the path followed by electromagnetic waves is not a strictly rectilinear trajectory. This causes the radio horizon to recede (Figure 4.8).

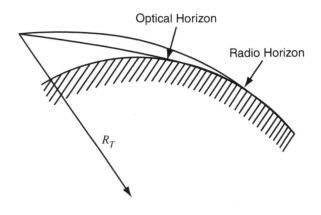

FIGURE 4.8 RADIO HORIZON

In order to take this into account, we assume that radar waves move in a straight line and that the Earth's radius is greater than its real value by a factor of approximately 4/3. The line-of-sight range for a target at altitude h for a radar at altitude H is

$$R_H = \sqrt{2Rh} + \sqrt{2RH},$$

where R equals the Earth's radio radius:

$$R = \frac{4}{3}R_T$$

See Figure 4.9.

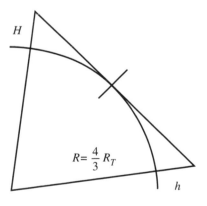

FIGURE 4.9 RADAR HORIZON

REMARK

Given that the radius of curvature of the wave path is greater than that of the Earth, in comparison with flat ground, the waves appear to be deviated upwards. We can therefore define a modified index n', whose variation with h is positive in a standard atmosphere.

4.3.2 ABNORMAL PROPAGATION

The true characteristics of the atmosphere are often quite different than those of the standard model described above. Under specific climatic conditions, an area of inversion of the temperature gradient and the modified refractive index is produced close to ground level (see Figure 4.10) or at altitude.

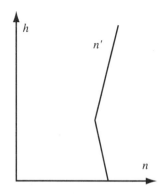

FIGURE 4.10   INVERSION LAYER

In this case, the propagation trajectory curves towards the ground and the waves are trapped in a duct between the inversion layer and the ground (see Figure 4.11). The formation of this duct can result from surface evaporation over the sea (evaporation duct). Radar range is therefore high for targets within the duct. However, targets located at higher altitudes go undetected.

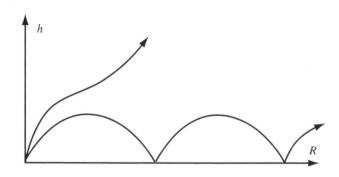

FIGURE 4.11   ABNORMAL PROPAGATION

Whenever the duct is situated at altitude, the wave is trapped between two low index layers surrounding a higher index area (a phenomenon similar to propagation in optical fibers).

4.3.3 ATMOSPHERIC ABSORPTION

As in any refractive medium, the atmosphere absorbs part of the energy transmitted. This absorption is influenced by numerous factors, as shown in Figure 4.12.

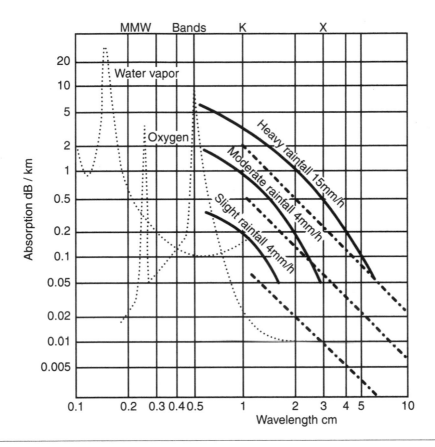

FIGURE 4.12 ATMOSPHERIC ABSORPTION

Water (in the form of rain or fog) considerably increases this absorption measured in dB/km. In the X- and Ku-bands, and above, this phenomenon assumes major importance. It imposes an upper limit on the frequency band used for any given application.

Chapter 1 showed how choosing a high frequency helps increase antenna gain. A trade-off must be reached based on the desired application:

- For ground-based radars or air-surveillance radars on large platforms, you can use large antenna. You can therefore stay in S- or L-band (10 or 23 cm), especially as the long range increases the chances of encountering rain or fog.
- For aircraft nose cone radar, antenna size is limited. You can find a compromise solution around X-band (λ = 3 cm).
- For missile seekers, the antenna size is even smaller and should be in Ku-band (compatible with shorter range).
- Finally, should you need to increase frequency for specific applications (missiles, detection of power lines), you should choose transmission windows (34 GHz, 94 GHz), located between the absorption lines of the atmospheric components.

Note: *This absorption is taken into account in the radar equation by the microwave loss term l (see Chapter 3). The product $2\alpha R$ should be added to l (in dB), where α is the absorption coefficient (in dB/km) and R is the size of the cloud (rain or fog).*

4.4 OTHER PHENOMENA

Other physical phenomena influence wave propagation, leading to radar applications such as surface wave propagation or ionospheric propagation. These phenomena are used in over-the-horizon radar, for example, and do not directly concern airborne radars.

5

NOISE AND SPURIOUS SIGNALS

5.1 INTRODUCTION

The detection of a target signal is hindered and limited by the presence of a variety of unwanted signals, both internal and external, artificial or natural. All these signals, which are not the expected target signal, will be considered noise, although some of them do not have the random nature generally associated with noise. Moreover that which is considered a disturbance (noise) in one application may well be the useful signal in another. One such example is ground returns (or atmospheric echoes), which constitute noise for air target detection radar but are useful signals for terrain mapping radar (or weather radar).

Noise consists of the following:

- internal noise, in particular thermal noise
- natural external noise such as radiometric noise
- artificial external noise such as jamming signals—known as Electronic Counter-Measures (ECM)—and interference from other radars. Part II deals specifically with these sources of electromagnetic pollution.
- spurious echoes created by the reflection of waves transmitted by the radar itself onto natural reflective surfaces around the target (the ground, rain, etc.)

5.2 THERMAL NOISE

As with any electronic receiver system, the ultimate limit on detection of a useful signal depends on the internal noise of the radar receiver. This is known as thermal noise, because of its thermal origin. This subject is more than adequately covered in specialist works on reception. We shall therefore limit ourselves to a reminder of its main characteristics.

5.2.1 THE CHARACTERISTICS OF THERMAL NOISE

Thermal noise is created by thermal agitation of the electrons in the various elements that make up the receiver. Its characteristics therefore depend mainly on the temperature of these elements. Thus, for a passive dipole, the spectral power density of noise available across its terminals is

$$b = kT_0,$$

where k is Boltzmann's constant ($k = 1,38.10^{-23} J/°K$) and T_0 is the absolute temperature of the dipole in degrees Kelvin. Generally spectral density for a dipole equals $b = kT$, with T representing the noise temperature of the dipole.

The spectral power density of noise is constant (independent of frequency). This noise is said to be white.

It is a Gaussian noise as it is produced by a combination of many independent sources (the electrons). Its components I and Q obey normal (Gaussian) independent functions:

$$p(x) = \frac{1}{\sqrt{2\pi}\sigma}e^{-\frac{x^2}{2\sigma^2}} = \frac{1}{\sqrt{\pi N}}e^{-\frac{x^2}{N}},$$

where $x = I$ or Q and $N = 2\sigma^2$ noise power.

The magnitude $\rho = \sqrt{I^2 + Q^2}$ obeys a Rayleigh function:

$$p(\rho) = \frac{2\rho}{N}e^{-\frac{\rho^2}{N}}.$$

Power $P = \rho^2$ obeys a Laplace function:

$$p(P) = \frac{1}{N}e^{-\frac{P}{N}}.$$

5.2.2 DEFINITION OF THE NOISE FACTOR

Take an amplifier that is characterized by power gain G and that does not introduce any reduction in bandwidth. For an input signal of power S_i, the output power is $S_o = GS_i$. If, however, N_i is the input noise, the output noise is $N_o > GN_i$. The amplifier has added internal noise to the input noise, thus reducing the contrast between signal and noise (the signal to noise ratio, S/N).

This degradation is measured by comparing the amplifier S/N ratios at input and output:

$$F = \frac{(S/N)_i}{(S/N)_o} = \frac{S_i/N_i}{S_o/N_o}, \tag{5.1}$$

where F is the amplifier noise factor.

Note that F is not an intrinsic characteristic of the amplifier, as it depends on input noise power, N_i.

When several amplifiers are placed in cascade, the overall noise factor of the chain is

$$F = F_1 + \frac{F_2 - 1}{G_1} + \frac{F_3 - 1}{G_1 G_2} + \cdots + \frac{F_i - 1}{G_1 G_2 \cdots G_{i-1}} + \cdots \qquad (5.2)$$

where F_i and G_i are the noise factor and gain of the ith element of the chain.

This shows that the gain of one element reduces the influence of noise from the elements that follow it. The noise factor of the first element predominates.

5.2.3 Noise Factor in a Reception Chain

A typical radar receiver chain consists of the following elements (see Figure 5.1):

- an antenna that picks up energy radiated in the receiving *bandwidth* by different sources in the space surrounding the radar. These sources include radiometric noise, jammers, and interference. In current applications, excluding jammers and interference that are dealt with separately, noise temperature available on input to the receiver (after microwave losses) equals the surrounding temperature $T_0 = 300$ K (see Section 5.3)
- a microwave amplifier with gain G_M and noise factor F_M
- a frequency converter that brings the signal into a lower frequency band by mixing with a local oscillator (heterodyne receiver). This converter is characterized by conversion loss l_c (gain $G_c = 1/l_c$) and noise factor F_C
- the rest of the amplification chain, characterized by noise factor F_{IF}

The noise factor of the entire chain is

$$F = F_M + \frac{F_C - 1}{G_M} + \frac{F_{IF} - 1}{G_M G_C} = F_M + \frac{F_C - 1}{G_M} + l_c \frac{F_{IF} - 1}{G_M}. \qquad (5.3)$$

Figure 5.1 shows typical values.

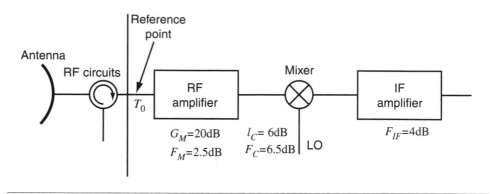

Application of Equation 5.3 gives the receiver noise factor F = 2.7 dB. This is very similar to that of the microwave amplifier, which confirms its importance.

REMARK

A practical way of overcoming the problem of the different gains in the reception chain is to calculate the S/N ratio at the receiver input (known as the reference point), supposing that all noise generated within the chain is referred to this point. The spectral density of thermal noise at the reference point is $b = kT_0F$. The total thermal noise power is

$$N = kT_0FB,\tag{5.4}$$

where B is the receiver bandwidth.

5.3 RADIOMETRIC NOISE

Even when no signal is being transmitted, the radar antenna picks up all the signals radiated by the environment in its reception bandwidth, such as the following:

- cosmic radiation. Except for the solar radiation (when the antenna beam is pointed in that direction), this has very little influence
- industrial radiation due to human activity. This generates spectral components within the radar bandwidth. Apart from signals from other radar or equipment fitted onboard the platform, categorized as part of the interference studied in Chapter 12, interference from industrial radiation is rare and generally negligible for the radar covered by this study
- ground thermal radiation, the main subject of this section

- power radiated by the ground in reception bandwidth B depends on its physical temperature θ, its emissivity, its reflectivity, and the temperature of the sky. It is characterized by an apparent noise temperature T_A. This temperature is approximately 200 K to 300 K (100 K for a stretch of water reflecting the sky)

The radar integrates the energy received, in bandwidth B, via the main lobe and all the spurious lobes of the antenna. For homogenous ground (constant temperature and emissivity), the received power is independent of the antenna pattern

$$(\int_{4\pi} G(\vec{u})d\Omega = 1 \text{ for an antenna with no losses})$$

and is equal to $P = kT_A B$.

If, however, we wish to obtain a thermal image of the ground (radiometry), antenna directivity (gain) is important in differentiating between small objects or plots of land at slightly different temperatures.

In reality, because the radar antenna and microwave circuits placed before the receiver have ohmic losses (a few dB), a large proportion of the radiometric noise is hidden by noise produced by these elements. The noise temperature at the receiver input is practically the same as surrounding temperature T_0.

5.4 SPURIOUS ECHOES AND CLUTTER

Discriminating between target echoes and unwanted echoes from sources surrounding these targets is one of the major difficulties associated with airborne radar. The properties of these echoes and the means of eliminating them vary depending on their source (ground, sea, or atmosphere).

5.4.1 CLUTTER AND GROUND CLUTTER

Reflective surfaces at ground level—be they natural surfaces such as the ground itself or vegetation, or man-made such as buildings—scatter signals back to the radar. Their power depends on their range, their radar cross section, and the gain of the antenna in their direction. Those signals received at the same time as a target at range R (excluding range ambiguity), and which thus hinder target detection, are located within an area $\mathfrak{D}$ (Figure 5.2). This is centered around the vertical projection of the target onto the ground and limited by circles with a radius $(R - r/2)\cos\theta_E$ and $(R + r/2)\cos\theta_E$, where θ_E is the elevation angle and $r = c/2B$ the range resolution of the radar (see Chapter 6).

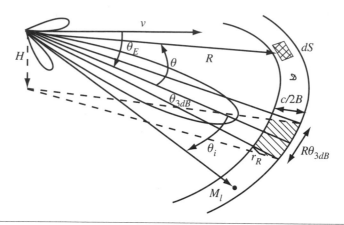

FIGURE 5.2 GROUND CLUTTER CALCULATION

An isolated reflector M_i with a radar cross section σ_i within this area $\mathfrak{D}$ produces an echo of power

$$P_i = \frac{P_t G^2(\theta_i)\lambda^2 \sigma_i}{(4\pi)^3 R^4 l}.$$

The total ground power at range R is equal to $P = \sum_i P_i$, spread over the entire area $\mathfrak{D}$.

5.4.1.1 HOMOGENOUS CLUTTER

For an element with a *homogenous* surface area dS within $\mathfrak{D}$ (e.g., part of a wheat field), the resulting power is proportional to the geometric area dS. One can calculate the proportionality coefficient for this type of area as $\sigma_0 = d\sigma/dS$, where $d\sigma$ is the radar cross section of the element with surface area dS.

The coefficient σ_0 is known as the backscattering coefficient. It is dimensionless (m^2/m^2) and characterizes the type of terrain in question. The signal corresponding to such echoes is known as ground clutter or homogenous clutter.

If this type of clutter exists throughout area $\mathfrak{D}$, the power of the clutter received at distance R is

$$P = \int_D d\sigma = \frac{P_t \lambda^2 \sigma_0}{(4\pi)^3 R^4 l} \int_D G^2(\theta)dS \tag{5.5}$$

Because radar antennas are tapered for optimal reduction of spurious lobes, the approximation

$$\int_D G^2(\theta)\,dS \approx G_0^2 S_r$$

is valid (where G_0 is the antenna gain along the axis), and

$$S_r = \frac{c}{2B}\theta_{3dB}R$$

is the surface area of the radar resolution cell in terms of range and angle, as shown by crosshatching in Figure 5.3).

The ratio in Equation 5.5 thus becomes

$$P = \frac{P_t G_0^2 \lambda^2 \sigma_S}{(4\pi)^3 R^4 l},$$

where σ_S is the radar cross section of the cell S_r given by

$$\sigma_S = \sigma_0 S_r = \sigma_0 \theta_{3dB} Rc/2B. \qquad (5.6)$$

5.4.1.2 MEASUREMENT OF THE BACKSCATTERING COEFFICIENT

The statistical values of σ_0 for different types of terrain are calculated using measurements obtained by calibrated radar. Starting with P, the value of σ_0 can be estimated using Equations 5.5 and 5.6.

The value of σ_0 depends on numerous parameters, such as

- type of terrain, humidity, season, etc.
- the angle at which the ground is observed, or grazing angle α
- wavelength and polarization

These measurements are the subject of numerous publications.

Figure 5.3 shows an example of the measurement of σ_0 as a function of the angle α for an average terrain (in a rural area) in the X-band.

Generally speaking, the dependence is of the $\sigma_0 = \gamma\sin\alpha$ type. Ground backscattering is characterized only by the parameter γ.

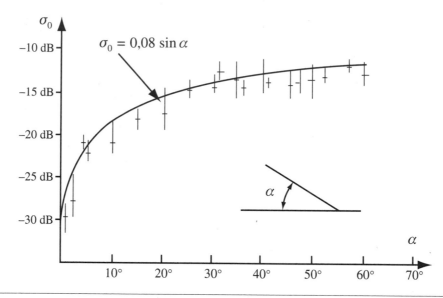

FIGURE 5.3 σ_0 MEASUREMENTS AND MODEL

5.4.1.3 GROUND CLUTTER MODELS

Evaluating radar performance means modeling ground returns (as in the case of targets). A widely accepted model is the one for homogenous clutter with constant γ, where $\gamma = 0.15$. It is fairly representative of actual situations that a radar may encounter and you can use it to calculate detection probabilities under normal circumstances.

In some cases, however, this model is insufficient and must be improved. The hypothesis of homogenous clutter can only be satisfied when the resolution sector is sufficiently large to contain many reflectors, giving a high and statistically constant average value for the radar cross section σ_S (several thousand square meters). When resolution is increased, the area of the cell decreases, as does the value of σ_S, and powerful point echoes can predominate. For those cells containing these echoes, the radar cross section is no longer proportional to S_r. These echoes must therefore be considered separately.

A more comprehensive model (Lacomme 1989), is composed of homogenous clutter with constant γ, onto which are superimposed point echoes spread randomly in radar cross section and in location, with distribution densities determining the number of point echoes in a given radar cross section category per unit area.

Two examples of clutter models are shown below.

TABLE 5-1. EXAMPLES OF CLUTTER MODELS (RURAL AREA AND URBAN AREA)

Rural Area		Urban Area	
$\gamma = 0.08$		$\gamma = 0.25$	
100	100 m^2 echoes per NM^2	300	100 m^2 echoes per NM^2
15	1 000 m^2 echoes per NM^2	100	1 000 m^2 echoes per NM^2
1	10 000 m^2 echo per NM^2	15	10 000 m^2 echoes per NM^2
0.1	100 000 m^2 echo per NM^2	0.3	100 000 m^2 echo per NM^2

EXAMPLE OF APPLICATION

Consider a typical example of ground clutter calculation in the case of a fighter flying at altitude H = 20 000 feet and trying to detect a low-level target of RCS $\sigma_T = 5m^2$, located at range R = 50 km from the radar. Assuming the different radar parameters are θ_{3dB} = 50 mrd (= 3°) and B = 1 MHz (r = 150 m), and that the clutter can be considered homogeneous with $\gamma = 0.15$, the ground clutter RCS is

$$\sigma_S = \sigma_0 S_r = \sigma_0 \theta_{3dB} R c / 2B \approx \gamma \frac{H}{R} \theta_{3dB} R c / 2B = 6750 \text{ m}^2.$$

This is very high compared to the target RCS $\sigma_T = 5m^2$ (more than 40 dB) and shows how low-flying target detection will be difficult.

5.4.1.4 GROUND CLUTTER SPECTRUM

A point ground echo M_i seen at angle θ_i compared with the velocity vector $\tilde{v}$ of the platform (see Figure 5.2) is received at Doppler frequency

$$f_D = \frac{2v}{\lambda} \cos \theta_i .$$

Ground return is spread across the spectrum as shown in Figure 5.4.

In the absence of frequency ambiguity, this spectrum has a very high level due to the echoes received by the antenna main lobe and located around

$$f_P = \frac{2v}{\lambda} \cos \theta_E \cos \theta_{Az},$$

where θ_e equals the antenna elevation and θ_{Az} the azimuth (or bearing). The spectral width of this zone equals

$$\Delta f = \frac{2v}{\lambda} \cos \theta_E \sin \theta_{Az} \theta_{G3dB} .$$

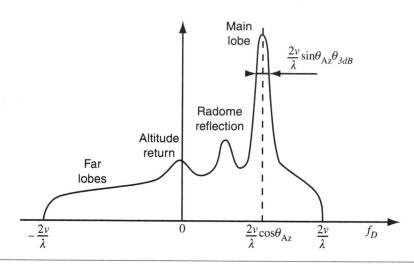

FIGURE 5.4 GROUND CLUTTER SPECTRUM

Because cos θ_e is generally close to one, the spectral width is essentially proportional to $v\sin\theta_{Az}$.

REMARK

Terrestrial vehicle echoes, with their own velocity, are mixed with the (static) ground echoes with a shift in Doppler frequency. As a result, the total spectrum of echoes received by the main lobe is considerably extended.

5.4.2 SEA CLUTTER

Any stretch of water (lake, sea, ocean) behaves like an equivalent ground surface. For moderate sea states (sea < 4) and average range resolution ($r > 30$m), sea clutter can be considered homogenous.

The σ_0 of the sea, which has been the subject of numerous calculations, depends on the following:

- the *sea state* and, in particular, wind strength. The fine structure of the sea surface, linked to wind-driven capillary waves, is the main influencing factor on sea clutter, which is generally moderate compared with ground clutter
- the *radar look direction* in relation to wind direction. The σ_0 of the sea downwind is higher (10 dB) than the σ_0 upwind
- *Wavelength* and *polarization*. The σ_0 with horizontal polarization is on average weaker (7 to 8 dB) than with vertical polarization

However, when range resolution drops below a dozen or so meters, sea clutter is no longer homogenous (sea swell is range resolved by the radar), particularly with horizontal polarization. Clutter is then distributed as large spikes on the breaking waves. This results in false alarms and makes detection more difficult. Under these conditions, horizontal polarization is much less effective.

As well as the effects caused by platform motion, the sea clutter spectrum includes a component linked to sea swell motion and wind speed. This component shifts and greatly increases the spectrum. The time duration of wave crests is several seconds.

5.4.3 METEOROLOGICAL ECHOES (ATMOSPHERIC CLUTTER)

In addition to absorbing radio waves, as examined in Chapter 5, the atmosphere plays a role by backscattering waves onto reflectors such as weather clutter (rain, hail, clouds), birds, and insects.

Birds and insects have a small radar cross section (pigeon = 0.002m^2, wild duck = 0.01m^2), which can nevertheless be compared to that of stealthy targets (missiles, stealth planes). They occur as transient point echoes. Because they cannot be identified with total certainty, they are classed as angel echoes (elements that appear and disappear mysteriously), causing false alarms whenever velocity discrimination is impossible.

Meteorological radar echoes (mainly rain) are distributed in volume. As with ground clutter, it is possible to define a reflectivity coefficient η. This is the ratio $\eta = \sigma/V$ of the radar cross section of a portion of space and its volume.

The radar cross section of atmospheric clutter is

$$\sigma_A = \eta V = \eta \theta_{E3dB} \theta_{Az3dB} R^2\, c/2B\,,$$

where V is the volume of the resolution cell at range R (see Figure 5.5).

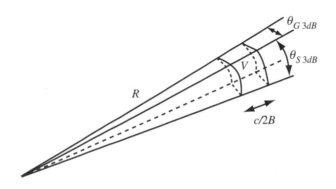

FIGURE 5.5 ATMOSPHERIC CLUTTER

The coefficient η (whose dimension is m^{-1}) depends mainly on the intensity I of the rain (in millimeters per hour—mm/h) and on f_0. An empirical ratio would be

$$\eta = 7.10^{-12} f_0^4 I^{1,6},$$

with f_0 in GHz.

EXAMPLE OF APPLICATION

$$I \quad = \quad 4\text{mm/h (moderate rainfall)}$$
$$f_0 \quad = \quad 10\text{GHz}$$
$$\theta_{E3\,\text{dB}} \quad = \quad \theta_{Az3\,\text{dB}} = 3°$$
$$B \quad = \quad 5 \text{ MHz}$$

At range R = 50 km, σ_A = 130m^2. The radar cross section of rain is therefore far greater than that of airborne targets and disturbs their detection if Doppler velocity discrimination is impossible.

In addition to factors dependent on platform velocity and common to ground clutter, the rainfall spectrum consists of a mean component, which is linked to wind speed, and a widening component due to turbulence and wind shears inside the cloud.

Rain spectrum can attain several hundred Hz in X-band.

6

DETECTION OF POINT TARGETS

6.1 INTRODUCTION

The basic problem of detecting a target swamped by surrounding noise can be considered in two phases:

- filtering, in the broadest sense of the term, in order to eliminate as much of the spurious noise masking the useful signal as possible, while at the same time maximizing the signal
- actual detection, which, using the signal received after optimal filtering, allows the radar operator to decide whether or not the target is present

The problem is alleviated in radar applications by the existence of relatively accurate information regarding the useful signal, or more precisely the form (to within a few parameters) of the signal reflected by a point target.

This signal is characterized by

- a plane wave perpendicular to the (unknown) direction of the target
- a temporal signal, which is a replica of the transmitted signal time shifted by the (unknown) two-way radar-target propagation delay and frequency, which are transposed by the Doppler effect linked to the (unknown) radial velocity of the target (see Chapter 3, Equation 3.5)

The useful return signal is thus defined by four unknown parameters: the angles Elevation and Azimuth, which determine its direction; its range; and its radial velocity. When in surveillance mode, with no information on these parameters, there will be a systematic search for each of their possible values in the investigation domain in which the target may be located.

There are several possible cases:

- The classic situation of *white noise* in which spurious signals are uniformly spread throughout the space-time-frequency domain. In this case, the receiver carries out processing *matched* to the useful signal
- The situation of *non-white* noise either in space (e.g., off-boresight jammer) or in time/frequency (ground clutter behind a formed beam)

The adaptive processing that optimizes the signal-to-noise ratio (to eliminate spurious noise) must then be found. The specific case of coupling between spatial and frequency noise. Here we shall look at space-time processing

6.2 THE OPTIMAL RECEIVER (WHITE NOISE)

6.2.1 DEFINITION OF PROCESSING

In this case the useful signal is characterized by

- a direction of arrival (normal to the wave plane) defined by two angles, e.g., Elevation and Azimuth (E and Az)
- a temporal signal expressed as

$$s(t) = Au\left(t - t_0\right)e^{-j2\pi f_D t}$$

where

A	=	propagation attenuation
$u(t)$	=	the transmitted signal
t_0	=	the delay caused by the distance R_0 from the target
f_D	=	the Doppler frequency linked to the radial velocity v_r of the target.

See Chapter 3.

This useful signal is swamped by noise that is

- omnidirectional in space (no preferred arrival direction)
- temporally (spectrally) white

This combination of signal plus noise is first spatially filtered by the radar antenna, whose radiation pattern maximizes the received signal in the direction of the desired signal while eliminating a maximum number of spurious signals coming from other directions. The required pattern maximizes the gain of the main lobe (and therefore its directivity) while it minimizes the side and far lobes (classic antenna pattern).

The signal sent by the antenna and picked up by the receiver therefore takes the form $x(t) = s(t) + n(t)$, where $n(t)$ is white noise.

This signal persists throughout observation time T_e of the radar in the direction in question (E and Az). T_e is a fraction of the time needed to explore the search area, that is, T_{exp}, imposed by operational considerations (search update rate).

When no other information is available, T_e is given by the basic equation

$$T_e = T_{\exp} \frac{\Delta\Omega}{\Omega} \, ,$$

where $\Delta\Omega$ is the solid angle illuminated by the antenna beam and Ω is the angular search domain.

In the specific case of a mechanical antenna scanning space at velocity ω, T_e is given by

$$T_e = \frac{\theta_{3dB}}{\omega} \, ,$$

where θ_{3dB} is the antenna aperture in the scanning plane.

The optimal receiver theory (Le Chevalier 1989), which maximizes the signal-to-noise ratio for Gaussian noise, is the correlator that performs time correlation over observation period T_e between the received signal $x(t)$ and the required signal with a form

$$u(t - \Delta t)e^{j2\pi\Delta f t} \, ,$$

where Δt and Δf are unknown parameters (see Figure 6.1). The time correlation is followed by envelope detection, whose role is to eliminate the unknown phase term, that is,

$$y(\Delta t, \Delta f) = |c(\Delta t, \Delta f)|^2 \, ,$$

with

$$c(\Delta t, \Delta f) = \int_{T_e} x(t)[u(t - \Delta t)e^{j2\pi\Delta f t}]^* dt = \int_{T_e} x(t)u^*(t - \Delta t)e^{-j2\pi\Delta f t} dt \quad . \quad (6.1)$$

Naturally, this operation should be carried out for each E, Az, Δt, and Δf value of the search domain.

REMARK

The theoretical term used in the reference is normalized by

$$\int_{T_e} |u(t)|^2 dt \, .$$

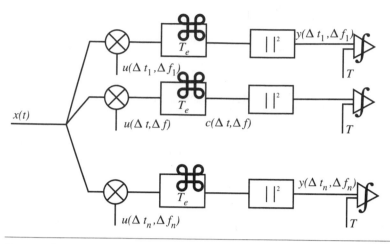

FIGURE 6.1 OPTIMAL RECEIVER (WHITE NOISE)

This normalization is usually taken into account directly by the false alarm regulation device, the Constant False Alarm Rate (CFAR), described in Chapter 8.

6.2.2 INTERPRETATION OF THE OPTIMAL RECEIVER

The optimal receiver described in the previous section carries out a correlation (linear operation) and can be considered equivalent to a filter.

With $u'(t) = u(-t)$, we can write

$$c(\Delta t, \Delta f) = \int_{T_e} x(t)u'^*(\Delta t - t)e^{-2\pi j \Delta f t}dt \ .$$

This is a convolution between $x(t)$ and $u'(t)$ and represents the output of a filter with a pulse response equal to

$$h(t) = u'(t)e^{-2\pi j \Delta f t} \ .$$

The transfer function of the filter is the Fourier transform of

$$u'(t)e^{-2\pi j \Delta f t} \ ,$$

that is

$$H(2\pi j f) = U^*(f - \Delta f), \tag{6.2}$$

where $U(f)$ is the Fourier transform of $u(t)$.

The optimal receiver is therefore based either on time processing (correlation) or frequency processing (optimal or matched filtering), depending on the implementation possibilities in the signal processing hardware.

In real-life situations, processing is usually broken down into successive linear operations carried out either by correlation or filtering.

6.2.3 SIGNAL-TO-NOISE RATIO AT THE OPTIMAL RECEIVER OUTPUT

Signal-to-noise ratio (S/N) is an important concept in radar applications (generally for detection). This ratio between the peak power of the desired signal and the effective noise power at the receiver output represents the contrast between target and noise. It can therefore be used to determine the capability of the radar to determine the presence of a useful signal.

Signal power and noise power at receiver output are calculated separately. (As the operator is linear, outputs can be calculated separately.)

CALCULATING SIGNAL POWER

Without the term $n(t)$,

$$c(\Delta t, \Delta f) = A \int_{T_e} u(t - t_0) e^{2\pi j f_E t} u*(t - \Delta t) e^{-2\pi j \Delta f t} dt .$$

This equation is an autocorrelation function.

It presents a global maximum (maximum superior to all the maxima) for the origin, for the parameters Δt and Δf such that $\Delta t = t_0$ and $\Delta f = f_D$ (i.e., for the choice of unknown parameters equal to the actual values of the target parameters).

This gives

$$c(\Delta t, \Delta f) = A \int_{T_e} |u(t - t_0)|^2 dt = \frac{1}{A} \int_{T_e} |s(t)|^2 dt ,$$

where

$$E_r = \int_{T_e} |s(t)|^2 dt$$

is the energy received by the radar during observation time.

$c(\Delta t, \Delta f)$, representing the filter output, is compatible with a voltage. The power at the filter output is

$$T = (E_r/A)^2.$$

CALCULATING NOISE POWER

Because noise, $n(t)$, is a random variable, it is described by its spectral density. Consider a matched receiver in the form of a transfer function filter:

$$H(2\pi jf) = U^*(f - f_D).$$

The power density at the filter output equals

$$b_s(f) = |H(2\pi jf)|^2 b = |U^*(f - f_D)|^2 b,$$

where b is the spectral density of input noise ($b = KT_0F$ white noise).

The noise power at the filter output is therefore

$$N = \int b_s(f)df = b\int |U^*(f - f_D)|^2 df.$$

The term

$$\int |U^*(f-f_D)|^2 df = \int |U(f-f_D)|^2 df = E_e = \frac{E_r}{A^2}$$

represents the energy, E_e, transmitted by the radar (during the processing period).

Consequently, the signal-to-noise ratio is

$$\text{T/N} = \frac{E_r^2/A^2}{(bE_r)/A^2} = \frac{E_r}{b} = R.$$

This is a fundamental relationship. It shows that the signal-to-noise ratio at the matched receiver output depends only on the ratio between the received energy and the spectral density of noise (known as the energy ratio and often written as R), and not on $u(t)$ itself.

In terms of signal-to-noise ratio, and therefore detection capability, radar performance does not depend on the form of $u(t)$ (known as waveform: pulse repetition frequency, pulse width, phase encoding), but only on the energy received during T_e i.e., mean power).

6.2.4 SIGNAL DETECTION IN WHITE NOISE

It has been shown (Le Chevalier 1989) that the previously described optimal receiver can be considered from two points of view:

- as the optimal estimator of the target parameters t_o and f_D (range and velocity), that is, the estimator that produces minimal quadratic error between actual values and measured values
- as the optimal detector, that is, the detector that maximizes quantity

$$\frac{P_{x/H_1}}{P_{x/H_0}},$$

known as the *likelihood ratio*, where P_{x/H_1} is the probability of having received the signal $x(t)$ in the hypothetical situation H_1, where a target is present, and P_{x/H_0} is the probability of having received $x(t)$ in the hypothetical situation H_0, where only noise is present. This likelihood ratio, compared with a decision threshold T, makes it possible to decide whether or not a target is present when the signal $x(t)$ is received.

No decision-making process is totally risk-free. Choosing the threshold T is therefore a trade-off between the two types of possible errors (see Figure 6.2):

- Deciding, if the threshold is exceeded by a noise peak, that there is a signal when in fact there is only noise. This type of error is characterized by the false alarm probability, P_{fa}, which is the proportion of time during which the signal output from the receiver exceeds the threshold. It is written as

$$P_{fa} = \int_T^\infty P_{H_0}(y)dy, \tag{6.3}$$

where $P_{H_0}(y)$ is the probability density of the output signal in the presence of noise alone.

- Failing to indicate the presence of a target when a target is effectively received. The probability of non-detection is complementary to the target detection probability P_D (proportion of time during which the signal exceeds the threshold), which is written as

$$P_D = \int_T^\infty P_{H_1}(y)dy, \tag{6.4}$$

where $P_{H_1}(y)$ is the probability density of the receiver output signal in the presence of a target superimposed on noise.

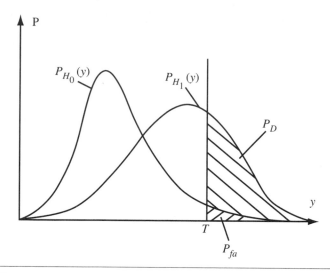

FIGURE 6.2 FALSE-ALARM AND DETECTION PROBABILITIES

These two types of errors do not have the same consequences for radar operation. A false alarm results in mobilization of resources (the operator is alerted, tracks are created, the fire control system is cued). These resources are quickly saturated if there are too many false alarms per unit of time. The system thus ceases to be operational.

This explains why a limit is set for P_{fa}; the threshold T is then set in accordance with this constraint, based on Equation 6.3.

REMARK

This calculation supposes that the probability density $P_{H_0}(y)$ is known. However, this is not usually the case, as the characteristics of noise are unknown.

In this case, the threshold is calculated by estimating noise based on the received signal using the CFAR (Constant False-Alarm Ratio), which we will examine in Part II.

CALCULATING THE DETECTION PROBABILITY P_D

Since the threshold is fixed by P_{fa}, Equation 6.4 can be used to calculate P_D for a target characterized by its power ratio R (its signal-to-noise ratio S/N).

The different P_D and S/N ratios for different P_{fa} are calculated numerically. Figure 6.3 shows these detection curves.

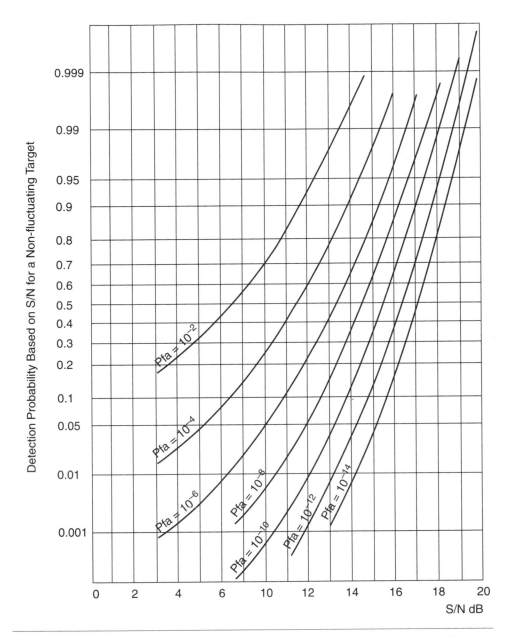

FIGURE 6.3 DETECTION PROBABILITY BASED ON S/N FOR A NON-FLUCTUATING TARGET

REMARK

In Chapter 3, we saw that the signal from an actual target fluctuates with time. Therefore, the S/N ratio, as well as P_D, vary from one observation to the next. The actual detection probability, measured over several observations, takes such variations into account. Figure 6.4 illustrates the situation for a target of type SW1.

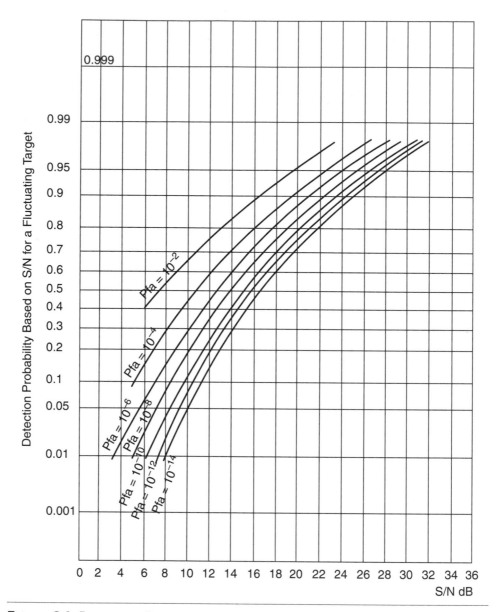

FIGURE 6.4 DETECTION PROBABILITY BASED ON S/N FOR A FLUCTUATING TARGET

Note that the detected target/non-detected target transition is more abrupt for a non-fluctuating target. The high P_D values ($>90\%$) are more difficult to attain for a fluctuating target. However, if a low P_D value ($<40\%$) is sufficient, a fluctuating target is preferable.

REMARK

In the case of a Rayleigh fluctuating target (Swerling I or Swerling II with no post detection integration) there is a simple relationship between the probability of detection, the probability of false alarm, and the signal-to-noise ratio, which is:

$$P_0 = P_{fa}^{\frac{1}{1 + S/R}}$$

6.3 OPTIMAL RECEIVER FOR KNOWN NON-WHITE NOISE

In this case, one of the noise parameters (space or time) is not uniform but presents preferential values (direction or frequency) at which power density is higher. Moreover, the statistical characteristics of this noise are known.

We shall examine the (temporal) case of a noise with non-uniform spectral density $S(f)$. This problem, developed in Le Chevalier in 1989, can be solved using two different methods, both of which give the same result:

- "whitening" the noise to recover the situation of the optimal receiver for white noise
- subtracting the non-white part of the noise

Given that both methods produce comparable results, we shall concentrate on the first one, which is equivalent to designing a filter whose transfer function $H_b(f)$ is such that

$$S(f).\left|H_b(f)\right|^2 = b,$$

where b is a uniform spectral density.

Filtering produces white noise. The previously described optimal receiver can therefore be used. However, because the useful signal has been changed by the whitening filter, the replica must be modified in the same way by the same filter. This gives the receiver design shown in Figure 6.5.

Given the "matched filter" aspect of the optimal receiver, it is modified, compared to Equation 6.2, to become

$$H(f) = \frac{U^*(f - \Delta f)}{S(f)}.$$

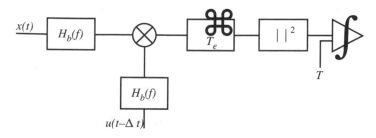

FIGURE 6.5 OPTIMAL RECEIVER FOR NON-WHITE NOISE

APPLICATION TO GROUND CLUTTER REJECTION

The ground clutter signal received by the antenna main lobe is spectrally non-white. Its position within the spectrum is known (see Chapter 4). In fact, noise power within the corresponding frequency zone is, for airborne radar, so high compared to the target signal that the whitening filter automatically eliminates the spectral components of the useful signal in this zone. The whitening filter is therefore replaced by a rejecting filter that cancels all the components of clutter and of the signal in the spectral zone of the main lobe.

6.4 ADAPTIVE RECEIVER FOR UNKNOWN NON-WHITE NOISE

As a general rule, noise is unknown. A receiver has to be designed to measure (estimate) noise characteristics using received signals. Then the receiver can use these measured parameters to process the signal as previously described for the optimal receiver for non-white noise.

In the simplest case, noise characteristics can be measured using an auxiliary reception channel that only receives external noise and supplies a "noise-only reference signal." This channel can be the main channel in the absence of a useful signal (e.g., transmission cutoff or dead time), or a non-selective channel that receives the useful signal without gain, and thus at a negligible level with respect to spurious signals.

The solution is then to subtract this noise-only reference signal from the main channel with the appropriate matched gain using an adaptive loop process, so as to cancel out spurious noise output from the channel processed.

In the absence of a noise-only reference signal, samples (spatial and/or temporal) containing both signal and noise are used.

In this case, for a given set of target parameters (direction, frequency), we

- assume that a target with such parameters is present
- perform filtering (weighted linear combination of the processed samples), which minimizes output energy, with the constraint that the useful signal is not affected by the filter (giving a gain constraint on the frequency and direction in question)

We shall examine both these methods in turn.

6.4.1 Adaptive Radar with a Noise-only Reference Signal

Taking the example of jammer cancellation (which represents the case of spatially non-white noise) using Side Lobe Suppression (SLS) gives us (see Figure 6.6):

- a main channel with high gain that receives the useful signal (at a significant level) and the jammers through the side lobes or far lobes
- one or more non-directive auxiliary channels (also known as ancillary channels) that receive the jammers. (The signal received without gain is negligible.)

In the simple case of a single jammer and a single auxiliary channel, the processing involves subtraction from the main channel $\vec{\Sigma}$ of the signal from auxiliary channel $\vec{A}$, weighted by a complex coefficient w that compensates for the difference in gain between the side lobes of the main channel and the auxiliary channel in the direction of the jammers:

$$\vec{\Sigma}' = \vec{\Sigma} - w\,\vec{A}\,.$$

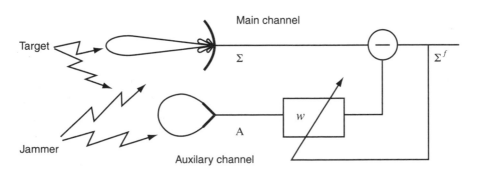

FIGURE 6.6 ANTIJAMMING WITH A NOISE-ONLY REFERENCE SIGNAL (SLS)

The coefficient w is calculated in such a way as to minimize output noise $\vec{\Sigma}'$. Residual noise is measured by correlating $\vec{\Sigma}'$ with the noise reference signal $\vec{A}$. Because the noise and signal are uncorrelated, the correlation component composes the error signal of the adaptive loop. (It measures the remaining jammer residue.)

One example of processing is the Widrow recursive algorithm, which makes w converge towards the optimal value using the equation

$$w_{n+1} = w_n + 2\mu \frac{\vec{\Sigma}'.\vec{A}}{E\left(|A|^2\right)},\tag{6.5}$$

where μ determines the convergence velocity.

This principle can be extended to N auxiliary channels and M jammers, provided that $M \le N$.

6.4.2 ADAPTIVE RADAR WITHOUT A NOISE-ONLY REFERENCE SIGNAL

The previous method assumes the presence of independent auxiliary channels in sufficient number to eliminate all potential jammers. The method also assumes enough gain to cover the main antenna side lobes, thus avoiding a fall-off in radar performance due to the addition of the internal noise of the auxiliary channel to the main channel, if $|w| > 1$.

Phased arrays and, in particular, active phased-array antennas, provide access to sources (sensors) or groups of sources (within an array).

Each sensor delivers a signal $x_i(t)$ comprised of in some cases, a useful signal, $s_i(t)$; and noise signals (noise + jammers), $b_i(t)$—that is, $x_i(t) = s_i(t) + b_i(t)$.

For a standard electronically scanned antenna, the signals x_i are summed using a complex weighting function, a_i, which allows the formation of a directive beam in the direction in which the target is being tracked (i.e., θ).

In the simple case of a linear array with N elements separated by distance d (Figure 6.7), the coefficient a_i has phase φ_i compensating for the difference in path between the sensors for a plane wave arriving from direction θ; that is,

$$\varphi_i = (i-1)\frac{2\pi d\sin\theta}{\lambda}.$$

(Assuming a narrow bandwidth, any delay is compensated for by phase shift.) The magnitude of a_i can be used to weight the antenna so as to reduce side lobes.

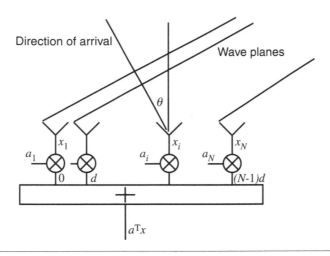

FIGURE 6.7 BEAMFORMING

This beamforming is written in vector form as:

$$d = a^{\mathrm{T}}.s, \tag{6.6}$$

where d is the useful signal, s is the signal vector $(s_1, s_2....s_i...s_N)$, and a^T is the transposed weighting vector $(a_1, a_2....a_i...a_N)$.

Access to the element-level signals gives enough degrees of freedom to optimize the global radiation pattern by forming "zeros" in the direction of the jammers.

These adaptive antijammer methods, known as Digital Beamforming (DBF), are dealt with extensively in the literature. We shall therefore limit ourselves to the main points.

The basic principle of adaptive DBF is to minimize the energy output from a spatial filter (linear combination of signals produced by elements x_i, with complex weighting coefficients w_i) while imposing a gain constraint in the observed direction θ.

In vector form, the processed signal is therefore $\hat{w}^T x$, where $\hat{w}$ is the optimal weighting vector and x is the received signal vector $(x_1, x_2...x_i...x_N)$.

Given that $d = a^T \cdot s$ is the required signal (with a gain constraint in direction θ), estimation error is written as

$$\varepsilon = a^{\mathrm{T}}s - \hat{w}^{\mathrm{T}}x.$$

Minimizing the output noise is equivalent to forcing ε to be orthogonal to the received signal, written as

$$\mathrm{E}[x\varepsilon^*] = 0 \ \text{ or } \ \mathrm{E}\left[\left(xw^T x - d = 0\right)\right],$$

that is,

$$R_{xx}\hat{w} - \mathrm{E}[xd^*] = 0 \ \text{ or } \ \hat{w} = R_{xx}^{-1}\mathrm{E}[xd^*],$$

where R_{xx} is the covariance matrix of signals received, x_i,

$$\text{(a matrix whose elements are } R_{ij} = \mathrm{E}[x_i x_j]).$$

$$\mathrm{E}[xd^*] = \mathrm{E}[xs^T a] = R_{xs}a = R_{ss}a,$$

where R_{xs} is the correlation matrix of vectors x and s, which can be reduced to the correlation matrix R_{ss} of s, as the components s (signal) and b (noise or jammers) of vector x are uncorrelated.

The optimal weighting coefficient is therefore

$$\hat{w} = R_{xx}^{-1} R_{ss}a = R_{xx}^{-1}\phi, \tag{6.7}$$

where ϕ represents the gain constraint in direction θ, given the characteristics of each sensor.

Calculating $\hat{w}$ requires knowledge of R_{xx}, which is a mathematical expectation that is, in theory, unknown. R_{xx} is estimated using time samples received by each sensor, supposing that noise is stationary (ergodicity hypothesis).

Recursive algorithms can also be used to converge towards $\hat{w}$ (still assuming that signals are stationary).

APPLICATION TO AIRBORNE RADAR

For airborne radar, the presence of powerful ground clutter scattered throughout the spectral domain and mainly dependent on the side lobe level creates additional problems:

- ground returns hinder calculation of $\hat{w}$
- the creation of *zeros* in the direction of jammers causes the side lobes or far lobes to increase in the other directions. It can also increase ground clutter received in these directions, thus diminishing visibility of the echoes in clutter

This problem can be solved either by adapting the previously described method (estimating $\hat{w}$ in "clutter-free" frequency or time zones or adding constraints to the radiation pattern), or by using a global method covering both time and space domains. These methods, known as space-time adaptive processing, are described in the sections that follow.

6.5 SPACE-TIME ADAPTIVE PROCESSING

The previous examples assume that space (angles) and time (Doppler frequency) dimensions are independent. Spatial (*nulling*) and temporal (*whitening*) adaptive processings are thus independent and can be implemented separately.

In the case of airborne radars, ground clutter is a noise with related time and space components. So, if θ is the angle between the velocity vector of the platform and the direction of the ground returns, the Doppler frequency of those returns is

$$f_D = \frac{2v}{\lambda}\cos\theta .$$

In this case, adaptive processing combines both the spatial dimension (N elements) and the time dimension (M temporal samples). The result is the generalized adaptive receiver, an extension of the description given in the previous chapter (the vectors having a dimension $N \cdot M$).

A number of works explain how to solve this problem (Klemm 1984, 1992, etc.). Figure 6.8 illustrates the principle.

There are N elements and M temporal samples (interpulse periods). For each direction, θ, and each Doppler frequency, f_D, the output signal is the weighted sum of the NM samples, x_i, (p being the sensor index and i the interpulse index). The weighting vector is given by Equation 6.7 in Section 6.4.2: $w = R_{xx}^{-1}\phi$, where R_{xx} is a matrix MM, with each element being a matrix NN whose general term

$$R_{ij}^{pq} = \mathrm{E}\left[x_i^p . x_j^q\right]$$

is the intercorrelation of the samples x_i^p and x_j^q from sensor p during interpulse period i, and from sensor q during interpulse period j. The constraint vector, ϕ, comprises $N \cdot M$ elements, forcing the gain in direction θ for Doppler frequency f_D of the required target.

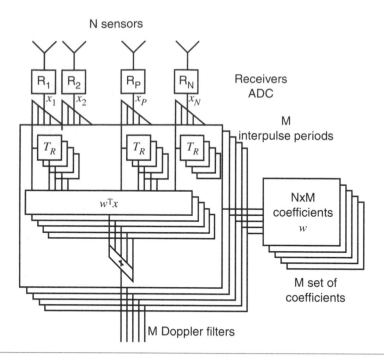

FIGURE 6.8 SPACE-TIME PROCESSING

PRACTICAL APPLICATIONS

The number of practical applications remains limited for two reasons. First, the amount of range-Doppler ambiguities means that in most cases it is impossible to associate one Doppler frequency with a single direction. Second, the calculations are extremely complex. However, in two specific cases this technique proves essential:

- airborne radars operating at low frequencies (VHF, UHF) to combat stealth targets. (In this case, ground clutter is not ambiguous.) The radiation pattern for these frequencies is of poor quality. Adaptive reduction of spurious lobes is therefore indispensable

- airborne radars used to detect targets moving at very low speeds (land vehicles) and whose Doppler frequency is mixed in with that of the ground returns received by the antenna main lobe. Chapter 10 covers this particular point, which is behind the notion of "Displaced Phase Center Antenna" (DPCA)

6.6 WAVEFORM AND AMBIGUITY FUNCTION

The signal $y(\Delta t, \Delta f)$, given by Equation 6.1, is available at the output of the matched receiver (optimal receiver for white or non-white noise) for each direction observed, θ.

This signal is the result of the correlation, over the observation time T_e, of the received signal $x(t)$ with a replica of the transmitted signal $u(t)$, with time delay Δt and frequency shift Δf. These signals can be modified using a whitening filter in the case of non-white noise. The signal $y(\Delta t, \Delta f)$ therefore appears as a surface that is dependent on two parameters, Δt and Δf. If a target is present, this surface has a global maximum for $\Delta t = t_0$ and $\Delta f = f_D$ corresponding to the actual position of the target. The amplitude of this maximum depends solely on the target energy ratio R and not on the form of $u(t)$. Comparison with the threshold T indicates whether or not a target is present. Radar detectability therefore depends on the signal produced by the energy transmitted during T_e, that is, the mean power of the transmitter. It is independent of the waveform $u(t)$.

In contrast, analysis of $y(\Delta t, \Delta f)$ shows that this surface, dependent on $u(t)$, can have several maxima when in the presence of a target (Figure 6.9). These maxima will also exceed the detection threshold, thus creating ambiguity as to the target position in the range-velocity plane ($\Delta t - \Delta f$).

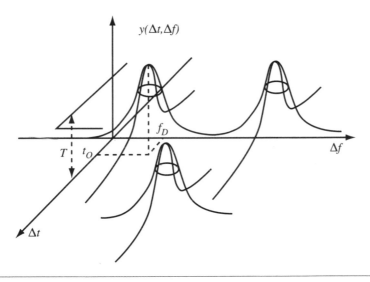

FIGURE 6.9 OUTPUT SIGNAL OF THE MATCHED RECEIVER

Moreover, if two adjacent targets in the $\Delta t - \Delta f$ plane are present simultaneously, the combined response of the receiver to these targets can produce a single correlation peak (Figure 6.10), leading to confusion between the two targets. Once again, the shape of the surface $y(\Delta t, \Delta f)$ will determine the radar discrimination capability. Of course, the narrower the correlation peaks, the easier it is to discriminate between the targets.

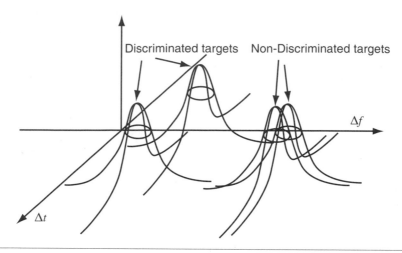

FIGURE 6.10 TARGET DISCRIMINATION

Finally, measurement of the target position in the range-velocity plane ($\Delta t - \Delta f$) is linked to evaluation of the maximum of $y(\Delta t, \Delta f)$, which also depends on the shape of this surface; the narrower the peak, the easier it is to determine the position of the maximum with precision.

The choice of the waveform $u(t)$, which influences the shape of the surface $y(\Delta t, \Delta f)$, thus determines the following characteristics:

- range-velocity ambiguity
- resolution capability
- the precision with which distance and velocity can be measured

6.6.1 AMBIGUITY FUNCTION

6.6.1.1 DEFINITION

In order to determine the intrinsic properties of the waveform, we shall study $y(\Delta t, \Delta f)$ in a situation where noise is negligible compared to the target signal, which is normalized

$$(A = 1, \ \int_{T_e} |u(t)|^2 dt = 1).$$

The received signal is thus reduced to

$$x(t) = u(t - t_0)e^{2\pi j f_D t},$$

and $y(\Delta t, \Delta f)$ becomes

$$y(\Delta t, \Delta f) = \left| \int_{T_e} u(t - t_0)u^*(t - \Delta t)e^{j2\pi(f_D - \Delta f)} dt \right|^2 \tag{6.8}$$

Following changes in the variables that bring the target back to its origin, ($t = t - t_0, \ldots \tau = \Delta t - t_0, \ldots v = \Delta f - f_D$), Equation 6.8 becomes

$$|x(\tau, v)|^2 = \left| \int_{T_e} u(t) u^*(t - \tau) e^{-j2\pi v t} dt \right|^2 \tag{6.9}$$

$|x(\tau, v)|^2$ is known as the ambiguity function of waveform $u(t)$. It represents the response of the filter matched to a target located at the origin at any point on the range-velocity plane (τ, v). Symmetrically it also represents the disturbance caused by a normalized spurious echo located at τ, v on output from a receiver matched to the reference target and placed at the origin. This function measures the imperfections of the matched filter, often known as the filter side lobes.

6.6.1.2 PROPERTIES OF THE AMBIGUITY FUNCTION

The ambiguity function has a dual symmetry with respect to the time axis, τ, and Doppler frequency axis, v.

The volume under the surface circumscribed by $|x(\tau, v)|^2$ and the origin plane is equal to one for normalized $u(t)$:

$$\iint |x(\tau, v)|^2 d\tau dv = 1 . \tag{6.10}$$

This is a very important property as it shows that the disturbance caused by an echo is constant throughout the range-velocity plane. The only available flexibility is to define $u(t)$ in such a way as to minimize this disturbance in certain areas of the plane.

6.6.1.3 EXAMPLES OF THE AMBIGUITY FUNCTION

THUMBTACK-TYPE AMBIGUITY FUNCTION

The ideal ambiguity function comprises a Dirac delta function at the origin (single peak without sidelobe). This requires an infinite transmitted bandwidth and an infinite illumination time. However, because the bandwidth of the transmitted signal (B) is limited, as is the illumination time (T_e), the variation domain of τ extends from $-T_e$ to $+T_e$, while that of v is from $-B$ to $+B$. The volume under the surface $|x(\tau, v)|^2$ is spread over a surface $4BT_e$. Given Equation 6.9, the minimum side lobe level is obtained via uniform diffusion of spurious energy over the surface $4BT_e$, that is, a pedestal of level $1/4BT_e$. This diffusion can be obtained, for example, by transmitting a noise (or a pseudo-random signal) from bandwidth B during time T_e (Figure 6.11).

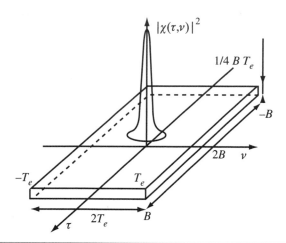

FIGURE 6.11 THUMBTACK-TYPE AMBIGUITY FUNCTION

KNIFE-EDGE-TYPE AMBIGUITY FUNCTION

A particularly interesting case is that of a continuous wave (constant amplitude) of duration T_e and with a fixed frequency or a linearly modulated frequency (frequency excursion ΔF). This waveform corresponds, for example, to a single pulse of duration T_e. The spectrum width of the transmitted signal is given by

- $B \approx 1/T_e$ for a fixed frequency
- $B \approx \Delta F$ for linear frequency modulation (and $\Delta F(T_e \gg 1)$)

Figure 6.12 shows the ambiguity function. It covers the domain $- T_e$, $+ T_e$ and $-B$, $+B$. The main lobe (correlation peak) looks like a narrow triangular knife-edge, which, in the case of frequency modulation, is aligned along the diagonal axis. Chapter 8 studies the specific properties of this function.

"BED OF NAILS"-TYPE AMBIGUITY FUNCTION

If the waveform consists of a series of repeated pulses with an interpulse period T_R (pulse repetition frequency $F_R = 1/T_R$), then the ambiguity function is periodic and looks like a group of identical peaks separated by time T_R along the time axis and by frequency F_R along the frequency axis (Figure 6.13).

6.6.1.4 RANGE-VELOCITY AMBIGUITY

For various reasons that we will explain later, the most common waveforms are made up of recurrent pulse trains. This gives an ambiguity function of the *bed of nails* type. Any echo located at t_0, f_D will produce a group of correlation peaks located at $t_0 + kT_R, f_D + k'F_R$ (integer k and k'), at the receiver output.

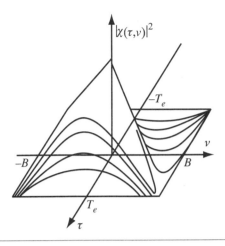

FIGURE 6.12 KNIFE-EDGE-TYPE AMBIGUITY FUNCTION

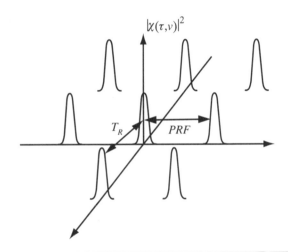

FIGURE 6.13 "BED OF NAILS"-TYPE AMBIGUITY FUNCTION

If the echo is strong enough to be detected, then all the correlation peaks will be detected. The exact location of the target will thus be ambiguous.

REMARKS

The non-ambiguous range is $R_a = cT_R/2$ and the non-ambiguous velocity is $v_a = \lambda F_R/2$. In the range-velocity plane, the non-ambiguous zone equals

$$R_a v_a = \frac{cT_R}{2}\frac{\lambda F_R}{2} = \frac{c^2}{4f_0},$$

depending on the carrier frequency, f_0, only.

Generally this surface is much smaller than the useful domain $R_{\max} v_{\max}$. Thus, for a fighter radar (in the X-band), the ratio $R_{\max} v_{\max} / R_a v_a$ is around 100, that is, a total of approximately 100 ambiguities (100 possible target positions) exist. The choice of F_R can only be used to disperse these ambiguities differently in either the velocity domain (Low Pulse Repetition Frequency radar, LPRF) or the range domain (High Pulse Repetition Frequency radar, HPRF), or in both the range and the velocity domains (Medium Pulse Repetition Frequency radar, MPRF). Chapters 7 and 8 examine the properties of these different radars.

Since detection of any echo leads to the detection of ambiguous echos—in particular an echo in the first ambiguity zone, $R_a v_a$—we can simply search for echos in this smaller domain, then measure the non-ambiguous position of the echo by means of ambiguous measurements provided by specific processing procedures.

6.6.2 RESOLUTION CAPABILITY

Two echoes are said to be resolved (or discriminated) when they produce two distinct detections. Figure 6.10 shows clear examples of resolved and non-resolved targets. However, the situation is not always so clear-cut; indeed, discrimination depends not only on the range of the targets and the form of the ambiguity function, but also on the relative level of these echoes, as shown in Figure 6.14, which illustrates various possible situations.

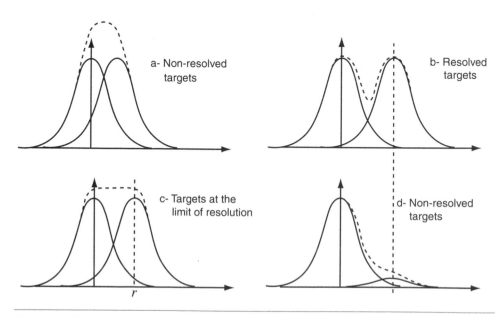

FIGURE 6.14 EXAMPLES OF RESOLUTION

The combination of signals produced by the two targets is shown by the dotted line. The targets are presumed to be separable if this signal has two maxima (Example b). Example c shows resolution limits. A distance r separates the targets, r being the resolution value for the dimension in question. In Example d, the targets are not discriminated, despite being separated by the same distance as in Example b. This is because the side lobes of the stronger target hide the weaker target.

Chapter 8 gives more precise definitions of resolution.

RANGE RESOLUTION

Range resolution is the minimum distance between two resolvable targets traveling at the same speed and with the same amplitude (Example c). This resolution depends only on the *transmitted bandwidth B* given by the equation

$$r \approx \frac{c}{2B}.$$

The corresponding time resolution is $r_t \approx \frac{1}{B}$ (Cook 1967).

It should be noted that range resolution is not directly linked to the duration of the transmitted signal T, except when the transmitted signal is a fixed-frequency pulse. The transmitted spectrum then has a bandwidth $B = 1/T$ and range resolution of $r = cT/2$ (time resolution is $r_t = T$).

VELOCITY RESOLUTION

This is the minimum velocity, r_v, separating two resolvable targets having the same amplitude and located at the same distance. This resolution is linked to the *coherent integration time* of the signal T_c via the equation

$$r_v \approx \frac{\lambda}{2T_c}.$$

Frequency resolution is $r_{fD} \approx \frac{1}{T_c}$.

In the case of an optimal receiver, this coherent integration time equals the illumination time; that is, $T_e = T_c$ (Rihaczek 1977).

6.6.3 PRECISION OF RANGE AND VELOCITY MEASUREMENT

We have seen how the optimal receiver is also the optimal estimator of range $ct_0/2$ and velocity $\lambda f_D/2$ of the target. The values measured are characterized by a measurement noise

- of zero average value (non-biased measurements)
- of standard deviation linked to the form of the correlation peak—the narrower the peak, the smaller the deviation—and by the energy ratio R with

$$\sigma_R \approx \frac{\sqrt{3}\, c}{2\pi B \sqrt{R}}$$

for range measurement and

$$\sigma_v \approx \frac{\sqrt{3}\, \lambda}{2\pi T_c \sqrt{R}}$$

for velocity measurement, where B is the equivalent transmitted bandwidth and T_C is the coherent integration time.

PART II
TARGET DETECTION AND TRACKING

FIGHTER RADAR UNDER TEST IN AN ANECOID CHAMBER

CLUTTER CANCELLATION

7.1 INTRODUCTION

Target detection and tracking effectiveness depend mainly on the choice of transmitted signal and how this signal is processed on reception. Whereas radar range over thermal noise depends on mean transmitted power and is independent of the waveform, all other factors being equal, range in the presence of ground clutter depends on the radar capacity to discriminate the wanted signal from spurious echoes, which is effectively a function of the form of the transmitted signal.

In this latter case, detection capability is determined by

- the waveform, chosen to suit the application in question
- imperfections in generation of the transmitted signal, as well as during reception and processing

First we shall examine the choice of the waveform for each operational detection situation. We shall then go on to study the influence of the above-mentioned defects on radar performance and the constraints imposed on equipment in order to meet the performance requirements in the various cases.

7.2 WAVEFORM SELECTION

7.2.1 CALCULATION OF GROUND CLUTTER RECEIVED BY THE RADAR

In Chapter 5, Part I, we examined the intrinsic physical behavior of ground clutter. The ground clutter signal effectively received depends not only on these properties but also on the radar characteristics, such as

- the terms used in the radar equation (Chapter 3, Part I)
- the ambiguity function of the waveform (Chapter 6, Part I)

Consider (Figure 7.1) a ground element with a geometric area dS located at the intersection of coordinates (X,Y) at range R_S from the radar, in direction

Φ in relation to velocity vector $\vec{V}$ of the platform, and forming angular deviation θ with respect to the antenna axis. If the backscattering coefficient of the ground is σ_0, the power reflected by this element is given by

$$dP = \frac{P_t G^2(\theta) \lambda^2 \sigma_0}{(4\pi)^3 R^4 l} dS .$$

(7.1)

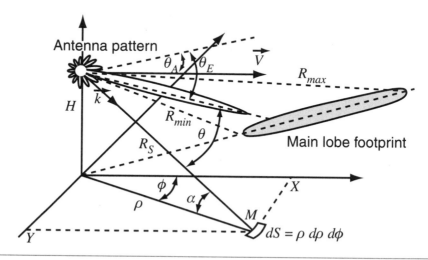

FIGURE 7.1 DEFINITIONS

For each ground element with coordinates (X,Y), we can determine a corresponding point (R, v) within the range-velocity plane, or a corresponding point (t_0, f_D) in the propagation delay-Doppler frequency plane.

The iso-range curves are parallels to the f_D axis in the (t_0, f_D) plane and circles centered on the vertical projection of the platform onto the (X,Y) plane.

The iso-velocity curves (or iso-Dopplers), v, are parallels to axis t_0 in the (t_0, f_D) plane and conics in the (X,Y) plane. These conics are formed by the intersection between the half-angle cones at summit β. Their axis of rotation is the velocity vector with the ground, with α given by $V \cos\beta = v$, where V is the platform velocity. For an aircraft in horizontal flight, these conics are hyperbolas whose axes are the projection of the velocity vector on the ground and its perpendicular (Figure 7.2a).

The range-velocity plane (Figure 7.2b) represents the map of ground return power

$$dP = \frac{P_t G^2(\theta) \lambda^2 \sigma_0}{(4\pi)^3 R^4 l} dS = I(t_0, f_D) dt_0 df_D$$

(7.2)

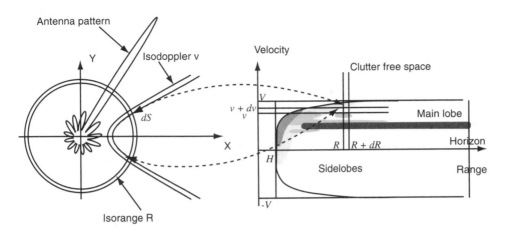

a) Projection along the coordinate system (X,Y)
network of isoranges and isodopplers

b) Map of ground clutter in the
range-velocity plane (time-frequency)

FIGURE 7.2 NON-AMBIGUOUS GROUND CLUTTER MAP

as it would be received by an ideal radar with a resolution, characterized by dR and dv ($dR = cdt_0/2$, $dv = \lambda df_D/2$), whose ambiguity function (Chapter 6, Part I) comprises a single peak with width $d\tau = 2dR/c$, $dv = 2dv/\lambda$ with no side lobe or ambiguity.

This map can be divided into several regions:

- a region of high levels corresponding to ground returns received by the antenna main main. Under normal operating conditions (low antenna elevation, $\cos\theta_E \approx 1$) the area illuminated by the main main is asymptotic to iso-Doppler $v = V\cos\theta_A$. This region is parallel to the range axis in the plane (R, v)

- a region situated between $-V$ and V in the velocity domain ($-2V / \lambda$ and $2V / \lambda$ in the Doppler domain) and between H (platform altitude) and the horizon in the range domain. Ground returns received by the antenna side and far lobes are located within this region

- a region located outside the above-mentioned region that receives no ground returns. This region is known as a *clutter-free zone*

- It should be stressed that under such reception conditions, only targets superimposed on main lobe echoes will be masked by ground clutter, as they are at a higher level elsewhere than the echoes of the side lobes (for normal, non-stealth targets)

In reality, we have seen that an ambiguity function cannot be composed of a single isolated peak and that this function,

$$|\chi(\tau,v)|^2,$$

is a surface representing the perturbation caused by an echo located at the origin (0,0) in the whole plane (τ, v). The parcel of land, dS, in R, v (i.e., t_0, f_D) returns the power $dP = I(t_0, f_D)dt_0 df_D$ to the radar. This generates, at the point (τ, v), a signal:

$$dP(\tau, v) = I(t_0, f_D)|x(\tau - t_0, v - f_D)|^2 dt_0 df_D \,. \tag{7.3}$$

The total ground power received in (τ, v) is therefore

$$P(\tau, v) = \iint I(t_0, f_D)|x(\tau - t_0, v - f_D)|^2 (dt_0) df_D \tag{7.4}$$

This is the convolution of the ground map (without ambiguity) with the ambiguity function, where the ambiguity function depends on the form of the transmitted wave. The amount of ground clutter superimposed onto a target located at point (τ, v) is therefore directly related to the ambiguity function. The following sections examine how waveform influences ground clutter cancellation.

7.2.2 GENERAL CLUTTER CANCELLATION

Ground clutter can be eliminated using

- antenna spatial (angular) selectivity
- range discrimination (temporal separation)
- velocity discrimination (Doppler filtering)

We shall now examine the different ways of eliminating clutter.

If D_{BT} is the domain limited by the ambiguity function (i.e., $-T_c, +T_c$ and $-B$, $+B$, where T_c is the coherent processing time and B is the transmitted bandwidth), the total disturbance introduced by ground returns can be written as follows:

$$P = \iint_{D_{BT}} P(\tau, v)d\tau dv = \iint_{D_{BT}} \iint_{D_S} I(t_0, f_D)|x(\tau - t_0, v - f_D)|^2 dt_0 \cdot df_D \cdot d\tau \cdot dv \,.$$

Given

$$\iint_{D_{BT}} |x(\tau, v)|^2 d\tau dv = 1 \text{ (see Chapter 6, Part I)},$$

we have

$$P = \iint_{D_S} I(t_0, f_D)dt_0 df_D \iint_{D_{BT}} |x(\tau - t_0, v - f_D)|^2 d\tau dv = \iint_{D_S} I(t_0, f_D)dt_0 df_D \tag{7.5}$$

The total power, P, of this disturbance, integrated over the entire D_{BT} domain, is independent of waveform. In contrast, the waveform directly influences the way in which this disturbance is spread throughout the domain.

If we take a radar transmitting a power P_e in order to detect a target with an RCS of σ at range R, the power received from the target is given by

$$P_c = \frac{P_t G^2 \lambda^2 \sigma}{(4\pi)^3 R^4 l}.$$

(The target is presumed to be located in the antenna main beam with a gain G.)

The ground return power, based on Equations 7.2 and 7.5 and by writing

$$R_S^2 = \rho^2 + H^2 \Rightarrow dS = \rho \, d\rho \, d\phi = R_S \, dR_S \, d\phi,$$

is given by Equation 7.6:

$$P_S = \int\int\limits_{D_E} I(t_0, f_D) dt_0 df_D = \int\int\limits_{D_E} \frac{P_t G^2(\theta) \lambda^2 \sigma_0 dS}{(4\pi)^3 R_S^4 l} \tag{7.6}$$

$$= \frac{P_t \lambda^2}{(4\pi)^3 l} \int_0^{2\pi} \int_H^{horizon} \frac{G^2(\theta)(d\phi \sigma_0) dD_S}{R_S^3}$$

Assuming the ground is flat (a justifiable assumption, as only ground in the immediate vicinity is involved) and that this ground is homogenous and therefore independent of ϕ, with respect to the function $\sigma_0 = \gamma \sin\alpha = \gamma(H/R_S)$, the contrast between clutter and target is given by

$$C = \frac{P_S}{P_C} = \frac{\gamma H R^4}{\sigma G^2} \int_0^{2\pi} \int_H^{horizon} \frac{G^2(\theta) d\phi dR_S}{R_S^4}. \tag{7.7}$$

To give an idea of the order of magnitude, we shall examine two typical situations: a combat aircraft (fighter) radar and an airborne early warning radar (AEW). Typical values are given in Table 7.1.

TABLE 7.1. Typical Parameters for Combat Aircraft and AEW

Combat Aircraft	AEW
$V = 300 \text{ ms}^{-1}$	$V = 200 \text{ ms}^{-1}$
$H = 1\,000$ m	$H = 10\,000$ m
$\lambda = 0.03$ m	$\lambda = 0.1$ m
$\theta_A = 3°$	$\theta_A = 1°$
$\vartheta_{A3dB} = 50$ mrd	$\vartheta_{A3dB} = 15$ mrd
$\theta_E = 3°$	$\theta_E = 10°$
$R = 100$ km	$R = 400$ km
$\sigma = 1 \text{ m}^2$	$\sigma = 1 \text{ m}^2$
$r = 100$ m	$r = 100$ m
$r_v = 3 \text{ ms}^{-1}$	$r_v = 3 \text{ ms}^{-1}$

7.2.2.1 TOTAL GROUND POWER WITH NO SELECTIVITY

In this case, $G(\theta) = 1$, therefore

$$C_1 = \frac{P_{S1}}{P_{C1}} \approx \frac{2\pi \gamma R^4}{3\sigma H^2}. \tag{7.8}$$

Ground/target contrast is 129 dB for the fighter and 141 dB for the AEW.

7.2.2.2 ANGULAR SELECTIVITY

In the general case of a directive antenna (Figure 7.1), the level of the ground echo is reduced by

- elevation angular selectivity, as the integral in range of Equation 7.7 is bounded by ranges R_{min} and R_{max} and of the antenna footprint
- Azimuth angular selectivity, as the angle integral of Equation 7.7 is practically limited to the azimuth aperture, that is, θ_G

With $R_{max} \approx \infty$ and G antenna gain, Equation 7.7 becomes

$$C_2 = \frac{\gamma H R^4 \theta_G}{\sigma} \int_{D_{min}}^{\infty} \frac{dR_S}{R_S^4} = \frac{\theta_G \gamma H R^4}{3\sigma R_{min}^3}. \tag{7.9}$$

Applied to the two previous examples, this gives

- fighter: $C_2 = 83$ dB
- AEW: $C_2 = 92$ dB

Despite being considerably reduced by angular selectivity, the ground/target contrast remains very high (>80 dB). Range and velocity selectivities are necessary.

7.2.2.3 RANGE SELECTIVITY

Achieving the effect of range selectivity alone (without ambiguity) means limiting the integral range of Equation 7.7 to that part of the range that corresponds to target R at range resolution r, that is,

$$C_3 = \frac{\gamma H R^4}{\sigma} 2\pi \int_{D}^{D+r_D} \frac{dR_S}{R_S^4} = \frac{2\pi\gamma Hr}{\sigma}. \tag{7.10}$$

C_3 = 50 dB for the combat aircraft and C_3 = 60 dB for the AEW.

If angular selectivity is added to range selectivity, the angle integration of Equation 7.6 is limited to the sector defined by θ_G, that is,

$$C_4 = \frac{\theta_G \gamma Hr}{\sigma} = \frac{\gamma \sin\alpha\,\theta_G Dr}{\sigma} = \frac{\sigma_0 \Delta S}{\sigma}. \tag{7.11}$$

This produces the same equations as in Chapter 5, Part I: The RCS of ground clutter is the product of σ_0 and the area of the resolution sector. The power ratio equals the RCS ratio (without ambiguity).

Applied to the examples, this gives

- fighter aircraft: C_4 = 29 dB
- AEW: C_4 = 34 dB

7.2.2.4 VELOCITY SELECTIVITY

In this case, integration is over the zone defined by the iso-Doppler in question, v, at velocity resolution r_v. C is calculated using the range-velocity plane and by integrating Equation 7.6 between v and $v + r_v$.

We obtain

$$C_5 = \frac{\pi\gamma(V^2 - v^2)R^4 r_v}{2\sigma V^3 H^2}; \tag{7.12}$$

that is, C_5 = 112 dB for the fighter and C_5 = 118 dB for the AEW (for an average value of v/V = 0.5).

In practice, angular selectivity is added to velocity selectivity. This situation applies to radars with high duty factors, e.g., High Pulse Repetition

Frequency (HPRF) radars without velocity ambiguity and with little or no range selectivity. There are two possibilities:

- The iso-doppler v is illuminated by the main beam. This occurs when $v = V \cos \theta = V \cos \theta_E \cos\theta_A$, that is, where v is the velocity of the main beam echoes (meaning that the inherent radial velocity of the target is zero). The ground/target contrast is then determined mainly by angular selectivity and by the ratio between the resolution and the velocity spread of the ground clutter, that is

$$C_6 \approx C_2\, r_v /V\theta_{A3dB} \sin\theta_A \cos\theta_E \;. \tag{7.13}$$

For an average pointing direction, $\theta_A = 45°$, the calculation results in $C_6 = 77\text{dB}$ for the fighter and $C_6 = 92\text{dB}$ for the AEW. Detection is impossible with such values.

- The inherent radial velocity of the target is such that iso-Doppler v is not illuminated by the antenna main beam. Ground clutter superimposed on the target is received either by the side lobes or far lobes of the antenna if $v < V$. The detection capacity is thus determined by the antenna quality and by the frequency rejection capability of the radar. If $v > V$, there are no ground returns at the Doppler frequency of the target, and detection is limited by rejection capability only.

The rejection capability is defined by both the ambiguity function and the spectral purity of the transmission-reception system. It must be greater than C_6 and will be examined in Section 7.3.

7.2.2.5 COMMENTS ON GROUND CLUTTER CANCELLATION METHODS

First, we recall the selectivity options, which are listed in Table 7.2.

TABLE 7.2. SELECTIVITY OPTIONS

		Fighter Aircraft	AEW
No selectivity	C_1	129 dB	141 dB
Angular selectivity	C_2	83 dB	92 dB
Range selectivity	C_3	50 dB	60 dB
Range + angular selectivity	C_4	29 dB	34 dB
Velocity selectivity	C_5	112 dB	118 dB
Velocity + angular selectivity			
Radial target zero	C_6	77 dB	92 dB
$v>V$ (clear zone)	C_6	depends on rejection capability	depends on rejection capability
$v<V$ (radial target¼0)	C_6	depends on rejection capability and antenna side lobes	depends on rejection capability and antenna side lobes

The following observations can be made:

- The total energy backscattered by ground returns is far higher than that received from the target: 130 to 140 dB
- Because it reduces the very close ground returns, angular selectivity produces high gain: 50 dB
- Because it eliminates ground clutter at ranges other than target range, range selectivity (range unambiguous) considerably reduces ground-target contrast. Combined with angular selectivity, this contrast is no higher than 30 dB for the examples studied. In certain applications, such as air-to-sea modes (sea clutter, high range resolution, high target RCS), this contrast can even be reversed, thus enabling detection without velocity selectivity
- Velocity selectivity has little effect when used alone, as very close ground returns can superimpose themselves on a distant target if $v < V$
- Only velocity selectivity, given by the Doppler effect, combined with angular selectivity enables target detection for the applications studied, when target radial velocity is not zero (i.e., $v = v \neq V \cos \theta_E \cos \theta_A$). If $v > V$ no ground clutter is seen at the same velocity as the target. The only limit to detection is the frequency-rejection capability of the radar. If $v < V$ (and $v \neq V \cos \theta_E \cos \theta_A$), ground clutter superimposed on the target is received by the antenna side lobes. Antenna quality influences the detection capabilities of the radar
- The rejection capability must be greater than the ground-target contrast (so that the *filter residue* is below target level), that is, 80 to 90 dB, which is highly restrictive
- When angular, range, and velocity selectivities are combined, the calculation is made even more complex by the presence of ambiguities. This will be examined in later chapters. However, whenever there is a lot of range ambiguity, the ground-target contrast before velocity selectivity is practically proportional to τ/T_R and is written as

$$C_7 = C_6 \tau / T_R \qquad (7.14)$$

For example, where $\tau/T_R = 1/10$ and target radial velocity is zero, the contrast is

- fighter: $C_7 = 67$ dB
- AEW: $C_7 = 82$ dB

These conclusions are identical to those reached in the previous paragraph, except for this variation in contrast.

7.2.3 CLUTTER CANCELLATION AND WAVEFORM SELECTION

We shall now determine the waveform best suited to the elimination of clutter in the different situations already encountered.

7.2.3.1 WAVEFORM WITH A THUMBTACK-TYPE AMBIGUITY FUNCTION

In theory, a *thumbtack*-type waveform (see Chapter 6, Part I), from a signal with bandwidth B and duration T_c, coupled to a directive antenna, enables a combination of angular, range, and velocity selectivity without ambiguity. However, it would cause the disturbance to be spread evenly throughout the range-velocity plane over surface $4BT_c$, at a power level far greater than that of the targets in the most commonly encountered situations. This is because the clutter to target power ratio is very high (> 80 dB). To obtain a ground residue at the same order of magnitude as the target, $4BT_c$ must be $> 10^8$, which would need many hypotheses on the target (range, velocity, acceleration, speed of acceleration) and require power processing far beyond the limits of current technology.

The chosen technique involves concentrating ground energy in the correlation peaks associated with a waveform made up of recurrent pulses with an ambiguity function of the *bed of nails* type (see Chapter 6, Part I).

7.2.3.2 PULSE RADAR

The ambiguity function of a waveform composed of recurrent pulses with period T_R is of the *bed of nails* type. Practically all the energy is concentrated in the correlation peaks. If we ignore the level between these peaks (Figure 7.3)—we will describe the techniques used to this effect later—we can say that

$$\left|\chi(\tau,v)\right|^2 = 1$$

for

$$\forall \tau \in \left[kT_R, kT_R + r_\tau\right] \text{ and } \forall v \in \left[k'F_R, k'F_R + r_v\right],$$

where k and k' are integers, r_τ and r_v are resolutions ($r_\tau = 1/B$, $r_v = 1/T_c$, and

$$\left|\chi(\tau,v)\right|^2 = 0 \text{ elsewhere.}$$

Based on Equation 7.4, we obtain

$$P(t_0, f_D) = \sum_k \sum_{k'} I(t_0 + kT_R, f_D + k'F_R) r_\tau \, r_v, \qquad (7.15)$$

where r_τ and r_v are small compared with variations in $I(t_0, f_D)$.

The term $I(t_0 + kT_R, f_D + k'F_R)r_\tau r_v$ represents the power reflected by the terrain of surface area $\Delta S = (cr_\tau/2) \times ((\lambda r_v)/2) = r \times r_v$ (where r and r_v are the range and velocity resolutions) at the intersection of iso-range $R = c(t_0 + kT_R)/2$ and iso-Doppler $v = \lambda(f_D + k'F_R)/2$ (Figure 7.3).

The ground clutter signal superimposed on a target located at point (t_0, f_D) is therefore the sum of the ambiguous ground returns and the target, to within the range-velocity resolution (Figure 7.3).

We can then define an ambiguous map of ground returns. We can obtain this by dividing the non-ambiguous map $I(t_0, f_D)r_\tau r_v$ into rectangles whose sides measure T_R and F_R and by superimposing all these rectangles. Figure 7.4 shows the result.

Depending on its position within the R, v plane, and on pulse repetition frequency F_R, the target will be either detectable (C_1 in Figure 7.3) or non-detectable if it is folded over on clutter either in velocity or range (C_2 in Figure 7.3).

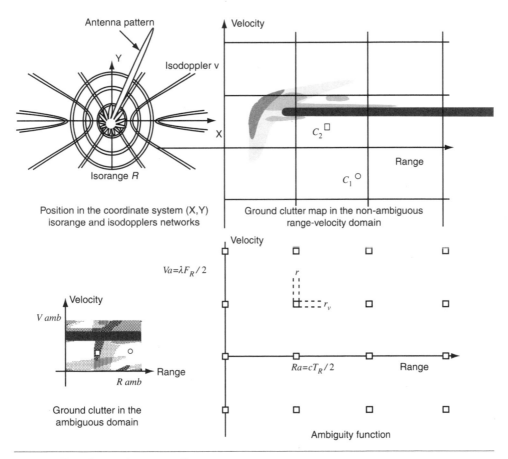

FIGURE 7.3 PULSE DOPPLER RADAR

We shall now examine the various possibilities.

7.2.3.3 Low Pulse Repetition Frequency Radar (LPRF)

These radars are non-ambiguous in range and their velocity ambiguity is therefore very high for usual carrier frequencies (L- to Ku-band). In this case, only velocity foldover occurs, as shown in Figure 7.4.

Except for certain relatively high pulse repetition frequencies, which Chapter 8 will examine, velocity foldover is practically equivalent to projection along the range axis. There are two possible situations:

- *look-up:* The antenna beam does not illuminate the ground. (This is the case when the target tracked is at a higher altitude than the platform.) The received ground returns come through the antenna side or far lobes and are at the same range as the target (no range ambiguity). For normal (non-stealth) targets, this clutter does not generally limit detection, as it is less powerful than the target (Figure 7.4)

- *look-down:* The antenna beam illuminates the ground, and the spectral width of ground clutter is high with respect to the pulse repetition frequency. Velocity selectivity is impossible and the ground-target contrast (see Section 7.2.2.5) is such that target detection is generally impossible except over a very short range (Figure 7.4)

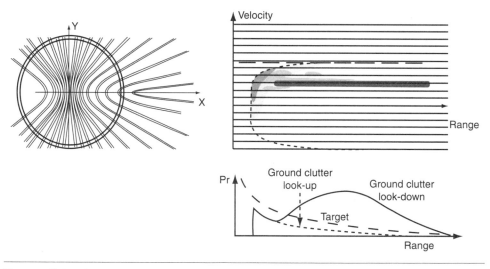

FIGURE 7.4 LPRF RADAR

This waveform is well suited to look-up but gives mediocre results in look-down.

7.2.3.4 High Pulse Repetition Frequency Radar (HPRF)

This expression refers to radars that have no velocity ambiguity but are highly range-ambiguous at the usual carrier frequencies (L- to Ku-band).

Foldover is thus equivalent to projection along the velocity axis (Figure 7.5), and we can compare the radar to a continuous wave (CW) radar without range selectivity (see Section 7.2.2.5):

- When target radial velocity is zero (i.e., when $v = V \cos \theta_E \cos \theta_A$), ground clutter received by the main beam has the same velocity as the target (same Doppler frequency). Detection is impossible
- If $v > V$, no ground returns are seen at the same velocity as the target. The only limit to detection is thermal noise, provided that the frequency rejection capability of the radar is sufficient (Section 7.3 covers this constraint)
- If $v < V$ (and $v \neq V \cos \theta_E \cos \theta_A$), ground returns superimposed on the target are received by the antenna side or far beams. Antenna quality is a determining factor in the detection capability of the radar. Although these echoes are considerably attenuated by angular selectivity, they can be located at very close range—particularly if the aircraft is flying at low altitude—and the propagation function in R^{-4} means that they are more powerful than a distant target. Moreover, their power considerably limits radar range, given that their RCS is generally much greater than that of targets

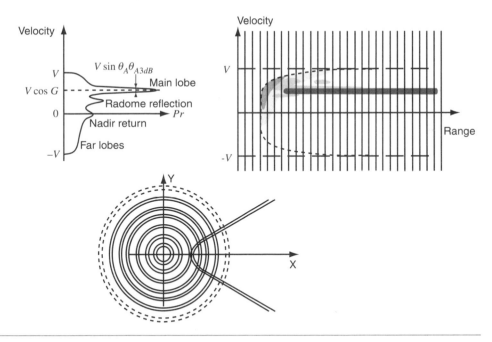

FIGURE 7.5 HPRF RADAR

This waveform is well suited to look-down target detection, where $v > V$, that is, total radial velocity is greater than platform velocity, because the target is in a clear Doppler zone. This configuration occurs in front sector presentation (head on). In a tail attack situation (where the platform is tracking the target), target and platform radial velocities cancel each other out and $v < V$, thus limiting detection.

7.2.3.5 MEDIUM PULSE REPETITION FREQUENCY RADAR (MPRF)

In order to reduce the effect of range foldover and assist target detection when $v < V$, range ambiguity must be increased. This means reducing pulse repetition frequency F_R. Section 7.2.3.2 and Figure 7.3 describe the resulting situation: target detection depends on its *range and velocity* on the one hand, and pulse repetition frequency F_R on the other. If no *a priori* information is available regarding the range-velocity position of the targets, several different values must be used for F_R to ensure that for each target in the useful domain (domain observed) at least one F_R value can be used to fold this target into an area free from ground return signals.

Detection range is reduced in comparison with detection over thermal noise, as during the observation period in the target direction, certain values of F_R prevent detection (foldover on ground clutter). Energy transmitted is thus lost. This is the case for target T_1 in Figure 7.6, which would be in a clear zone in HPRF mode. However, this waveform enables detection of all targets independently of the respective positions and altitudes of the platform and the target (except, of course for targets with

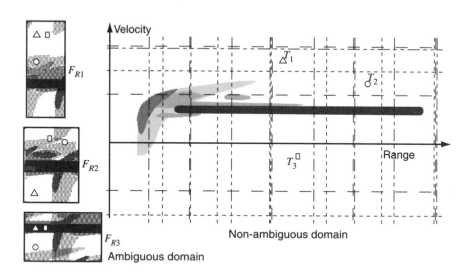

FIGURE 7.6 MPFR RADAR

zero radial velocity). Figure 7.6 illustrates this, showing the various possibilities for three targets and three F_R:

- T_1 is in a clear Doppler zone. It is not detected at F_{R3}—it aliases onto the main main—but is detected at F_{R1} and F_{R2}
- T_1 and T_3 are not detectable at LPRF (it folds over onto the main main), nor at HPRF (foldover onto the far lobes). However, T_2 is detectable over thermal noise at F_{R3}, and T_3 is detectable at F_{R1} and F_{R2}

This waveform is used when the previous waveforms have failed.

7.3 IMPROVEMENT FACTOR AND SPECTRAL PURITY

7.3.1 DEFINITIONS

SUBCLUTTER VISIBILITY

Subclutter-visibility, τ_V is the ratio of ground return power to the minimum target power that can be *detected in the presence of this clutter* and located at the same ambiguous range. It is determined by the contrast between ground clutter power and target power (calculated in Section 7.2.2). It is recommended that you add a coefficient k to take into account fluctuations in ground clutter (calculated as a mean value), that is, $\tau_V = kC_7 \approx 10C_7$.

IMPROVEMENT FACTOR (WITH NO TRANSMISSION-RECEPTION DEFECTS)

We have seen that only velocity selectivity (Doppler frequency selectivity), combined with spatial (angular) selectivity of the antenna, provides sufficient target/ground contrast for target detection. Ignoring the antenna pattern quality required in certain configurations (see Section 7.2.3), let us now look at the frequency rejection constraints imposed on the radar in order to meet this objective.

Consider ground clutter received by the antenna main lobe (Figure 7.2b). It is distributed parallel to the range axis. Its Doppler frequency is $v_0 = (2V/\lambda)\cos\theta_E \cos\theta_A$. In the ambiguous range cell in which the target is located, the level of ground clutter is $P(\tau,v_0)$.

The ratio

$$\tau_E = \frac{P(\tau,v_0)}{P(\tau,v)}$$

is known as the improvement factor. It represents the frequency rejection capability of the radar.

Given the waveforms that interest us (periodic pulse train),

$$|x(\tau, v)|^2$$

is of the "bed of nails" type and can be written as

$$|x(\tau, v)|^2 = f(\tau)g(v).$$

Given that $I(T_0, f_D) = 0$ if $f_D \neq v_0$, Equation 7.4 takes the form

$$P(\tau, v) = P_S(\tau)|x(0, v - v_0)|^2.$$

The improvement factor is given by

$$\tau_E = \frac{P_S|x(0, 0)|^2}{P_S|x(0, v - v_0)|^2} = \frac{1}{|x(0, v - v_0)|^2}. \qquad (7.16)$$

This must be such that ground return filtering residue

$$P(\tau, v) = P_S(\tau)|x(0, v - v_0)|^2$$

does not impede target detection. Two situations call for closer examination:

- Residue is *random* in nature (noise). It combines with thermal noise and *degrades radar sensitivity*. This reduction in sensitivity is considered negligible if the total noise (thermal noise + residue) increases by only 1 dB. This corresponds to a level of residue of approximately 6 dB below thermal noise (Figure 7.7a)
- Residue is *deterministic* (spurious lines, sidelobes). It has the same processing gain as the targets and produces false detections, thus *increasing the number of false alarms*. This increase is generally acceptable if residue is 5 dB below thermal noise (Figure 7.7b)

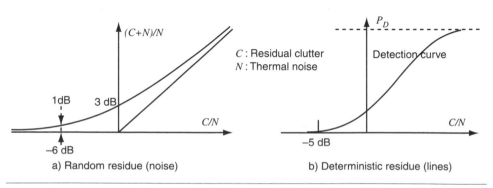

a) Random residue (noise)   b) Deterministic residue (lines)

FIGURE 7.7 TOLERABLE RESIDUE LEVEL

Generally, because the minimum detectable signal is 5 to 10 dB above thermal noise, the improvement factor must be at least 10 dB greater than the required visibility ratio, that is, $\tau_E = k'\tau_V > 10\tau_V$.

The improvement factor depends on the ambiguity function and thus on the form of the transmitted signal, $u(t)$. It also depends on the quality of $u(t)$ and its reference signal $u(t - \Delta t)e^{j2\pi\Delta ft}$. So far we have presumed that the terms $u(t)$ and $u(t - \Delta t)e^{j2\pi\Delta ft}$ are known, without error, which is obviously not the case in real-life situations. The following sections describe the elements that determine the radar improvement factor.

7.3.2 SPECTRAL PURITY

7.3.2.1 MODELING

Figure 7.8 shows a simplified configuration applicable to most radars. The signal to be transmitted, $u(t)$, modulates a microwave carrier, $s_0(t) = Au(t - t_0)e^{j2\pi f_D t}$, produced by a reference source (the frequency oscillator of the exciter unit). This same source generates the different waves or local oscillators (LO), enabling the transposition of the received signal from the microwave to the different intermediate frequencies (IF), then to the baseband (video frequency signals), in order to produce the received signal, $x(t)$, to be processed. The successive changes in frequency are equivalent to a single change that directly transposes the microwave signals to baseband, as shown in the simplified configuration (Figure 7.8).

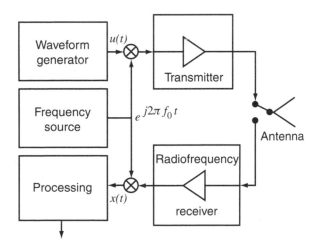

FIGURE 7.8 SIMPLIFIED RADAR BLOCK DIAGRAM

The stability of the frequency source is characterized by slow drifts, spurious lines, and a phase noise spectrum, £(f). Section 7.3.3.2 describes the effects.

DEFINITIONS

The voltage produced by a frequency source is

$$v(t) = v_0[1 + a(t)]\sin[2\pi f_0 t + \varphi(t)],$$

where $a(t)$ and $\varphi(t)$ are random processes of zero average, varying slowly in comparison with $2\pi f_{0t}$. The term $a(t)$ represents amplitude noise, and $j(t)$ represents phase noise.

The spectral power density of phase noise $\varphi(t)$ is

$$S_\varphi = \int_{-\infty}^{+\infty} \Re_{\varphi\varphi}(\tau)e^{j2\pi ft}d\tau . \qquad (7.17)$$

The spectral noise is characterized by £(f), defined by

$$£(f) = \frac{2\int_{f-1/2}^{f+1/2} S_v(u)du}{\int_{-\infty}^{+\infty} S_v(u)du} ,$$

where

$$S_v = \int_{-\infty}^{+\infty} \Re_{vv}(\tau)e^{j2\pi ut}d\tau . \qquad (7.18)$$

$£_{(f)}$ is the ratio of signal power in the bandwidth of one Hertz, shifted by f with respect to the central frequency f_0, to total signal power produced by the frequency source.

7.3.2.2 TRANSMITTER AND RECEIVER BEHAVIOR

The frequency oscillator generates a signal with high spectral purity at a frequency f_0, written in complex form as

$$v(t) = v_0[1 + a_0(t)]e^{j[2\pi f_0 t + \varphi_0(t)]} \approx v_0[1 + a_0(t) + \varphi_0(t)]e^{j[2\pi f_0 t]} . \qquad (7.19)$$

This signal is modulated and then amplified (P) by the transmitter, which adds its own noise ($a_e(t)$, $\varphi_e(t)$) and compresses (α), the amplitude noise of the oscillator. The transmitted signal is therefore

$$v_e(t) \approx Pv_0 u(t)\Big[1 + a_e(t) + \alpha a_0(t) + j\big(\varphi_e(t) + \varphi_0(t)\big)\Big]e^{j2\pi f_0 t} .$$

At the receiver input, the transmitted signal has been attenuated by the factor k, delayed by t_0, and Doppler-shifted by f_D:

$$v_r(t) = kv_e(t - t_0)e^{j2\pi f_D t}.$$

It is transposed by multiplication with the reference signal from Equation 7.15; that is,

$$s(t) = v_r(t)v_0^*(t) \approx A \left\{ \begin{matrix} 1 + a_e(t - t_0) + \alpha a_0(t - t_0) + \beta a_0(t) \\ + j[\varphi_e(t - t_0) + \varphi_0(t - t_0) - \varphi_0(t)] \end{matrix} \right\} u(t - t_0)e^{j2\pi f_D t}.$$

As a general rule, the amplitude noise of the reference signal has negligible influence. (Moreover, it is compressed by the transmitter on transmission and by the mixer on reception.) The signal actually received is given by

$$s(t) = [1 + b(t)]Au(t - t_0)e^{j2\pi f_D t} \quad \text{(see Figure 7.9)}$$

with

$$b(t) = a_e(t - t_0) + j[\varphi_e(t - t_0) + \varphi_0(t - t_0) - \varphi_0(t)] = a(t) + j\varphi(t).$$

The received signal is therefore the sum of a deterministic term representing the noiseless signal, $s_0(t) = Au(t - t_0)e^{j2\pi f_D t}$, and a noise term, $b(t)s_0(t)$, carried by the signal.

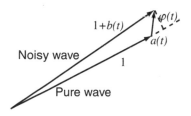

FIGURE 7.9 AMPLITUDE AND PHASE NOISE

7.3.2.3 SPECTRAL NOISE PROPERTIES

PRELIMINARY REMARK

In reality, signal $s(t)$ only exists during observation period T_e. We should speak in terms of *energy density*. In fact we can consider that the signal is transmitted and received from $-\infty$ to $+\infty$ and that signal duration is limited to T_e by the receiver. This method has the advantage of enabling calculation of *power densities* that are easier to handle and measure. Indeed, these are the conditions under which spectral purity is measured (permanent signals).

REFERENCE SOURCE NOISE (LO)

Note that noise caused by the reference source (LO) is introduced by the term $\varphi(t) = \varphi_0(t - t_0) - \varphi_0(t)$, whose autocorrelation is given by the equation

$$\Re_{\varphi\varphi}(\tau) = \int_{-\infty}^{+\infty} \left[\varphi(t - t_0) - \varphi(t)\right]\left[\varphi(t - t_0 - \tau) - \varphi(t - \tau)\right]dt$$

or

$$\Re\varphi\varphi(t) = 2\Re\varphi_0\varphi_0(\tau) - \Re\varphi_0\varphi_0(\tau + t_0) - \Re\varphi_0\varphi_0(\tau - t_0);\qquad(7.20)$$

that is, by applying Equation 7.16 to Equation 7.20, we get

$$S(f) = 4S_0(f)\sin^2 \pi f t_0.\qquad(7.21)$$

The noise produced by demodulating the signal transmitted and received by the reference source is filtered by the transfer function

$$\left|H(j2\pi f)\right|^2 = 4\sin^2 \pi f t_0,\qquad(7.22)$$

which is that of a two-pulse canceller of period t_0 (see LPRF Doppler radar in Chapter 8).

The LO noise, $\pounds(f)$, shown in Figure 7.10a, falls rapidly from the value of the platform frequency f_0 to reach a plateau of $\pounds_0$ at several kiloHertz from f_0.

Noise is considerably reduced for values of f close to that of the carrier frequency, particularly at short ranges (Figure 7.10a). Note that for $t_0 = 0$, the LO noise has no influence. This creates a problem when we try to measure it in a laboratory, as a propagation delay is necessary.

For a distant echo, the spectrum modulation introduced by Equation 7.21 is extremely rapid and has no effect, given the usual filtering values (Figure 7.10b).

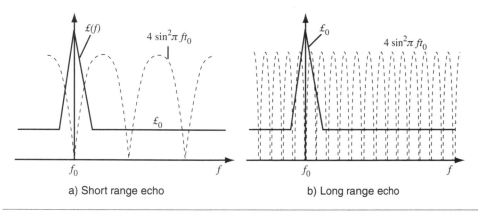

a) Short range echo   b) Long range echo

FIGURE 7.10 REFERENCE SOURCE NOISE (LO), INFLUENCE OF RANGE

SPECTRAL DENSITY OF RECEIVED NOISE (THERMAL NOISE EXCEPTED)

Let us now calculate the correlation function of received noise $x(t) = b(t)s_0(t)$, knowing that $b(t)$ and $s_0(t)$ are independent:

$$\mathfrak{R}_{xx}(\tau) = \mathfrak{R}_{bb}(\tau)\mathfrak{R}_{ss}(\tau)$$

with

$$\mathfrak{R}_{bb}(\tau) = \mathrm{E}\big[b(t)b(t-\tau)\big] \text{ and } \mathfrak{R}_{ss}(\tau) = \int_{-\infty}^{+\infty} s_0(t)s_0(t-\tau)dt .$$

Based on Equation 7.16, spectral power density is

$$S_x(f) = \int_{-\infty}^{+\infty} \mathfrak{R}_{xx}(\tau)e^{-j2\pi f\tau}d\tau = \int_{-\infty}^{+\infty} \mathfrak{R}_{bb}(\tau)\mathfrak{R}_{ss}(\tau)e^{-j2\pi f\tau}d\tau ;$$

that is,

$$S_x(f) = \int_{-\infty}^{+\infty} S_b(v)S_s(f-v)dv , \tag{7.23}$$

where S_b and S_s are obtained from $\mathfrak{R}_{bb}$ and $\mathfrak{R}_{ss}$ by applying Equation 7.17.

The spectral power density of the noise carried by an echo is equal to the convolution of the transmission-reception chain noise with the spectrum of the echo signal.

For a waveform composed of pulses with spectrum B and repetition frequency F_R, the spectrum is made up of lines spaced at intervals F_R (Figure 7.11a). Convolution results in noise spectrum foldover (Figure 7.11b). The noise level between the components that are multiple of F_R thus increases to

$$\pounds(f) = \pounds_0 T_R B,$$

that is,

$$\pounds(f) = \pounds_0 T_R B. \tag{7.24}$$

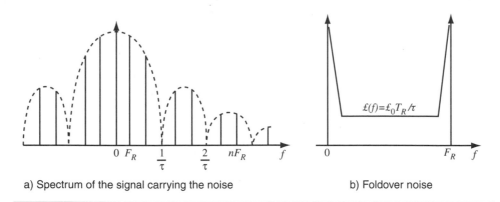

a) Spectrum of the signal carrying the noise   b) Foldover noise

FIGURE 7.11 FOLDOVER OF LO NOISE

7.3.3 CONSTRAINTS LINKED TO CLUTTER CANCELLATION

7.3.3.1 IMPROVEMENT FACTOR AND SPECTRAL PURITY

Equation 7.15, which defines the improvement factor, is based on the ideal receiver concept examined in Part I. In reality, the transmission-reception-processing assembly is different. Design faults such as the previously mentioned amplitude and phase noise must be taken into consideration, as must processing modifications such as

- the presence of temporal weighting, *w(t),* of the received signal. Equation 7.15 shows that rejection capability is limited by the *frequency side lobes* of the ambiguity function
- limitation of the integration time linked to the inherent spectrum of the target. The next chapter examines this. This integration time (or coherent processing time), T_c, determines the velocity resolution ($r_v = 1/T_c$) and is generally less than target illumination time T_e

This means we must replace the ambiguity function in Equation 7.15 with

$$|\chi'(\tau,v)|^2 = \left|\int_{T_c}(1+b(t))u(t)w(t)u*(t-\tau)e^{j2\pi vt}dt\right|^2;\qquad(7.25)$$

that is, considering that *b(t)* and *u(t)* are independent and that $b(t) < < 1$,

$$\tau_E \approx \frac{\left|\int_{T_c}|u(t)|^2 w(t)dt\right|^2}{\left|\int_{T_c}|u(t)|^2 w(t)e^{j2\pi(v-v_0)t}dt\right|^2 + \left|\int_{T_c}b(t)|u(t)|^2 w(t)e^{j2\pi(v-v_0)t}dt\right|^2}\qquad(7.26)$$

where the numerator is a normalization term close to one (and represents weighting loss), and the denominator consists of a deterministic term linked to waveform and processing and a random term linked to spectral purity. *The improvement factor is determined by the greatest of these two denominator terms.*

Now let us examine the limits imposed by these two terms.

7.3.3.2 INFLUENCE OF PROCESSING AND WAVEFORM

The first term of the denominator represents residue from the *filtering of a spurious echo of unit power located at Doppler frequency v_0* by means of the Doppler filter centered on the desired echo at frequency v (Figure 7.12). In the absence of weighting (tapering), this residue is given by the ambiguity function

$$|\chi(0, v - v_0)|^2 ,$$

whose form is described in Chapter 6, Part I, for a waveform consisting of a series of repeated pulses (Figure 7.12).

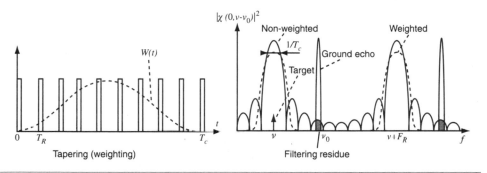

FIGURE 7.12 INFLUENCE OF WAVEFORM AND PROCESSING (TAPERING)

The level of the ambiguity peaks located at $v - v_0 = kF_R$ is close to one. The improvement factor for these values is very low, thus eliminating any chance of target detection. These values correspond to blind speeds if $v = v_0$.

The side lobes on the frequency axis between these ambiguity peaks decrease as $\sin x / x$ decreases. This decrease, which is very slow from v_0 onward, leads to masking of the desired target by ground return residue from the main main received in the same ambiguous range cell. The $\sin x / x$ form is due to the Fourier transform of measurement window T_c. (The processed signal can be seen as the product of an infinite pulse train by this window.)

Amplitude tapering (or weighting) of this window causes a reduction in the side-lobe level of its Fourier transform (Figure 7.12), as well as processing loss (tapering loss due to a mismatched receiver) and a widening of the correlation peak (degradation of velocity resolution). The article by Harris (1978) describes the characteristics of the various weighting functions.

COMMENT 1

The improvement factors required in the above examples (Section 7.2.2.5) are

- Fighter: $\tau_V = kC_7 \approx 75$ dB $\tau_E \approx k'\tau_V \approx 85$ dB
- AEW: $\tau_V = kC_7 \approx 85$ dB $\tau_E \approx k'\tau_V \approx 95$ dB

The weighting factors required to achieve such rejection directly would cause unacceptably heavy losses, so it is preferable to use a rejection filter for the main main clutter (clutter notch) before the matched receiver (Fourier transform), with this filtering acting as a whitening filter. The role of this filter (a two-pulse or three-pulse canceller) is to reduce ground return dynamic range before Doppler processing, thus limiting tapering losses.

COMMENT 2

The above equations assume that signal reception and processing are *linear*. *Non-linearity* within the chain causes *harmonics* and *intermodulations* between the signals, reducing frequency rejection capability and therefore the improvement factor (see Section 7.4).

7.3.3.3 INFLUENCE OF SPECTRAL PURITY

The second term of the denominator in Equation 7.26 represents filtering by processing of the phase and amplitude noise carried by ground echoes (mainlobe) located in the (ambiguous) range cell in question (see Figure 7.13).

The filtering residue is given by

$$ p_S = \pounds(f)\frac{1}{T_c}P_S = \pounds_0 \frac{1}{T_c}\frac{T_R}{\tau}P_S . $$

The improvement factor is given by

$$ \tau_E = \frac{P_S}{p_S} = \frac{T_c}{\pounds_0}\frac{\tau}{T_R} = \frac{\tau}{\pounds_0 r_v T_R} . $$

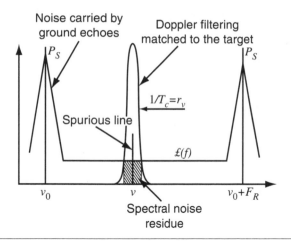

Noise carried by ground echoes

Doppler filtering matched to the target

Spurious line

$£(f)$

Spectral noise residue

FIGURE 7.13 AMPLITUDE AND PHASE NOISE FILTERING

This ratio can be used to determine the spectral purity constraint. We have seen that

$$\tau_E = k\tau_V = kk\,'C_7 \approx kk'\,C_6\,\tau/T_R$$

and can deduce that

$$£_0 = \frac{T_c}{kC_6} = \frac{1}{kC_6 r_v}.$$

$£_0$ does not depend on the waveform. The spectral purity constraint is the same whatever the pulse length and pulse repetition frequency. This conclusion holds true with regard to amplitude and phase noise, except for spurious lines, and assumes that the radar is range ambiguous.

Let us now calculate these constraints for the two previously mentioned applications, with T_C = 0.01 s (r_v – 100 Hz) and kk' = 100:

- fighter aircraft: $£_0 = -115$ dB$_c$ / Hz
- AEW: $£_0 = -130$ dB$_c$ / Hz

These are extremely severe constraints.

COMMENT

The ratio between target and thermal noise power is given by the energy ratio

$$\frac{P_C}{N} = \frac{S}{N} = \frac{P_m T_c}{kT_0 F},$$

where P_m is the mean power of the target signal.

The ground residue-to-noise power ratio after processing (filtering) is

$$\frac{p_S}{N} = \frac{p_S}{P_S}\frac{P_S}{P_C}\frac{P_C}{N} = \pounds_0 \frac{T_R}{\tau}\frac{1}{T_c}C_6\frac{\tau}{T_R}\frac{P_m T_c}{kT_0 F} = \frac{\pounds_0}{kT_0 F}C_6 P_m. \qquad (7.27)$$

This shows that in terms of processing, the spectral noise carried by ground returns behaves in the same way as thermal noise and is not dependent on waveform. Pulse compression in particular does not enable reduction of the ground-target contrast, as the random nature of this noise does not produce the range resolution gain obtained with deterministic signals.

In contrast, spurious lines carried by ground returns are processed with the same gain as wanted signals. The constraints associated with these spurious lines depend on waveform and the pulse compression factor.

7.4 DYNAMIC RANGE AND LINEARITY

Saturation or non-linearity in the reception or processing circuits leads to the creation of harmonics or intermodulations if powerful echoes are received. These spurious signals produce interference with the Doppler spectrum and make it more difficult to detect wanted targets. Let us examine the constraints imposed on radar *dynamic range* by the rejection of ground echoes (linear signal processing range).

Figure 7.14 shows the variation in ground echoes and target power as a function of range for a fighter radar in the HPRF mode, where the antenna beam remains pointed at the target. The difference between the two curves represents the ground/target contrast C_7 calculated in Section 7.2.2.

Figure 7.14 also shows that total dynamic range (the maximum ground level-to-minimum target level ratio) is high (> 110 dB). The higher (< 65 dB) instantaneous dynamic range (contrast at a given range) is obtained at maximum detection range.

In addition, Figure 7.14 shows the relative levels of the different signals (ground, ground residue, minimum target, thermal noise) and the dynamic range constraints imposed at the different stages of reception:

- The instantaneous dynamic range is $C_7 = 65$ dB
- The subclutter visibility is $\tau_V = kC_7 \approx 75$ dB (taking into account the peak value of ground echoes)
- The ground return-to-thermal noise ratio at filter output is ≈ 80 dB.
- The required clutter rejection is $\tau_E = k'\tau_V \approx 85$ dB
- The ratio of ground return to thermal noise before Doppler filtering, which represents the dynamic range before analog-digital conversion

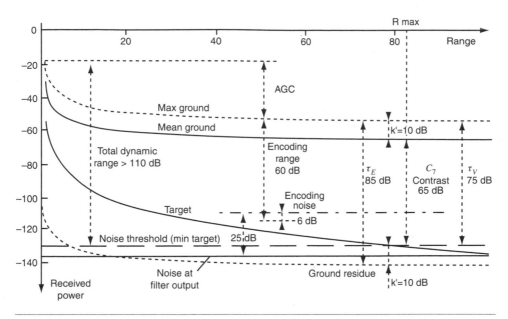

FIGURE 7.14 POWER AND DYNAMIC RANGE OF THE PROCESSED SIGNALS

(ADC), is $80 - 10 \log(512) \approx 55$ dB for Doppler filtering over 512 interpulse periods

- Given that thermal noise must be encoded using an LSB (Least Significant Bit) equivalent at most to half of the RMS noise, minimum *quantization dynamic range* is $55 + 6 > 60$ dB, which corresponds to at least 10 bits plus the sign, or 11 bits

This quantization dynamic range assumes that the receiver has automatic gain control (AGC), which adjusts gain so it is centered on the maximum level of ground clutter. Given that AGC is sensitive to jamming and its adjustment is a delicate operation, it is advisable to allow some margin for dynamic range; and quantization is generally carried out using a 12 to 14 bit ADC.

8

AIR-TO-AIR DETECTION

8.1 INTRODUCTION

Air-to-air detection, that is, the search for an air target by an aircraft, is undoubtedly one of the most difficult missions a radar has to perform, especially when the target is flying at low altitude and is mixed with strong ground clutter returns.

In the previous chapters, we saw the importance of the choice of the waveform in the performance of radars operating in the presence of strong ground echoes, even if the waveform does not affect the power budget over thermal noise. Depending on the operating conditions (velocities, altitudes, relative angle of view), the most suitable waveform is either LPRF (range-unambiguous) coherent or non-coherent, HPRF (without velocity ambiguity), or MPRF (ambiguous in range and velocity). These various waveforms have led to the development of increasingly complex systems as technology has advanced. They will be successively analyzed in the order listed.

8.2 NON-COHERENT LOW-PRF MODE

The first radars to be developed were of the non-coherent, range-unambiguous type. They were classic radars, as described in Chapter 1.

The introduction of the pulse-compression technique enabled an increase in range without changing the radar design and without being constrained by technological limits or degrading other performances.

Finally, thanks to recent technological progress, coherence has been added to radars that are range-unambiguous, making it possible to discriminate expected echoes from spurious ground returns and resulting in a significant enlargement of their field of application.

We shall now look at these three generations of radars that are range-unambiguous.

8.2.1 WAVEFORM AND THEORETICAL PROCESSING

A non-coherent LPRF radar transmits a waveform composed of pulses of peak power P_c, duration T, and interpulse period T_R, such that $cT_R/2 > R_{max}$, where R_{max} is the maximum range of desired echoes received. These pulses can be frequency modulated. First, let us assume that this is not the case (pulse without frequency modulation) and then apply the results to the pulse-compression case. Figure 8.1 represents signal $u(t)$.

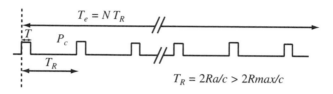

FIGURE 8.1 LOW-PRF WAVEFORM

A transmission tube of the self-oscillating type (Magnetron) is generally used, and the phase of the transmitted signal is not controlled from one interpulse period to the next. It is not the purpose of this radar to use the target Doppler frequency that is assumed to be zero.

For a search radar, the received signal is composed of the train of N pulses received during time T_e when the antenna beam illuminates the target, with $N = T_e/T_R$.

The receiver *matched* to this signal *would* be the correlator

$$c(\Delta t) = \int_{T_e} x(t)u(t - \Delta t)dt \, ,$$

that is

$$c(\Delta t) = \sum_{i=0}^{n-1} \int_{iT_R}^{iT_R + T} x(t)u(t - \Delta t)dt = \sum_{i=0}^{n-1} x_i(\Delta t) \, ,$$

where $x_i(\Delta t)$ is the output of a filter matched to the unit pulse of interpulse period i, centered on $iT_R + \Delta t$ (see Figure 8.2).

The receiver is an *integrator* that, from one interpulse period to the next, sums the average of the received signal $x(t)$ in a "range gate" whose duration is T and is located at $iT_R + \Delta t$.

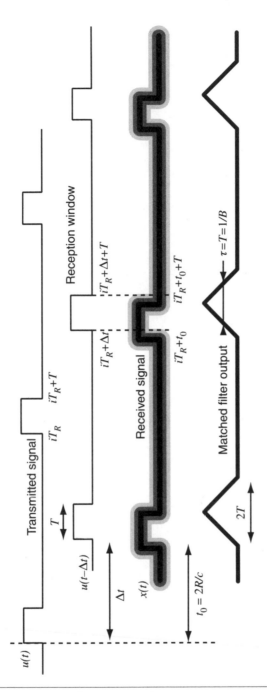

FIGURE 8.2 SIGNAL RECEIVED IN LOW PRF

In fact, as no information about the phase (or the Doppler velocity) of the transmitted signal is available, the phase of $x_i(\Delta t)$ remains unknown, and $E[x_i(\Delta t)] = 0$ so $c(\Delta t) = 0$.

The average of the receiver output signal is zero. Therefore, it is impossible to use a matched receiver as previously defined; the matched receiver is replaced by a simpler receiver composed of the following elements:

- a *filter matched* to the single pulse that calculates $x_i(\Delta t)$. At this stage, we assume that the radar is transmitting a single pulse and, considering the relatively short duration of T (< few μs), that the coherence of the entire system is sufficient from the point of view of the interpulse period
- an *envelope detector* that eliminates the carrier (and the phase problems)
- a *post-detection integration* system (or non-coherent integration), that accumulates the signal detected during the N interpulse periods of illumination time T_e

That means that the radar calculates the term:

$$c'(\Delta t) = \sum_{i=0}^{N} |x_i(\Delta t)|$$

which gives the block diagram described in the next sub-section.

8.2.2 NON-COHERENT RADAR BLOCK DIAGRAM

This radar is composed of the following:

- an antenna and duplexer
- a *transmitter;* a magnetron is generally used, whose oscillation frequency—which is the resonant frequency of its cavity—is relatively imprecise
- a microwave signal *mixer,* which converts the received signal frequency to a lower IF frequency band (some tens of Megahertz) where it can easily be amplified
- a *local oscillator,* which generates the local signal for the superheterodyne receiver. This oscillator frequency is locked on the transmitter frequency (shifted by the IF frequency) by means of an AFC (Automatic Frequency Control), allowing the received signal to remain centered in the IF band, independently of the frequency drift of the transmitter
- an *IF receiver* and the *matched filter;* the IF receiver amplifies the received signal to a level enabling envelope detection. In general, it also acts as a *filter matched to the pulse.* Section 8.2.2.1 discusses this function in detail
- an *envelope detector,* which eliminates the carrier and provides a monopolar signal
- *a CFAR* (Constant False Alarm Rate); this device regulates the noise level so that the false alarm remains constant (Section 8.2.2.2)

- a *post-detection integration*, which performs non-coherent post-detection integration of the data. Section 8.2.2.3 examines this
- a *detection device,* where a comparison is made with a threshold

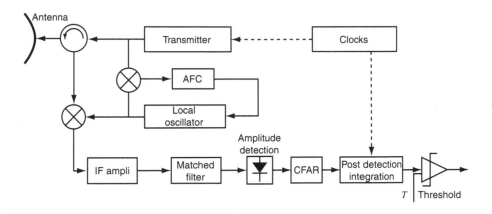

FIGURE 8.3 NON-COHERENT LPRF RADAR BLOCK DIAGRAM

8.2.2.1 FILTER MATCHED TO UNIT PULSE

As previously stated, in a standard radar the duration of the coherent processing is limited to the duration of the interpulse period.

Let us consider a radar operating only during the ith interpulse period. It transmits a single pulse of width T, ($u_i(t) = 1$ for $iT_R < t < iT_R + T$). The receiver matched to this radar performs the calculation

$$c_i(\Delta t) = \int_{T_R} x(t)u_i(t - \Delta t)dt = \int_{iT_R + \Delta t}^{iT_R + \Delta t + T} x(t)dt \cdot$$

In the case of a search radar, this computation must be carried out for each value of Δt of the range domain (with $n = T_R/\tau$ values by sampling the signal at the rate $\tau = r_t \approx 1/B \approx T$). This is implemented by means of the device described in Figure 8.4, representing a set of n correlators, for example.

Each correlator input is connected to $x(t)$ during time interval $[p\tau, (p + 1)\tau]$, and $c(p\tau)$ are available at the n correlator outputs.

However, as the number n of range cells is very high (several thousand), there is another way to proceed, which is to consider the matched receiver as a matched filter.

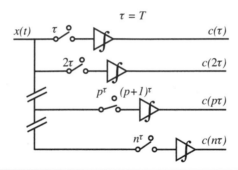

FIGURE 8.4 THEORETICAL MATCHED FILTER

The transfer function of such a filter is (see Chapter 6)

$$H(2\pi jf) = U^*(f),$$

where $U(f)$ is the Fourier transform of transmission modulation $u(t,)$ which is a rectangular pulse of duration T. At the filter output, the pulse has a triangular shape with duration $2T$ at the base. (It is the autocorrelation function of a pulse of width T whose maximum value is t_0, or more precisely, $t_0 + T$, for a real filter).

The maximum of the $R = S/N = $ SNR ratio is obtained for $\Delta t = t_0$. The ratio corresponds to $R = E/b$, where E is the energy received by the target during time T_R. This means that $E = P_{mr}T_R$, where P_{mr} is the received mean power. (E is also the energy of a single pulse $E = P_{cr}T$, where P_{cr} is the peak power, $P_{cr} = P_m T_R/T$.)

APPLICATIONS

The matched filter can be placed on video signal $x(t)$ (i.e., for the I and Q signals output from the demodulators), but it is usually more suitable to carry out filtering in the IF receiver.

This is a filter whose theoretical transfer function is

$$\left|H(2\pi jf)\right| = \left|\frac{\sin \pi Tf}{\pi Tf}\right|.$$

This filter is approximated by a bandpass filter whose bandwidth is $\Delta f \approx 1/T \approx B$, which results in better cancellation of the spurious signals located outside the bandwidth.

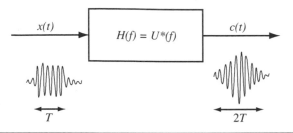

FIGURE 8.5 IF MATCHED FILTER

COMMENT 1

It should be remembered that SNR is *maximum* at this filter output.

This ratio, equal to $R = E/b$, is a known relationship that we can also determine in another way:

- The maximum signal received is the received peak power S = P_{cr}
- The noise power N is the product of noise spectral density b and filter bandwidth $\Delta f \approx 1/T \approx B$
- This results in

$$S/N = \frac{P_{cr}}{b\Delta f} = \frac{P_{cr}T}{b} = \frac{E}{b} \tag{8.1}$$

COMMENT 2

The matched-filter bandwidth, $\Delta f = 1/T$, is broad (several Megahertz) compared to the usual Doppler frequencies (some ten Kilohertz). So, a signal with a *non-zero Doppler frequency* will not be affected much by filtering. *The filter matched to zero velocity echoes is also appropriate for targets of any velocity.*

8.2.2.2 THE CFAR

The purpose of this device is to *normalize* the noise superimposed on clutter signals in order to be able to determine the detection threshold independently of the characteristics of this noise.

We have seen that the false-alarm rate is a very important feature of the radar specification, due to operational requirements. If the noise properties (its density function and its spectrum) were known, it could be possible to determine *a priori* the threshold needed to fix the false-alarm rate. As this is not the case in real life, we have to find the best value of this threshold. One solution would be to use some "gain margin."

This gain margin is not suitable in radar because it introduces unacceptable loss of detection, as shown on Figure 8.5a. In this case we face unexpected high-level noise superimposed to the thermal noise (weather clutter localized in range, for example). The threshold fixed by thermal noise

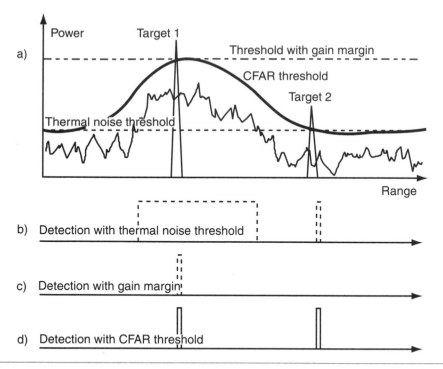

FIGURE 8.6   CFAR THRESHOLD

conditions (Figure 8.5b) enables you to detect with maximum sensitivity target 2, but the very high false-alarm rate in the clutter region prevents any detection of target 1, even if its power is much higher than the surrounding noise.

A gain margin (higher threshold, Figure 8.5c) avoids false alarms in a clutter region and enables the detection of target 1, but target 2 is lost.

The CFAR tries to locally adapt the threshold to the surrounding noise level (Figure 8.5d). It allows you to detect a target as soon as its level is sufficiently higher than the local noise. This threshold is calculated for each range bin from an estimation of the noise proprieties. It can be considered a "standardization" of the noise.

This standardization (a nonlinear operation that converts the noise-probability density into a known density, such as a Rayleigh function), assumes that the statistical properties of this noise are defined (function, mean value, standard deviation, etc.). Because, in practice, these properties are unknown, they are *evaluated from a set of samples* representative of the noise present in the cell in question (tested cell).

This set of samples is taken from the range cells in the vicinity of the tested cell. Its size has to be big enough to enable correct estimation of the noise properties but limited so it can take into account the nonstationary nature of noise, because the reference samples must remain close to the tested cell.

An ordinary device estimates the mean value from eight range cells and normalizes by dividing the signals by this mean value (see Figure 8.7).

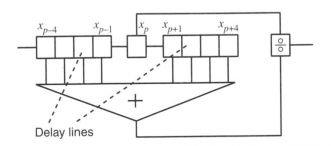

FIGURE 8.7 CFAR WITH SLIDING MEAN VALUE

The CFAR introduces into the processing procedure losses (some dB) resulting from the imperfect evaluation of the noise characteristics. False-alarm regulation performance depends on the nature and stationary aspect of the present noise. Moreover, using CFAR can desensitize the radar to nearby targets.

8.2.2.3 ENVELOPE DETECTION AND POST-DETECTION INTEGRATION

At the matched-filter output, all the useful phase information has been exploited and the envelope detector eliminates the subcarrier.

Then the receiver performs a non-coherent integration known as post-detection integration, whose function is to accumulate the energy in each range cell (for each value of Δt) from one interpulse period to the next during the illuminating time T_e (during the $N = T_e/T_R$ interpulse periods). See Figure 8.8.

APPLICATIONS

In modern radars, which can process a large amount of data by digital techniques, post-detection integration is performed by a sampled filter that performs a true integration from one interpulse period to the next, for each range cell (several thousand).

By contrast, with analog devices it was impossible to carry out this integration directly, and cumulative energy was obtained from the phosphor coating of a cathode ray tube on which the electrons accumulate,

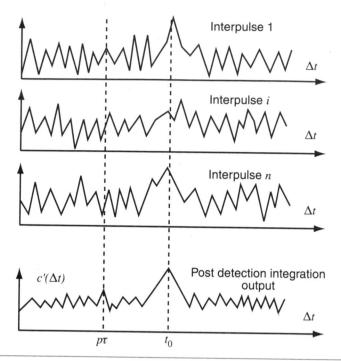

FIGURE 8.8 POST-DETECTION INTEGRATION

resulting in an enhancement of the plot brightness from one interpulse period to the next.

Obviously, this post-detection integration was not a mathematical summation of the information, and performance mainly depended on the tube settings that were available to the operator at that time.

In addition to detector losses, post-detection integration losses occurred, partly compensated for by the "correlation effect," which gave a typical "banana" shape to the useful plots.

8.2.2.4 TARGET DETECTION

Target detection is obtained by comparing the output signal from post-detection integration with the threshold, T, fixed by the imposed false-alarm probability.

Using post-detection integration results in losses not found in optimum processing. To evaluate these losses, the two processing procedures are compared, with all other conditions being equal (see Figure 8.9). In the optimum receiver, ratio R_0 is required to enable detection under the imposed conditions (P_D, P_{fa}). Optimum processing can be divided into two steps:

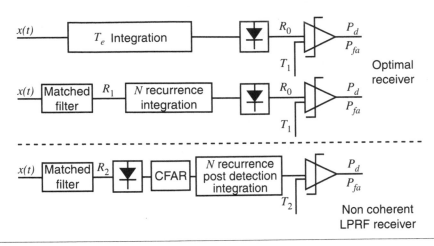

FIGURE 8.9 LPRF RADAR PERFORMANCE

- matching a filter to the pulse
- coherent integration on the N pulses received during time T_e

At the matched-filter output, this SNR is $R_1 = R_0/N$, where the coherent integration gain is N.

In order to obtain the same conditions (P_D, P_{fa}), the non-coherent LPRF receiver requires a greater SNR at the matched-filter output, R_2, depending on the number of post-detection integrated interpulse intervals, N.

Since the processing procedures are identical up to the matched-filter output, the ratio $l_t = R_1/R_2$ represents the processing loss introduced by post-detection integration. This loss can be expressed as a function of N and (P_D, P_{fa}).

The ratio $G(N) = R_0/R_2$ is the processing gain resulting from post-detection integration (or post-detection integration gain), which also depends on N and (P_D, P_{fa}). It takes the form $G(N) = N \cdot l_t$.

PROBABILITY OF DETECTION

The $G(N, P_D, P_{fa})$ function cannot usually be expressed analytically, but it can be computed digitally. Figure 8.10 shows some examples of detection curves representing P_D as a function of R_2, parametered in P_{fa}, for $N = 30$.

For usual values of (P_D, P_{fa}), it is convenient to use the empirical formula $G(N) = 2.3\sqrt{N} - 1.3$, which shows that post-detection integration gain varies asymptotically with respect to $\sqrt{N}$ instead of N for coherent integration.

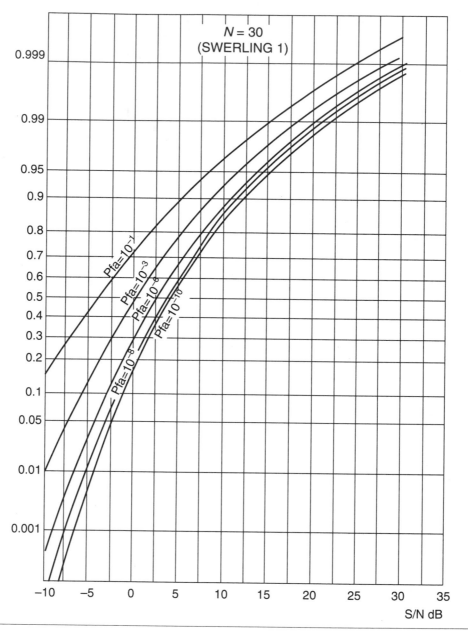

FIGURE 8.10 PROBABILITY OF DETECTION WITH POST-DETECTION INTEGRATION, N=30, SWERLING CASE I

Radar detection performance is calculated by reading the value of R_2 on the detection curves, and then applying these relationships:

$$R_2 = E_i/b = P_{mr}T_R/b = P_{cr}T/b$$

in order to obtain the P_{mr} or P_{cr} power required for reception.

NOTE ON DOUBLE THRESHOLD DETECTION

In certain applications, the post-detection integration can be replaced by a double threshold detection composed of (see Figure 8.11) the following:

- a first detection (threshold T_1), with, at the output, the probability of detection being p_{d0} and the probability of false alarm p_0
- a detection count, for each range cell, over N interpulse periods (with K being the result)
- a subsegment detection in which K is compared with a threshold M: If $K > M$, a target is detected. The probability of detection is P_d and the probability of false alarm is P_{fa}. The terms p_0, P_{fa} and p_{d0}, P_d, are related by a binomial function:

$$P_{fa} = \sum_{i=M}^{i=N} C_N^i p_0^i (1 - p_0)^{N-i} \approx C_N^M p_0^M \, ,$$

$$P_d = \sum_{i=M}^{i=N} C_N^i p_{d0}^i (1 - p_{d0})^{N-i} \, . \tag{8.2}$$

The calculations show that $p_0 >> P_{fa}$ and $p_{d0} < P_d$. The SNR required before the first detection threshold is therefore smaller than would be necessary for a single pulse. *Double threshold detection (or M-out-of-N detection) behaves like post-detection integration,* and, for usual values, the *equivalent SNR gain* is very close to that achieved by ordinary post-detection integration.

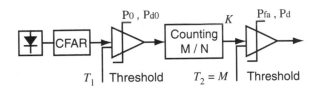

FIGURE 8.11 DOUBLE THRESHOLD DETECTION

8.3 PULSE-COMPRESSION RADAR

8.3.1 DEFINITION

The only difference between this radar and the previous one is the form of the transmitted pulse modulation.

We have seen that range resolution is determined by the transmission spectral bandwidth only. Therefore, by transmitting frequency- or phase-modulated pulses of reduced peak power but long duration, we can obtain range resolution equivalent to that of a radar transmitting signals at much higher peak power, with the mean power being equal.

A compression ratio can be defined: $N = BT$. This is the product of the transmission spectrum bandwidth and the pulsewidth, or the ratio between the pulsewidths before and after compression.

8.3.2 PULSE-COMPRESSION RADAR BLOCK DIAGRAM

The block diagram shown in Figure 8.12 differs from the previous one in two ways:

- The transmitter is controlled (and not self-oscillating).
- It is a microwave amplifier that amplifies a wave *dispersed* in a device that will be dealt with in the next section.

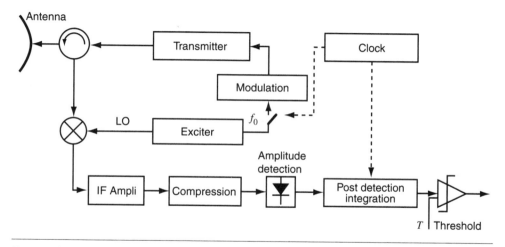

FIGURE 8.12 BLOCK DIAGRAM OF LPRF PULSE-COMPRESSION RADAR

On reception, the filter matched to the unit impulse acts as a transfer function,

$$H(2\pi jf) = U^*(f),$$

which is the conjugate transform of the transmitted pulse, therefore has a more complex form than a simple passband, and has the same function as the matched filter in Section 8.2.2.

The role of all the other components is comparable to that of a non-coherent radar, and performance calculation is carried out in the same way. The only difference would be that, if we were talking in terms of peak power, we would have to introduce the gain (N) of the matched filter (compression device).

8.3.3 PULSE-COMPRESSION SYSTEMS

The two mains classes of system examined in this section are

- linear frequency modulation (FM) compression
- phase-coding compression

8.3.3.1 LINEAR FM

Chapter 6 described this kind of compression. It consists of linear modulation of the transmitted frequency during the entire pulsewidth (see Figure 8.13). The instantaneous frequency varies between f_0 and $f_0 + \Delta F$, during T. This type of modulation can be obtained by means of a dispersive line that delays signals in different ways depending on their frequency. (A signal with bandwidth ΔF is dispersed by the line in a long pulse of duration T. The resulting bandwidth is then $B \approx \Delta F$.)

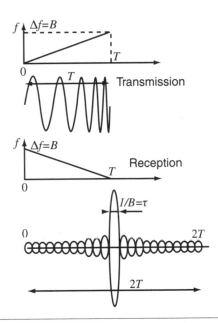

FIGURE 8.13 LINEAR FREQUENCY MODULATION

On reception, the signal is fed into a dispersive delay line whose delay-frequency characteristic is the inverse of that used for transmission.

As a result, all the frequency components are delayed globally (both on transmission and reception) by the same quantity T and added together to produce a narrow peak of duration $1/B$, accompanied by side lobes that should be reduced by weighting.

8.3.3.2 PULSE COMPRESSION BY PHASE CODING

In this case, pulse T is divided into N time intervals of equal duration $\tau = T/N$. For each interval (or subpulse), the phase is modified in accordance with a pre-established code. This code can be binary-phase (0 or π) or polyphase. At the receiver input, the signal is *correlated* in phase with a replica of the transmitted signal and performs the sum of the subpulses readjusted in phase. This type of processing is perfectly suited to the digital techniques used in modern radars (see Figure 8.14).

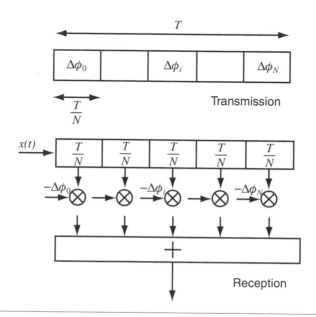

FIGURE 8.14 PULSE COMPRESSION BY PHASE CODING

The advantage of phase coding is that the waveform can easily be modified, having no link with a physical device such as the Surface Acoustic Wave (SAW) lines used in compression by frequency modulation.

BINARY-PHASE CODING (BARKER CODES)

The main characteristic of this waveform is that it can be coded (0 or π) and decoded with great simplicity. (A phase difference of π is obtained by modifying the signal sign.)

The ratio between the amplitude of the compression side lobe and that of the main lobe can be limited to $1/N$ if an adequate code is chosen. These codes are referred to as *Barker codes;* they exist for $N = 2, 3, 5, 7, 11, 13$ subpulses.

The limitations of these codes are

- the use of low compression ratio. Higher compression ratio can be obtained by interlacing Barker codes, but if this is done, some spurious lobes will be greater than $1/N$
- their sensitivity to echo Doppler frequency f_D, when f_D can no longer be neglected compared with $1/T$

POLYPHASE CODING

A polyphase code is a digital approximation of linear FM when the continuous quadratic phase variation is replaced by discrete phase steps. There are various types of codes (Franck codes, P_1, P_2, P_3, P_4 codes) detailed in the work of Kretschner and Lewis (1983).

Although the implementation of these codes is more complex than that of Barker codes, polyphase codes are increasingly used because

- they have no length limitation
- like linear FM, they are virtually insensitive to the echo Doppler frequency

8.4 LOW-PRF DOPPLER RADARS (MTI)

8.4.1 DEFINITION

Low PRF and non-coherent radar gives good results as long as clutter level is low with respect to the targets. As soon as ground (or sea) echoes are in the same range resolution cell as the wanted echoes, that is, as soon as the antenna beam scans the ground, clutter becomes predominant and masks the target echoes.

In particular, if the target is at low altitude or is ground-based, the antenna beam will not be able to separate it from the clutter, and the velocity difference (Doppler frequency) of these echoes will have to be used to discriminate them.

The LPRF Doppler radar is a coherent system, range-unambiguous, that cancels the clutter received by the antenna main lobe by Doppler filtering.

8.4.2 COHERENT LOW-PRF RADAR THEORETICAL ANALYSIS

The transmitter is coherent. The signal received by a target with (t_0, f_D) as parameters is

$$x(t) = Au(t - t_0)e^{j2\pi f_D t} + n(t),$$

where $u(t)$ (the transmitted signal) is a set of consecutive pulses of duration T and period T_R (Figure 8.15).

As for non-coherent radars, we shall first use a receiver matched to a unit pulse.

At interpulse period i, the operation performed by the receiver is:

$$x_i(\Delta t) = \int_{iT_R + \Delta t}^{iT_R + \Delta t + T} x(t)dt \ .$$

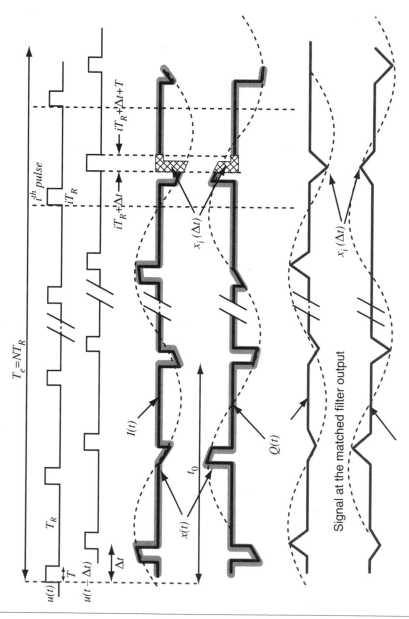

FIGURE 8.15 TRANSMITTED AND RECEIVED SIGNALS IN A COHERENT PULSE RADAR

Considering the useful signal only and assuming that, in most cases, $Tf_D \ll 1$ (which means that the phase variation $2\pi f_D t$ is negligible in integration interval T), we can write

$$x_i(\Delta t) = \int_{iT_R + \Delta t}^{iT_R + \Delta t + T} A u(t - t_0) e^{j2\pi f_D t} dt \approx e^{j2\pi f_D(iT_R + \Delta t)} A \int_{iT_R + \Delta t}^{iT_R + \Delta t + T} u(t - t_0) dt$$

and

$$x_i(\Delta t) \approx e^{j2\pi f_D(iT_R + \Delta t)} c_i(\Delta t) ,$$

where

$$c_i(\Delta t) = A \int_{iT_R + \Delta t}^{iT_R + \Delta t + T} u(t - t_0) dt$$

is the signal output from the filter matched to the pulse mentioned in Section 8.2.

As previously stated, it is a triangular signal.

This signal is multiplied by the phase term: $\varphi_i = 2\pi f_D(iT_R + \Delta t)$.

If the target speed is zero ($f_D = 0$), the phase remains constant and the target echo is identical from one interpulse period to the next. If a moving target is involved, the signals received at the instant $\Delta t = t_0$ is modulated by a sinusoid with frequency f_D (modulo F_R). It will therefore be possible to discriminate moving targets from fixed targets using simple high-pass filtering, which cancels targets whose speed equals 0 modulo F_R.

This technique is known as Moving Target Indication (MTI).

The filter used for clutter rejection is the *Doppler filter.*

At the Doppler filter output, the signal is processed in the same way as in a non-coherent and LPRF radar. The next section shows the block diagram of a typical MTI system.

8.4.3 MTI BASIC BLOCK DIAGRAM

The fundamental differences between the MTI block diagram (see Figure 8.16) and that of the non-coherent radar are in

- the *transmitter,* which is of the coherent type: it is a microwave amplifier whose noise, phase, and amplitude characteristics must be compatible with the expected improvement ratio

- the *exciter*: this provides the transmitted signal (f_0) that is modulated onto pulses, and the transposition (or demodulation) signal(s), whose spectral purity must also be compatible with the improvement ratio
- the *Doppler filter*, which eliminates echoes with low Doppler frequency. The next section studies its characteristics

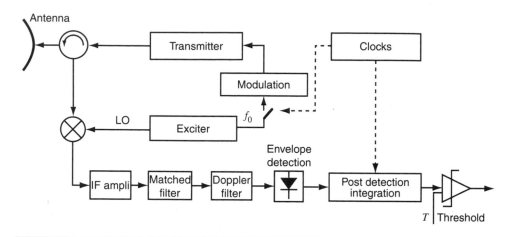

FIGURE 8.16 COHERENT LPRF BLOCK DIAGRAM (MTI)

Chapter 7 explained the very severe constraints imposed on the transmission-reception chain.

Apart from these differences, the MTI processing procedure is the same as that of a non-coherent radar, that is, it consists of

- *filter matching* to the elementary pulse T (performed in IF or video on the I and Q channels)
- *amplitude detection*, which eliminates the carrier (or the subcarrier) at the Doppler filter output
- *a CFAR device* that normalizes the noise statistical properties
- *post-detection integration*, which accumulates the detected signal from one interpulse period to the next
- a *comparison with a threshold T*, calculated to guarantee the false-alarm probability, P_{fa}

DOPPLER FILTER (CLUTTER CANCELLER)

The function of this device is to eliminate low Doppler-frequency signals.

The operation performed by the simplest filter is a subtraction of x_{i-1} (t_0) from $x_i(t_0)$, as shown in Figure 8.17.

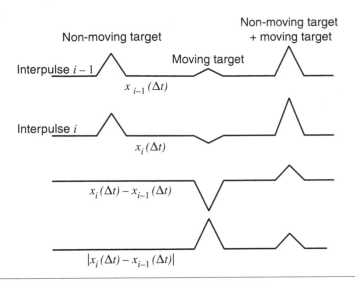

FIGURE 8.17 TWO-PULSE CANCELLER SIGNALS

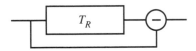

Theorical block-diagram

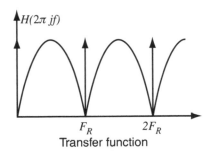

Transfer function

FIGURE 8.18 TWO-PULSE CANCELLER TRANSFER FUNCTION

At the receiver matched to T output, the signal is fed into a delay line of duration T_R. On output, the signals of the i and $i-1$ interpulse periods are subtracted.

Components of *zero* frequency, which are constant from one interpulse period to the next are eliminated. This Doppler filter is called a two-pulse canceller.

The transfer function is given by the Z transforms:

$$S(Z) = (1-Z^{-1})\, e(Z)$$
$$H(2\pi jf) = 2\sin\pi T_R f \text{ (see Figure 8.18)}$$

This is a periodic filter, or comb-filter, which cancels all the kF_R components of a signal of zero frequency.

In general, this filter does not have sufficient frequency-rejection bandwidth to efficiently cancel ground echoes that have a specific spectral width (such as wind-blown vegetation), or which are induced by platform motion (see Section 8.3.5). Its complexity is increased when it acts on several consecutive interpulse periods: three interpulse periods for a three-pulse canceller (two single cancellers in cascade), or more for a Doppler filter capable of rejecting a large frequency bandwidth.

8.4.4 ADDITIONAL MTI CONSIDERATIONS

BLIND SPEEDS

It has been explained that the Doppler filter not only cancels a frequency bandwidth centered around $f_D = 0$, but, because of the many frequency ambiguities, it also eliminates all the multiple frequencies of F_R (i.e., $f_D = kF_R$).

The corresponding target speeds are known as *blind speeds*.

To overcome this difficulty, the value of F_R is changed periodically; that is, F_R is staggered.

8.4.5 AIRBORNE MTI (AMTI)

When the radar platform is moving with respect to ground echoes, the clutter signals have nonzero radial velocity relative to the radar. Chapter 5 showed that the ground clutter spectrum corresponding to the echoes received by the main lobe is shifted, in mean value terms, by

$$f_p = \frac{2v}{\lambda} \cos\theta_E \cos\theta_A,$$

and its spectral width is

$$\Delta f = \frac{2v}{\lambda} \cos\theta_E \sin\theta_A \theta_{0A}.$$

Clutter cancellation is achieved by

- centering the rejection bandwidth of the Doppler filter around f_p. It is often preferable to transpose the entire spectrum of the received signal by f_p in order to bring the ground echoes down to around zero
- increasing the rejection area, which means selecting a relatively high F_R value. This in turn means a reduction in the non-ambiguous range domain and, therefore, in radar range (by definition, range-unambiguous)

Although different techniques exist for reducing the effect of the spectrum enlargement (Displaced Phase Center Antenna techniques, discussed in Chapter 10), this last point severely limits airborne MTI systems applications, and other waveforms have to be used for airborne applications in look-down air-to-air detection.

8.5 HIGH-PRF RADAR

The previous section explained that radars without ambiguity perform satisfactorily as long as no ground clutter is received by the antenna main lobe or the clutter spectrum remains small compared to the repetition frequency (as in ground-based radars, for example).

When the radar platform is moving with respect to the ground, however, there is a spread of the clutter spectrum, and radar repetition frequency F_R needs to be increased in order to discriminate desired echoes from clutter by Doppler filtering. This increase gives rise to range ambiguities. New difficulties appear along with these ambiguities:

- Due to range folding, it may be that the clutter and the target echoes are received in the same resolution cell, even if the clutter is much closer to the radar. In this case, even if the clutter echoes are not located in the antenna main lobe, they can reach a level that competes with that of the target, the $[R^{-4}]$ function of the radar equation acting in their favor and compensating for antenna gain attenuation
- As range is one of the major pieces of information a radar has to provide, some way has to be found to eliminate these ambiguities

The previous chapter demonstrates that clutter spectrum is limited to the frequency region $[-2V/\lambda, 2V/\lambda]$ and *no ground clutter can exist* outside this region if F_R is sufficiently high. As a result, in many important operating configurations (head-on target), the target can be separated from ground clutter.

These radars are called High Pulse-Repetition Frequency (HPRF) and they are used in modern interceptor aircrafts.

We shall first study the Continuous Wave (CW) radar, which of course has no velocity ambiguity (at least, as long as its transmission frequency is not modulated), and a direct derivative of which is the 0.5-duty cycle HPRF radar.

8.5.1 Continuous Wave (CW) Radar

8.5.1.1 Definition and Processing

The waveform employed in continuous wave (CW) radar is a continuous wave that, as a first stage, will be assumed to have fixed frequency f_0.

If T_e is the illumination time (> > a few ms), the ambiguity function has a lobe with frequency bandwidth $1/T_e$ and range width T_e.

The frequency resolution, $1/T_e$, may be excellent, but the range resolution, $cT_e/2$, is practically zero; and a radar with such a configuration *will not be able to deliver useful range information*. Therefore, only the frequency domain is of interest. For each value of Δf of the *frequency domain*, the optimum receiver will perform the following calculation (see Chapter 6):

$$c(\Delta f) = \int_{T_e} x(t) e^{-j2\pi\Delta ft} dt \qquad (u(t) = 1)$$

This corresponds to the *Fourier transform* of the received signal, which is equivalent to a bank of contiguous filters whose bandwidth is $\Delta F = 1/T_e$, that is,

$$N = \frac{F_{max}}{\Delta F}$$

filters, where F_{max} is the target frequency domain.

Each filter bank output is compared with a detection threshold that depends on the probability of false alarm, P_{fa}.

In practice, this type of radar, which gives no range indication and requires two antennas (*simultaneous transmission and reception with the same antenna is impossible due to the dynamic range involved*), would have a few direct applications. Frequency modulation would be introduced to give the radar the range measurement capability.

8.5.1.2 Modulated CW Radar (Range Measurement)

If frequency f_0 is linearly modulated, and, provided the transmitted spectrum has broad bandwidth B, excellent range resolution $r = 2c/B$ can be obtained; but confusion can arise between the velocity and the range measurements. If range t_0 of the echo is known, the Doppler frequency can be deduced. Conversely, knowledge of the echo Doppler frequency enables the measurement of range t_0.

As an example, in the case of fixed targets ($f_D = 0$), it is possible to define the echo range by measuring the frequency of the received signal (the filter number indicating the presence of the target). This property is implemented in certain altimeters.

Both f_D and to can be measured by reversing the slope of the frequency modulation (see Figure 8.19).

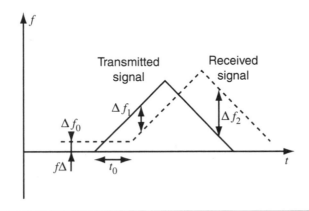

FIGURE 8.19 RANGE MEASUREMENT BY FREQUENCY MODULATION

Frequency Δf_1 is measured in one direction and Δf_2 in the other.

Assuming $t_0 = 0$, the difference between the transmitted and received frequencies is due to the target Doppler frequency, f_D, and $\Delta f_1 = \Delta f_2 = f_D$.

If $f_D = 0$, this difference is caused by transmission delay t_0 and $\Delta f_1 = \Delta f_2 = k t_0$ (where k is the modulation slope).

In most cases, the following relationships can be written as

$$f_D = \frac{\Delta f_1 + \Delta f_2}{2} \text{ and } t_0 = \frac{1}{k}\frac{\Delta f_1 - \Delta f_2}{2}.$$

These relationships provide both range and velocity measurements (but with less accuracy because illumination time T_e is divided by 2).

8.5.2 0.5-DUTY CYCLE, HIGH-PRF RADAR

8.5.2.1 DEFINITION

We can consider 0.5-duty cycle, high-PRF radar a CW radar whose antenna is multiplexed between transmission and reception.

Consequently, the waveform is a continuous wave (CW) with a duty cycle of 0.5, with half of the period being dedicated to transmission and the other to reception.

The modulation frequency (repetition frequency) is F_R, the period is $T_R = 1/F_R$, the pulse length is $T = T_R/2$ and the reception duration is T_e.

The transmitted (and received) signal is therefore *sampled* at frequency F_R, which must be sufficiently high to obey the sampling theorem; that is, $F_R > F$, where $F = f_{max} - f_{min}$ is the domain of the *frequencies* to be received (usually, $F_R > 100$ kHz).

Under these conditions, the radar has no velocity ambiguity, and the structure of this radar is similar to the CW radar (see Figure 8.20).

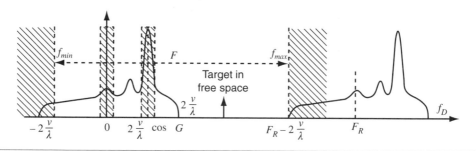

FIGURE 8.20 HIGH-PRF MODE SPECTRUM

COMMENT

The illumination time, T_e, generally produces a very narrow Doppler filter bandwidth, $\Delta F = 1/T_e$, at least compared to the *target's own spectrum* (which has been considered negligible in all the previous calculations). It follows that target energy is spread over several adjacent filters, and the receivers are no longer matched to the signal. To avoid this, (and to limit the number of Doppler filters), ΔF is defined in accordance with the target spectrum (some hundred Hertz), resulting in a reduction of coherent processing time T_c. At the Doppler filter output, processing continues with envelope detection followed by post-detection integration, as in MTI systems.

8.5.2.2 0.5-DUTY CYCLE, HIGH-PRF RADAR

Figure 8.21 shows the block diagram for 0.5-duty cycle, high-PRF radar. Note the following:

- The transmitter is coherent, driven by a pulse-modulated wave with a duty cycle of 0.5 supplied by the exciter
- The same antenna is used for transmission and reception

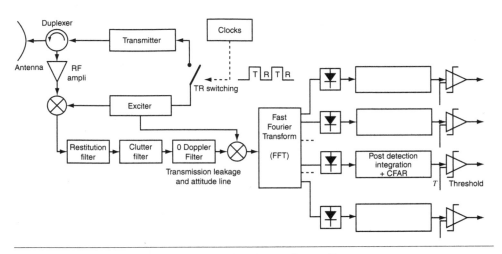

FIGURE 8.21 0.5-DUTY CYCLE, HIGH-PRF RADAR BLOCK DIAGRAM

- At the receiver input, the signal is switched so it is in phase opposition to transmission pulse
- The IF receiver has a set of filters that will be examined in Section 8.5.2.3
- The signal is demodulated in phase and quadrature
- The bank of N filters (of ΔF elementary bandwidth) covers frequency bandwidth

$$F\left(N = \frac{F}{\Delta F}\right)$$

- Its correlation time is

$$T_c = \frac{1}{\Delta F}$$

- It is smaller than the total illumination time, T_e. At the Doppler filter output

$$h = \frac{T}{T_c}$$

- *Decorrelated* (independent) samples are available and will be processed as in a LPRF radar, that is, with post-detection integration on h samples, carried out after envelope detection. These two processes result in processing losses that do not occur in the coherent processing that would be performed if $\Delta F = 1/T_e$
- Signal detection is achieved by a comparison with a threshold at each post-detection integration output

8.5.2.3 IF FILTERS

The receiver includes a set of filters whose function is explained in Figure 8.18 and which represents the spectrum of the received signals.

CW Restitution Filter

This bandpass filter, with a bandwidth of F (Doppler range of expected echoes) m eliminates the *lines produced by sampling* around $f_0 + kF_R$ (where k is a positive or negative integer, not zero).

At this filter output, waves are continuous and the signal is the one we would have obtained with a CW radar transmitting line f_0 only. The signal output from this filter is processed in exactly the same way as in a CW radar.

Transmission Leakage and Altitude-line Rejection Filter

This bandpass filter centered on f_0 cancels direct leakage from the transmitter to the receiver (f_0)—through the duplexer and the reception switch—and altitude-line.

This function could be performed by the filter bank (by eliminating the filter for f_0), but it is preferable to reduce the dynamic range of these spurious signals by pre-filtering (noise whitening), which facilitates the design of the filter bank.

Main Lobe Rejection Filter

The purpose of this filter is to reject the very strong clutter signals received by the main lobe. In this frequency zone, centered around

$$f_0 + \frac{2V}{\lambda}\cos\theta,$$

(where θ is the angle between the velocity vector and the radar-target axis), clutter signals are much stronger than the desired returns, and detection is impossible. It is preferable to eliminate this zone before the filter bank, for the same reasons as previously explained (dynamic range reduction, constraints on the Doppler filter).

8.5.2.4 Performance Calculation, Processing Losses

Consider a target with Doppler frequency

$$f_D > \frac{2V}{\lambda}.$$

In *free space*, and assuming that the spectral purity of the transmission-reception chain is sufficient (see Chapter 7) target detection is limited by the thermal noise only.

ECLIPSE-FREE PROCESSING

If the target is located at $t_0 = kT_R + \tau = \left(k + 1/2\right)T_R$, reception of the return signal coincides exactly with the period in which the receiver is open. There is no eclipse loss, and the overall reflected energy is processed by the receiver (Figure 8.22). The possible presence of a post-detection integration should be considered.

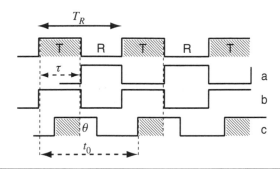

FIGURE 8.22 ECLIPSES

If the coherent time is T_e, which means there is no post-detection integration, we are dealing with the optimum receiver. At Doppler filter output Δf_i, centered on the target (f_D), the signal-to-noise ratio is

$$R = \frac{S}{N} = \frac{E}{b},$$

where E is the energy received during T_e ($E = P_c T_e/2$), and b is the noise spectral density.

In fact, we know that in order to obtain a sufficient rejection ratio, *tapering* is required (see Chapter 7), which introduces non-negligible weighting losses, *lp*.

In most cases, post-detection integration (and CFAR) is present. As previously stated, the entire reception duration of signal T_e (the time during which the target is illuminated by the antenna beam) is relatively long and would lead to a filter bandwidth of $\Delta F = 1/T_e$, which is too narrow in comparison with the target's own spectrum.

This is why, in this type of radar, the coherent processing time is deliberately limited to T_c (some milliseconds), where $T_c \ll T_e$ is the correlation time of the filter whose bandwidth is $\Delta F = 1/T_c$ (a few hundred Hertz).

Several coherent and independent measurements are therefore available during illumination time T_e, where $h = T_e/T_c$.

To improve the detection characteristics, the h measurements are post-integrated following envelope detection and prior to comparison of the signal with the detection threshold, as in LPRF systems.

This post-detection integration of h samples causes losses in noncoherent processing l_t (see Section 8.2.2.3).

Calculations are carried out in the same way as for LPRF radar: As h, P_d and P_{fa} are known, the value of R *before envelope detection,* and therefore at *Doppler filter output,* can be determined by means of the detection curves.

ECLIPSING

For $t_0 = kT_R$ (see Figure 8.22b), signal reception is impossible, as the echo is in phase with the transmission. The corresponding ranges are known as *blind ranges.* This is the *eclipse* phenomenon.

Apart from these blind ranges, the level of the received signal varies in accordance with the value of t_0, and partial eclipses occur (Figure 8.22c).

The loss in average power with regard to the maximum value is

$$l_{e'} = \int_0^1 \left(\frac{\theta}{T}\right)^2 d\frac{\theta}{T} = \frac{1}{3}.$$

Eclipsing results in an average loss of power of 5 dB. In fact, the real loss has to be estimated by calculating the SNR increase necessary to restore the probability of detection.

In addition, this eclipse loss is increased by the presence of *dead times* preceding and following the transmission pulse. The *blanking* of the receiver surrounding this pulse reduces θ by the same amount.

8.5.3 RANGE MEASUREMENT

8.5.3.1 FREQUENCY MODULATION

In general, range measurement is achieved by modulation of transmission frequency f_0, as in CW radars.

The illumination time, T_e, is divided into two or three periods (Figure 8.19). During these periods, detection, range, and speed measurement is performed as presented in Section 8.5.1.2.

Obviously, as range extraction requires *several detections* during T_e, performance is *reduced* to less than that when using the operating mode without range measurement (known as *velocity search*).

8.5.3.2 FSK MODULATION

We have shown (Section 3.5) that the received signal, $s(t)$, includes a constant phase term,

$$\varphi = -2\pi \frac{2R_0}{\lambda} = -2\pi \frac{2R_0 f_0}{c} .$$

which depends on target range R_0 and carrier frequency f_0 only. If two carrier frequencies (f_0 and $f_0 + \Delta f$) are used, the difference of their respective phases is

$$\Delta \varphi = 2\pi \frac{2R_0 \Delta f}{c} .$$

The measurement of Δf gives the non-ambiguous range R_0 if $\Delta \varphi < 2\pi$ or $\Delta f < c/2R_{max}$, where R_{max} is the maximum target range.

This principle can be implemented by interleaving f_0 and $f_0 + \Delta f$ from one pulse to the other.

8.6 PULSE-DOPPLER MODE (HIGH- AND MEDIUM-PRF)

8.6.1 DEFINITION

Radars referred to as *pulse-Doppler* are systems that transmit a train of coherent pulses (of pulsewidth T and period T_R) in order to detect moving targets masked in clutter by range-velocity selectivity, in accordance with the principles described in the previous chapter. In fact, this category embraces coherent LPRF radars, all the high-PRF radars, and medium-PRF radars.

The signal received during illumination time T_e is composed of N pulses, where $N = T_e/T_R$ (Figure 8.13).

It is generally assumed that the radar is ambiguous in both range and velocity.

We shall study the radar behavior with respect to a target echo located at t_0 and f_D, and assuming initially that the pulses are without phase or frequency coding.

8.6.2 IDEAL PULSE-DOPPLER RECEIVER

The ideal receiver calculates the term

$$C(\Delta t, \Delta f) = \int_{T_e} x(t) u^*(t - \Delta t) e^{-j2\pi \Delta f t} dt$$

(Chapter 6) for every possible echo position; that is, for every value of Δt and Δf of the range-velocity domain, the $C(\Delta t, \Delta f)$ global maximum must be sought in this domain.

However, Chapter 6 showed that range-velocity ambiguities result from the impossibility of determining the global maximum from all the maximum values of $C(\Delta t, \Delta f)$.

Consequently, at this stage of detection, we are obliged to limit our calculations to the ambiguous range-frequency domain (limited by T_R and F_R) and to look for the maximum of $C(\Delta t, \Delta f)$ in this domain.

Assuming that $r_f = 1/T_e = \Delta f$ is the resolution capability in frequency (in velocity) and $r_\tau \approx \tau \approx 1/B \,(= T)$ is the resolution capability in time (range), we can quantify the ambiguous domain in

$$m = \frac{F_R}{\Delta F} = \frac{T_e}{T_R} \ (=N)$$

frequency cells (or Doppler cells) and in

$$n = \frac{T_R}{\tau} \ (= \frac{T_R}{T})$$

time cells (or range cells).

$C(\Delta t, \Delta f)$ is determined in $m \cdot n = T_e/\tau$ points such that

$$\Delta t = p\tau \ \text{and} \ \Delta f = q\Delta F = q\frac{F_R}{m}.$$

The couple (p, q) represents a range-velocity resolution cell.

Thus

$$C(p, q) = \int_{T_e} x(t) u(t - p\tau) e^{-j2\pi q \Delta F t} dt \ .$$

Taking into account the form of $u(t)$ (Figure 8.13), this can be written as

$$C(p, q) = \sum_{i=0}^{m-1} \int_{iT_R + p\tau}^{iT_R + p\tau + T} x(t) e^{-j2\pi q \Delta F t} dt \quad ,$$

considering that

$$e^{-j2\pi q \Delta F t} \approx e^{-j2\pi q \Delta F (iT_r + p\tau)}$$

for

$$i T_r + p\tau \le t \le i T_r + p\tau + T$$

and

$$C(p, q) = e^{-j2\pi q \Delta F \tau} \sum_{i=0}^{m-1} e^{-j2\pi q \Delta F i T_r} \int_{iT_R + p\tau}^{iT_R + p\tau + T} x(t) dt ,$$

with

$$\Delta F T_R = \frac{T_R}{T_e} = \frac{1}{N} = \frac{1}{m}$$

and

$$C(p, q) = e^{-j2\pi p q \Delta F \tau} \sum_{i=0}^{m-1} e^{-j2\pi \frac{qi}{m}} x_{ip} ,$$

where

$$x_{ip} = \int_{iT_R + p\tau}^{iT_R + p\tau + T} x(t) \, dt$$

represents the correlation between $x(t)$ and the ith pulse of $u(t - p\tau)$.

x_{ip} is the sampling system output, at interpulse period i, of a sampler measuring signal $x(t)$ filtered by the filter matched to the unit impulse, in pth range gate $p\tau$.

The operation

$$\sum_{i=0}^{m-1} e^{-j2\pi q i} x_{ip}$$

is a discrete Fourier transform (DFT) of the x_{ip} samples.

Therefore, the *ideal quantified receiver* consists of the following:

- sampling the signal $x(t)$ through a series of n range gates $[p\tau, (p+1)\tau]$ for p ranging from 0 to $n-1$
- performing for each range cell p a discrete Fourier transform of the $m = N = T_e/T_R$ samples
- finding the maximum value from the $n \cdot m = T_e/\tau$ ambiguous range-velocity cells and comparing it with a detection threshold T

Note that, as with the 0.5-duty cycle, HPRF radar, coherent processing is not carried out over the entire duration of the measurement T_e (for target's own spectrum reasons). The coherent processing time is limited to T_c where $T_c \ll T_e$ (T_c = a few milliseconds).

The pulse train T_e is divided into $h = T_e/T_c$ bursts of $m = T_c/T_R$ interpulse periods ($m \cdot h = N$) (Figure 8.23).

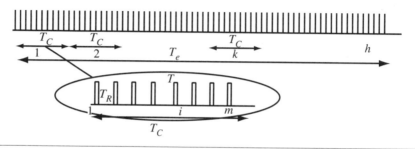

FIGURE 8.23 PULSE TRAIN IN A PULSE-DOPPLER RADAR

The coherent processing previously described (i.e., $C_k(p, q)$ for the kth burst) is carried out, with T_e *being replaced by* T_c *and* N *by* m. The resulting frequency resolution is $\Delta F' = 1/T_c$ (a few hundred Hertz).

Post-detection integration is then performed:

$$C'(p, q) = \sum_{k=1}^{h} |C_k(p, q)|,$$

which is identical to the HPRF radar post-detection integration.

$C'(p, q)$ is then compared with a detection threshold.

This operation produces non-coherent processing losses in contrast to the previous case, exactly like in LPRF radars (see Section 8.1).

8.6.3 PULSE-DOPPLER RADAR BLOCK DIAGRAM

Figure 8.24 presents a typical block diagram of a pulse-Doppler radar. The coherent transmission and reception part is classic: it supplies the I and Q signals output from phase-amplitude demodulators, representing the projections of $x(t)$ along orthonormal axes.

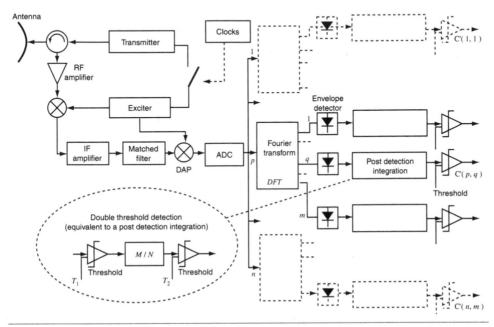

FIGURE 8.24 PULSE-DOPPLER RADAR BLOCK DIAGRAM

The signal is then sent to the filter matched to the pulse (or compression-pulse circuit) and to the sampling device that supplies the terms:

$$x_{ip} = \int_{iT_R+p\tau}^{iT_R+p\tau+T} x(t)\, dt$$

at interpulse period i.

More details of this process are given in the next section.

The x_{ip} terms are then analyzed by a device that performs a Discrete Fourier Transform (DFT) of the m samples, x_{ip}, of the k bursts. For each burst k and for each value of p, m, C_k (p, q) frequency samples are provided by the device (that is $m \cdot n$ terms).

Each term is then detected $(|C_k$ $(p, q)|)$ and post-integrated burst to burst, with k varying from 1 to h during observation time T_e.

Processing continues with the calculation

$$C'(p,q) = \sum_{k=1}^{h} |C_k(p,q)| .$$

Finally, these terms are compared with a detection threshold.

> **Note:** In the medium PRF modes, the eclipse elimination and the resolution of ambiguities result in a modification of the recurrence frequency, and the post-detection integration is consequently replaced by a double threshold detection (Sections 8.2 and 8.6.6).

8.6.4 RANGE GATE SAMPLING

The range gate sampling part of the processing involves the calculation of the terms

$$x_{ip} = \int_{iT_R+p\tau}^{iT_R+p\tau+T} x(t)\, dt ,$$

in which i and p are the interpulse period and the range cell numbers, respectively. This integration operation on received signal $x(t)$ during gate p ($p\tau$ to $p\tau + T$) has already been described. It can be seen from two standpoints:

- the *correlation* standpoint (Figure 8.4), which is the $x(t)$ integration in the range gate p
- the *matched filtering* standpoint (Figure 8.5), which has been mentioned for low-PRF radar and which represents a filtering of $x(t)$ matched to T followed by sampling at instant $p\tau$

Both operations gives the same results. When the pulses are not frequency-modulated, and the sampling rate is $\tau_e = \tau$, the response of the device relative to t'_0 (ambiguous range) varies in a sawtooth fashion between the maximum values (for $t'_0 = p\tau = pT$) and the mid-amplitude values (except for $t'_0 = 0 \bmod (T_R)$, where it equals zero).

Figure 8.25 illustrates the case $T_R = 4\tau$, when the sampling rate equals the pulse duration (i.e., when $\tau_s = \tau$).   The values $t'_0 = 0 \bmod (T_R)$ are the *blind ranges*, that is to say, the *eclipses* (see Section 8.5.2.4) produced when the receiver is blanked during transmission.

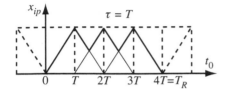

FIGURE 8.25 RANGE SAMPLING

The average sampling loss (range gate straddle range) is

$$l_{st} = \frac{1}{T_R} \int_0^{T_R} |x_i(t'_0)|^2 dt'_0 \ .$$

If

$$n = \frac{T_R}{T},$$

considering symmetries and periodicities, we can write

$$l_{st} = \frac{1}{n} \left[2\int_0^1 t^2 dt + 2(n-2) \int_0^{0.5} \left(\frac{1}{2}+t\right)^2 dt \right] = \frac{2}{3n} + 2\frac{n-2}{3n}\left(\frac{7}{8}\right) = \frac{1.75n - 1.5}{3n} \ .$$

This sampling loss tends towards 2 dB when n is high, and the matched filter is the ideal filter (in $\sin x /x$). Actually, the filter is usually replaced by a more specific filter (Section 8.2.2.1) that introduces approximately 1 dB of losses for the maximum value but allows some widening of the correlation peak, thus reducing sampling losses. Overall loss is of the order of 2 dB, and the matched filtering plus sampling must be considered as a whole in which the corresponding loss is 2 dB, regardless of the type of filter.

The matched filter + sampling loss could be reduced by oversampling such as $\tau_s = \tau/2$ (overlapping gate); in this case, the loss would become 1 dB (but computing power has to be doubled).

If the oversampling is further performed, sampling loss remains around 1 dB.

The values of matched filter + range sampling loss to retain are

$$\tau_s = \tau \qquad l_{st} \approx 2 \text{ dB}$$

$$\tau_s \le \tau/2 \qquad l_{st} \approx 1 \text{ dB}$$

8.6.5 FREQUENCY ANALYSIS

The operation that follows range sampling is the Discrete Fourier Transform (DFT):

$$C(p,q) = \sum_{i=0}^{m-1} e^{-j2\pi \frac{qi}{m}} x_{ip},$$

which is carried out on $m = T_c/T_R$ samples of x_{ip} (m interpulse periods).

It should be remembered that DFT, which is a circular convolution, is a bijection of the $m\, X_q$ terms on the $m\, x_i$ terms by the relation

$$X_q = \sum_{i=0}^{m-1} W^{qi} x_i,$$

where W is the mth root of the unit

$$W = e^{\frac{-j2\pi}{m}}.$$

Algorithms exist for calculation of this DFT, and they eliminate a considerable number of operations; the Fast Fourier Transform (FFT) is the most commonly used.

In order to lower the side lobes, signal $x(t)$ needs to be tapered (see the previous chapter). The corresponding calculation is

$$C(p,q) = \sum_{i=0}^{m-1} e^{-j2\pi \frac{qi}{m}} w_i x_{ip}.$$

This results in a *weighting loss*, linked to the widening of the Doppler filter, accompanied by a sampling loss in frequency (*filter straddle loss*) due to the fact that the peak value is not taken into account when $f_D \neq qF_R/m$. Overall losses are $l_f \geq 2$ dB.

8.6.6 ECLIPSE AND AMBIGUITY ELIMINATION

A pulse-Doppler radar with range and speed ambiguities consists of

- blind ranges (range eclipses) for $D = kcT_R/2$
- blind velocities for $v = k'\lambda F_R/2$
- a *Doppler notch* corresponding to the cancellation of the main lobe clutter
- range-velocity regions polluted by clutter signals received by the side lobes

By choosing an appropriate set of repetition frequencies, it is possible to ensure the detection of any target echo in a certain number of these frequencies (Chapter 7).

In addition, range and velocity ambiguities can be solved by periodic modification of repetition frequency F_R.

In fact, the measured characteristics p and q (the numbers of the range cell and the Doppler filter on the output of which a target is detected) are related to the true ranges and velocities by the expressions

$$t_0 = p\,\tau \bmod T_R \left(= p\,\tau + k\,T_R \right)$$

$$f_d = q\,\frac{F_R}{m} \bmod F_R \left(= q\,\frac{F_R}{m} + k'\,F_R \right),$$

where k and k' are integers.

When the value of T_R (and F_R) is modified, ambiguous ranges and velocities p and q vary.

If the consecutive values of T_R are properly chosen, the knowledge of p_i and q_i couples corresponding to the period T_{Ri} enables us to determine ambiguity levels k_i and k'_i and, therefore, unambiguous ranges F_R.

Figure 8.26 shows the case with two F_R. For F_{R1} the ambiguous positions of the target are marked by squares. For F_{R2} the corresponding positions are marked by circles. Due to ambiguities, the radar only knows the position of the target in the ambiguous domains (the hatched areas in Figure 8.24).

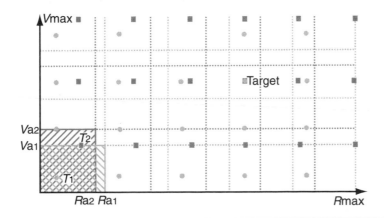

FIGURE 8.26 VELOCITY-RANGE AMBIGUITY RESOLUTION

From these values in ambiguity domain, we can deduce all the possible positions in the unambiguous domain (unfold positions) for F_{R1} and F_{R2}. If F_{R1} and F_{R2} are suitably chosen, superimposition of squares and circles occurs only for the correct position of the target and gives the unambiguous position of the target.

In MPRF radars where range and velocity ambiguities are always present, the presence of several coherent bursts, T_c, existing during illumination time T_e is used to change the repetition frequency from one burst to the next, in order to solve both eclipse and ambiguity problems. Typically eight F_R values are selected to guarantee at least three detections for any target in the range-velocity domain. Detection is achieved by locating, for each range-velocity cell in the unambiguous domain, the eight corresponding range-velocity cells in the ambiguous domain of each of the eight F_R values. If at least three of these F_R values have resulted in detection after CFAR and threshold operations, the target is said to be present in the corresponding range-velocity cell. The double threshold detection, 3/8, replaces post-detection integration.

8.6.7 DETECTION PERFORMANCE

We will illustrate calculating the performance of a pulse-Doppler radar by a typical example in the medium PRF mode.

The parameters are the following:

peak power	P_{ce} = 10 kW	wavelength	λ = 3 cm	
mean power	P_{me} = 200 W	antenna gain	G = 32 dB	
average	F_R = 20 kHz	RCS (SW1-type target)	σ = 5 m^2	
pulsewidth	τ = 1µs	microwave losses*	l_H = 5 dB*	
scan rate	ω = 60 °/s	aperture	$\theta_{3\,dB}$ = 3.6°	
noise factor	F = 3 dB			

    * these losses include propagation and radome losses

The required resolutions are r = 150 m in range and r_f = 150 Hz in frequency (r_v = 2.25 m/s in velocity).

The required probability of detection is P_d = 0.5 with a false-alarm rate of τ_{fa} = 1 /mn, in the range domain R_{max} = 150 km and the velocity domain V_{max} = 1500 m/s.

Ambiguity unfolding leads to the use of a 3/8 double threshold detection.

Note that the false-alarm rate is expressed as a *false-alarm rate* (number of false alarms per time unit) τ_{fa}, which corresponds directly to an operational need. Calculation of the false-alarm probability, P_{fa}, is required to define the value of the detection threshold.

8.6.7.1 CALCULATING FALSE-ALARM PROBABILITY

Consider the *number* (N_d) of decisions taken during the time unit (1 minute). If all the decisions have an equal risk of error, the P_{fa} is given by $P_{fa} = \tau_{fa}/N_d$.

During illumination time $(T_e = \theta_{3\ dB}/\omega)$, a decision (presence or absence of target) is taken for each unambiguous range-velocity cell; that is,

$$\frac{D_{max}}{r_D}\frac{V_{max}}{r_v}$$

decisions are taken, so there are

$$N_d = \frac{60}{T_e}\frac{D_{max}}{r_D}\frac{V_{max}}{r_v}$$

decisions per minute, and

$$P_{fa} = \tau_{fa}\frac{T_e}{60}\frac{r_D}{D_{max}}\frac{r_v}{V_{max}}.$$

$P_{fa} = 1.5\ 10^{-9}$ in the example in question.

As there is a double threshold detection (3/8), the probabilities of false alarm p_0 and of detection p_{d0} after the first threshold are given by the binomial functions Equations 8.2 and 8.3:

$$P_{fa} = \sum_{i=M}^{i=N} C_N^i p_0^i (1-p_0)^{N-i} \approx C_N^M p_0^M \tag{8.3}$$

and

$$P_d = \sum_{i=M}^{i=N} C_N^i p_{d0}^i (1-p_{d0})^{N-i}; \tag{8.4}$$

that is, $p_0 = 3\ 10^{-4}$ and $p_{d0} = 0.32$.

8.6.7.2 DETERMINATION OF THE REQUIRED SIGNAL-TO-NOISE RATIO

At the coherent processing output, which is prior to envelope detection, the signal-to-noise ratio (SNR) is obtained from the detection curves (Figure 6.4).

For P_{d0} = 0.32 and P_{fa} = 3 10^{-4}, the curves give SNR≈ 8 dB.

The various processing losses have to be added to this value:

- matched-filter and range sampling losses ($\tau_s = \tau$) $l_{st} \approx 2$ dB
- weighting and velocity-sampling losses $l_{sv} \approx 2$ dB
- CFAR losses $l_t \approx 2$ dB
- beamshape loss (the antenna gain is not maximum during the entire illumination time, T_e) $l_l \approx 1.5$ dB

The overall losses are $l_T \approx 7.5$ dB.

The resulting SNR required for detection is

$$SNR = 8 + 7.5 = 15.5 \text{ dB.}$$

Note: This ratio corresponds to the ratio that would have been measured at the *coherent processing (FFT) output if all the losses were transferred downstream* (after envelope detection).

8.6.7.3 RADAR RANGE

Let us now calculate the range at which the target considered can be detected under the stipulated conditions. First, the minimal received power, P_r, enabling detection must be determined; P_r must be such that R = 15.5 dB at the FFT output, assuming that the overall losses are transferred downstream in the processing procedure. *The processing performed until now is thus completely coherent, and the receiver theory applies.*

This is given by

$$R = \frac{E}{b} = \frac{P_{mr}T_c}{b} = \frac{mP_{cr}T}{b}, \tag{8.5}$$

where E is the energy received during coherent processing, that is, the energy of the m pulses of the coherent unit on which the FFT is computed (P_{mr} and P_{cr} are the received mean and peak power), where b is the noise spectral density (white noise).

This power is related to the target range by the radar equation (Chapter 3)

$$P_r = \frac{P_t G^2 \lambda^2 \sigma}{(4\pi)^3 R^4 l_h} \quad \text{or} \quad R = \left[\frac{P_t G^2 \lambda^2 \sigma}{P_r (4\pi)^3 l_h}\right]^{\frac{1}{4}} . \tag{8.6}$$

These expressions, which can apply to either the peak or mean power, take the form

$$R = \left[\frac{P_{me} T_c G^2 \lambda^2 \sigma}{(4\pi)^3 (S/N) k T_0 F l_h}\right]^{\frac{1}{4}} = \left[\frac{P_{ce} T G^2 \lambda^2 \sigma N_{FFT}}{(4\pi)^3 (S/N) k T_0 F l_h}\right]^{\frac{1}{4}} , \tag{8.7}$$

where $N_{FFT} = m$ is the number of pulses of the coherent burst.

In the present example, this results in the following:

- $T_e = \theta_{3\ dB}/\omega = 60$ ms
- $T_c = 1/r_f = 6,7$ ms
- eight coherent bursts ($8T_c = 53$ ms) are received during the illumination time
- $N_{FFT} = m = T_c/T_R \approx 128$
- $R = 54$ km

9

AIR TARGET TRACKING

9.1 INTRODUCTION

So far, this book has dealt with search or surveillance radars designed to *detect* the presence of a target in a range, velocity, or angular resolution cell.

The surveillance or coverage domain (set of resolution cells) was sequentially explored by antenna scanning (e.g., in azimuth) and, as a result, radar observation time T_e for each target was limited.

When a target is detected, it is useful to *correlate* consecutive target detections and *merge* the scan-to-scan measurements, that is, to *track* the target. This tracking can be fairly elaborate, from simple *tracking* to *Track-While-Scan* (TWS) or *Single Target Tracking* (STT), in which the radar continuously monitors a single target.

Apart from improving the detection conditions by providing available *a priori* information on the target, tracking also improves the quality of certain target measurements, such as

- range
- velocity
- elevation and azimuth angles

We shall now look at

- Single-Target-Tracking (STT), used in trajectography radars (firing test ranges and satellite tracking) and in conventional fire-control systems
- plot tracking, used for the surveillance functions of modern radars
- Track-While-Scan (TWS), used in certain modern fire control systems

9.2 PLATFORM MOTION AND ATTITUDE— COORDINATE SYSTEMS

To determine target trajectory independently of platform attitude and motion, you must perform calculations using an absolute system of coordinates.

As measurements are carried out using a radar-related system of coordinates (and therefore using a platform-related system), the coordinate systems have to be changed. In addition, target location is given by the radar in the form of two directions and range, while calculations must be performed using a Cartesian coordinate system.

The main coordinate systems used are

- the terrestrial coordinate system carried by the aircraft x_0, y_0, z_0 (Figure 9.1): this system is centered on the center of gravity, O, and defined by the horizontal plane containing x_0 (i.e., pointing North) and y_0 (pointing East). The z_0 axis is vertical and directed downwards. In this coordinate system, a direction (line of sight) is defined by azimuth angle θ_A and depression angle θ_D (which is also called elevation angle θ_E)

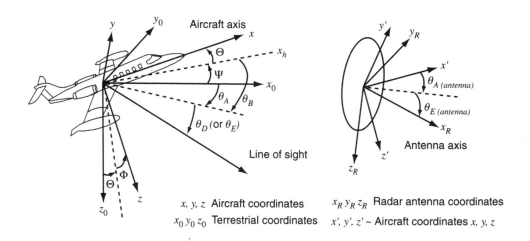

FIGURE 9.1 COORDINATE SYSTEMS USED IN AIRBORNE RADARS

- the aircraft-related terrestrial coordinate system x_h, y_h, z_0 (Figure 9.1): The x_h axis is in the vertical plane containing the platform axis. In this coordinate system, a direction is determined by bearing angle θ_B and depression angle θ_D, which usually are named azimuth angle θ_A and elevation angle θ_E
- the aircraft coordinate system x, y, z, (Figure 9.1): this is centered on O. The x axis represents the longitudinal axis of the aircraft, the y axis

is in the wing plane (to the right), and the z axis is perpendicular to this plane, pointing downwards. This coordinate system is deduced from the x_0, y_0, z_0 coordinate system by successive rotations:

- yaw Ψ, which transposes the x_0, y_0, z_0 coordinate system to the x_h, y_h, z_0 coordinate system
- pitch Θ, which transposes the x_h axis to the aircraft axis x
- roll Φ, which places y in the wing plane

The radar coordinate system x', y' z' differs from this aircraft coordinate system only in the radar alignment

- The radar antenna axis coordinate system x_R, y_R, z_R: in the case of a mechanically scanned antenna, this coordinate system is related to the antenna. Axis x_R is perpendicular to the surface of the antenna, and axes y_R and z_R are the main antenna axes. In the case of an electronic scanning antenna, the coordinate system depends on pointing direction x_R. It is related to the radar coordinate system (without roll stabilization) by antenna azimuth angle θ_A and antenna elevation angle θ_E. Target direction is measured in this coordinate system; it is defined through the antenna angular difference in azimuth ΔA and in elevation ΔE

9.3 SINGLE-TARGET TRACKING (STT)

9.3.1 DEFINITION

This kind of tracking (monotarget) is characterized by lock-on to a single target.

The antenna is continuously slaved in the direction of the target, and only the range-velocity domain containing the target is explored. However, *automatic target detection* or *acquisition* is performed, usually before the actual tracking phase, in a limited zone centered around a *marker* that gives the approximate location of the target. This marker is positioned either manually by the operator or automatically from data provided by tracking or by a search radar.

The various phases of tracking are

- acquisition
- lock-on of range and then velocity loops
- lock-on of angular loops

We will examine these three aspects in turn.

9.3.2 ACQUISITION—PRESENCE

This initialization phase of tracking consists of determining the range-velocity-elevation-azimuth cell containing the designated target in which the measurements are to be carried out.

This range gate is sought in a limited domain (correlation window) whose dimension depends on the accuracy of the designation (manual or automatic).

In practice, the presence of echoes is detected in a range *gate bank* centered on the estimated range and in a *filter bank* around the estimated frequency of the echo. Following acquisition, the banks are shifted so that the target remains in the *central gate*, in which the measurements are performed (Figure 9.2).

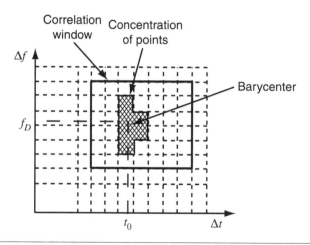

FIGURE 9.2 CORRELATION WINDOW

When the echo disappears (due to target fluctuation phenomena, for example), the position of the bank is extrapolated from the velocities measured and stored before the echo vanishes. At the next detection, a shift in the position of the echo relative to the central gate may occur; this shift is determined from the position of the target in the window.

Should the echo not reappear after a few seconds, its loss is confirmed and the tracking is abandoned.

9.3.3 GENERAL STRUCTURE OF TRACKING LOOPS

Assuming that x is the element that defines the target being tracked (range, velocity, elevation, or azimuth angles). Its estimate, $\hat{x}$, is achieved by a tracking loop whose general structure is illustrated in Figure 9.3. This estimate consists of

- a *discriminator* that measures the difference, Δx, between the value and its estimation
- a *filter* that determines the transfer function of the loop. This filter is composed of *two integrators* such that the estimation error is zero for a value varying at constant speed. The first integrator input is proportional to second derivative $\ddot{x}$ (acceleration). This integrator output (the second integrator input) is proportional to derivative $\dot{x}$ (velocity). Platform velocity $\dot{x}_p$ is injected at this point. When the first integrator input goes through zero, the velocity memory is introduced; $\hat{x}$ then evolves at constant speed on the basis of the most recently validated data
- a *range marker* that, from the estimated value $\hat{x}$, positions the device enabling measurement of Δx (range gates, frequency filters, or antenna positions)

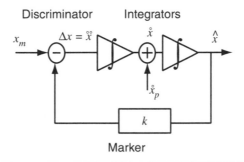

FIGURE 9.3 TRACKING LOOP

9.3.4 RANGE TRACKING

In the previous chapter, we showed that range processing (matched filter and range sampling) is performed before velocity processing, and, in addition, range and velocity need to be known in order to validate target angular data. The range-tracking loop includes

- the discriminator
- the Loop Transfer Function
- the range marker

9.3.4.1 THE DISCRIMINATOR

The discriminator enables measurement of the difference between the range of the echo, measured by the delay t_0, and the position of the central gate, measured by Δt_0.

This delay is determined by means of a *differential measurement* between two consecutive range gates (Figure 9.4).

As previously mentioned, measurement of a signal in a range gate (Figure 9.4a) can be replaced by a measurement obtained by sampling the signal output from the matched filter (Figure 9.4b).

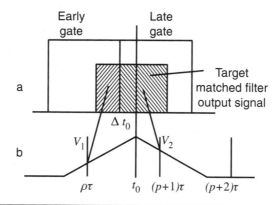

FIGURE 9.4 RANGE MEASUREMENT

The term

$$\Delta_D = \frac{(V_1) - (V_2)}{(V_1) + (V_2)}$$

gives a measurement of $t_0 - \Delta t$ independently of the power of the echo.

9.3.4.2 THE LOOP TRANSFER FUNCTION

The loop is composed of two integrators that ensure a *zero-delay error* for a target of *constant radial velocity*.

The second integrator input is proportional to the radial velocity; if the radar platform is moving, injection of the platform velocity at this point limits the workload of the loop.

The first integrator input is proportional to *target acceleration* (which is zero in the case of a target of uniform velocity).

The bandwidth of the loop must meet two requirements:

- it must enable rapid acceleration when acquisition occurs (velocity goes from 0 to v, which is a significant acceleration)
- it must *filter* the position difference measurement

Since these two objectives are incompatible, two bandwidths are often necessary: a wide bandwidth during acquisition and a narrower one (of the order of 1 Hz) following confirmation.

9.3.4.3 THE RANGE MARKER

The purpose of this device is to translate the range term calculated by the loop into a delay, Δt, for the sampling gates (or sampling gates preceding and following the received pulse).

It provides a ranging marker that releases the sampling or range-selection pulses.

9.3.5 DOPPLER VELOCITY TRACKING

The aim of this type of tracking is to slave the central position of the velocity filter bank to the Doppler frequency of the target.

The velocity difference is given by a differential measurement on output from two adjacent Doppler filters.

You can compare the structure of the loop to that of Section 9.3.3.

REMARK

Range tracking enables determination of velocity v and, thus, the velocity-tracking loop should not be necessary. In fact, the accuracy of Doppler velocity tracking is far better than the measurement deduced from range tracking. Similarly, it is possible to obtain the value of R by integration of f_D, but the value of R is then known to within one constant.

9.3.6 ANGLE TRACKING

In this kind of tracking, the antenna beam remains pointed at the target. As the antenna pointing direction is defined by two independent parameters (elevation and azimuth angles), tracking is generally composed of two independent and identical loops.

The structure of each loop is similar to the general configuration (Figure 9.2).

This is a conventional servosystem with two integrators controlling the antenna position by means of a servomechanism (bandwidth $\approx$ 1 Hz).

In modern radars, angular position difference is obtained by using the monopulse measurement as described in Chapter 3, Part I, Section 3.6.1.

Signal $\vec{\Sigma}$ is used for detection; $\vec{\Sigma} = x(t)$.

In the case of a noiseless point source, $\Delta \approx \Sigma \tan q\alpha$, where α is the angular deviation from the antenna maximum-gain angle. Signals $\vec{\Sigma}$ and $\vec{\Delta}$ are of the same nature; they are *processed in the same manner* in identical parallel channels.

Signals $\vec{\Sigma}_{p,q}$ and $\vec{\Delta}_{p,q}$, sampled on output from Doppler filter q and range cell p, which are free of spurious echoes (see Chapter 2), are used to calculate angular difference:

$$\Delta = \frac{\vec{\Sigma}_{pq} \cdot \vec{\Delta}_{pq}}{\left|\vec{\Sigma}_{pq}\right|^2}.$$

9.4 PLOT TRACKING

9.4.1 DEFINITION

This tracking mode enables scan-to-scan correlation of plots in search radars (or in the search mode for fire-control radars).

This operation, which was a manual operation in early radars, is performed automatically by the radar processing unit that presents useful information only (confirmed plot, trajectory configuration, velocity vector) to the operator (or to the system). The operator work load (in search radars) is lightened, and, in the case of fire-control radars, the information provided enables preparation of target take-over by the system. This is known as *plot tracking*.

Plot tracking involves two types of operations:

- trajectory estimation
- track creation and maintenance

9.4.2 TRAJECTORY ESTIMATION

Consider the simple case of tracking in which target *position* and *velocity* are estimated by filter α, β each time the antenna beam illuminates the target. As an example, we shall consider one of the coordinate systems, x (Figure 9.5), with the following definitions:

- x_i is the *actual* position of x during illumination i
- $\hat{x}_i$ is the position *predicted* before illumination i
- x_{im} is the position *measured* during illumination i
- $\bar{x}_i$ is the *corrected* position after illumination i
- v_i is the *corrected velocity* after illumination i

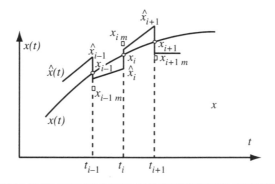

FIGURE 9.5 TRACKING

Just before the *i*th illumination, the predicted value of x is $\hat{x}_i$.

The position measured during illumination i is x_{im}; the difference $(x_i - x_m)$ is the real measurement error (due to glint or thermal noise, for instance). In fact, x_{im} is measured as a function of the predicted value, $\hat{x}_i$, and the signal processing circuits give the difference $\Delta x_i = x_{im} - x_i$.

After illumination i, the following corrections are made:

- position of $\hat{x}_i$, with $\bar{x}_i = x_i + \alpha(x_{im} - x_i) = x_i + \alpha \Delta x_i$

- velocity v_{i-1}, with $v_i = v_{i-1} + \beta \dfrac{\Delta x_i}{T_p}$,

where T_p is the time between two consecutive illuminations.

Between illumination i and $i + 1$, v_i is assumed constant and the estimated location is $x = \bar{x}_i + v_i t$.

The coefficients α and β are correcting coefficients whose function is to *smooth the trajectory*. If α and β are small, the values of $\bar{x}_i$ and v_i depend essentially on measurements made prior to i (of the track history); the new measurement will not be very important, and the *noise will be filtered*.

However, if the target is attempting an evasive maneuver (such as an acceleration), the weight of the recent measurements will have to be increased, and α and β will need to be higher. These coefficients therefore have a significant value and are often adapted to suit the circumstances:

- $-\alpha$ and β are high at the start of tracking (when v is unknown) or during an evasive maneuver

- $-\alpha$ and β are low for a well-established tracking of a target with uniform motion or when there is significant measured noise (low SNR)

Fixed-coefficient filters (α-β trackers) have the advantage of simple implementation but they are efficient only if the target is basically on a straight-line trajectory and with $P_D = 1$. A variable gain sequence through Kalman filtering is preferable in complex situations when high accuracy is required.

The Kalman filter minimizes the mean-squared error as long as the target dynamics and the measurement noise are correctly modeled. The main advantages of the Kalman filter are as follows:

- The gain sequence is automatically adapted to target maneuver and measurement noise models. For example, as the target range decreases, angular measurement increases and angular dynamics become higher. The adaptive parameters are adjusted accordingly
- The Kalman gain takes into account the missing detections when P_D is low
- The Kalman filter provides an estimate of the tracking accuracy (through the covariance matrix) enabling you to determine the prediction window in which the correlation of the track is made

The Kalman filter is presented in a very large number of works to which the reader may refer.

9.4.3 TRACKING MANAGEMENT AND UPDATE

Consider a track that has been initialized. Before the ith scan, the estimated location (for the parameter x) is $\hat{x}_i$. At illumination i, a plot located at x_{im} (close to $\hat{x}_i$) is detected. Does this plot belong to this track or to another one? Or is it a new echo that will initialize a new track?

To answer these questions, a *prediction window* is created (Figure 9.2). Its dimensions depend on

- the quality of the measurements
- the history of the track
- possible evasive maneuvers

If the echo is in this window (and is the only one in the window), it is associated with this track. If there are several echoes, or if the echo belongs to several prediction windows (as the result of several convergent tracks), the most probable track will be chosen for each echo.

If the echo does not belong to any prediction window, it is a new target (or a false alarm), and we must wait until the next series of illuminations to confirm it and create the corresponding track.

If a track is not updated over several consecutive illuminations, it is interrupted.

9.5 TRACK-WHILE-SCAN (TWS)

This kind of tracking enables relatively accurate measurement of the target parameters (range, velocity, and azimuth and elevation angles) while the search function continues to operate.

The antenna therefore keeps operating in the scanning mode (as for a plot tracking) and the tracking information (monopulse measurements) is updated while the antenna beam is pointing at the target. The difference between plot tracking and TWS is in the difference measurements (in range, speed, azimuth and elevation angles) that are calculated in the same way as for STT. Compared with plot tracking, accuracy is better and elevation tracking is possible (which is generally not the case with plot tracking).

Acquisition, track update, and trajectory estimation are performed in the same way as in plot tracking.

10

GROUND TARGET DETECTION AND TRACKING

10.1 INTRODUCTION

The detection and tracking of ground targets (vehicles such as tanks, trucks, or buildings) differs from that of air targets because their relatively low speeds. Aircraft velocity is sufficient to enable elimination of the entire spectrum of clutter signals received by the main main by Doppler filtering without greatly reducing the detection domain of the target. But in the case of airborne radars, the Doppler spectrum of ground targets and clutter may overlap, making Doppler filtering more difficult.

In this chapter, we shall make a distinction between fixed ground targets that can only be detected and tracked if there is sufficient contrast with surrounding echoes, and moving targets that can only be discriminated from fixed echoes by Doppler filtering.

10.2 DETECTION AND TRACKING OF CONTRASTED TARGETS

Fixed ground targets can only be detected if they have sufficient RCS compared with the RCS of the environment. This is particularly true for targets such as ground installations or armored vehicles in a homogeneous background when the resolution cell is small. The problem of contrast enhancement is discussed in Chapter 15, on imaging radars.

10.3 DETECTION AND TRACKING OF MOVING GROUND TARGETS

10.3.1 LOW-SPEED AIRCRAFT (HELICOPTERS)

Since the velocity of expected targets is limited ($\approx 30 \mathrm{ms}^{-1}$), an *unambiguous waveform* can be used. With interpulse period $F_R = 2$ kHz, the unambiguous velocity domain is

$$|v_{amb}| = 30 \mathrm{ms}^{-1},$$

apart from the velocity sign (in X-band), and the unambiguous range is R_a = 75 km (which is a sufficient operational value). A LPRF MTI mode is therefore suitable.

In the case of a slow-moving platform (helicopter), an airborne MTI operating mode such as that described in Section 2.4.4 is satisfactory, and we will therefore refer to this section for detection of terrestrial moving targets from a low-speed platform.

This type of target tracking does not significantly differ from the airborne target tracking dealt with in Chapter 14.

For a high-speed platform (airplane), detection of slow targets (e.g., a few ms^{-1}) is hindered by the broadening of the clutter spectrum signals received by the antenna main lobe when it is pointing off-boresight (see Section 2.4.4), and the space-time processing techniques discussed below have to be used.

10.3.2 HIGH-SPEED AIRCRAFT (AIRPLANES)

The spectral width of the clutter signals is given by

$$\Delta f = \frac{2v_C}{\lambda}\cos\theta_E\sin\theta_A\Delta A \approx \frac{2v_C}{\lambda}\sin\theta_A\Delta A \; , \qquad (10.1)$$

where ΔA is the width of the main lobe in azimuth (Figure 10.1).

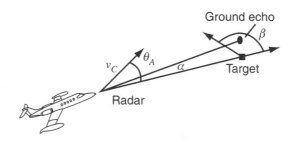

FIGURE 10.1 DEFINITION OF THE VELOCITY TERMS

If we take a numerical example, then

- v_C = 250 ms^{-1}
- λ = 3 cm
- θ_A = 60°
- ΔA = 6°
- Δf = 1450 Hz

This spectrum broadening prevents a correct target detection in a zone of $\Delta v = \pm 22$ ms^{-1}, which constitutes an important part of the velocity domain of the desired targets.

The Doppler frequency of a target depends on

- its inherent radial velocity:

$$f_T = \frac{2v_T}{\lambda} \cos \beta \qquad (10.2)$$

- the component linked to platform speed:

$$f_C = \frac{2v_C}{\lambda} \cos \theta_A \qquad (10.3)$$

(given the usual values of elevation angle θ_E, we can write $\cos \theta_E \approx 1$, and the off-boresight angle is similar to the azimuth angle).

Two targets with different speeds can therefore have the same Doppler frequency if they are viewed from different α angles (within the radar beam); this is why a target echo received in the beam axis can be masked by a clutter signal located at the angle α such that

$$\frac{2v_C}{\lambda} \cos \left(\theta_A + \alpha \right) = \frac{2v_C}{\lambda} \cos \theta_A + \frac{2v_T}{\lambda} \cos \beta .$$

The only way to discriminate between these echoes is to take into account the angle, α, formed by the return signals. This leads to the *space-time processing*—the simplest example of which is the *Displaced Phase Center Antenna* (DPCA)—described in Chapter 6.

10.3.2.1 Displaced Phase Center Antenna (DPCA)

Like all space-time processing techniques, DPCA tries to create a displacement of the antenna phase center in order to compensate for the forward movement of the platform, from one interpulse period to the next. With DPCA, this compensation is limited to two consecutive interpulse periods. As a result, the antenna is *immobilized*; the problem is like that of a fixed radar without spectrum broadening, which enables the detection of very low-speed targets.

The sum (Σ) and azimuth difference (Δ) signals received by the antenna monopulse channels are used. These signals are linked by the relationship

$$\vec{\Delta} = \vec{\Sigma} \tan q \alpha \approx q \alpha \vec{\Sigma},$$

where α is a small off-boresight angle. See Chapter 9.

Figure 10.2 illustrates the principle of DPCA. On the basis of the signals received at the i and $i+1$ interpulse periods on the sum and azimuth difference channels—that is $\vec{\Sigma}_i, \vec{\Delta}_i$ on one hand, and $\vec{\Sigma}_{i+1}, \vec{\Delta}_{i+1}$ on the other—the following *synthetic signals* are constructed:

$$\vec{\Sigma}'_i = \vec{\Sigma}_i + jk\vec{\Delta}_i \text{ and } \vec{\Sigma}'_{i+1} = \vec{\Sigma}_{i+1} - jk\vec{\Delta}_{i+1}$$

The value of k is calculated such that

$$\vec{\Sigma}'_{i+1} - \vec{\Sigma}'_i = 0$$

for all fixed echoes.

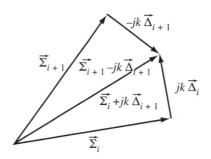

FIGURE 10.2 DPCA TECHNIQUE

For a ground return ($f_c = 0$), the Doppler frequency is

$$f_D = f_C = \frac{2v_C}{\lambda}\cos(\theta_A + \alpha) \approx \frac{2v_C}{\lambda}\cos\theta_A - \frac{2v_C}{\lambda}\sin\theta_A\,\alpha \ . \qquad (10.4)$$

The first term, independent of α, is compensated (aircraft velocity compensation), and the variation of residual phase between the two pulses is

$$\Delta\varphi = \varphi_{i+1} - \varphi_i = 2\pi\frac{2v_C T_R}{\lambda}\sin\theta_A\,\alpha \ .$$

The relationship between the signals received at the i and $i+1$ interpulse periods is

$$\vec{\Sigma}_{i+1} = \vec{\Sigma}_i e^{j\Delta\varphi} \text{ and } \vec{\Delta}_{i+1} = \vec{\Delta}_i e^{j\Delta\varphi} .$$

The value of k is given by

$$\vec{\Sigma}_{i+1} - \vec{\Sigma}_i = \vec{\Sigma}_i [e^{j\Delta\varphi}(1 - jkq \ \alpha) - (1 + jkq \ \alpha)] \approx \vec{\Sigma}_i(\Delta\varphi - 2jkq \ \alpha) = 0$$

(for $\Delta\varphi \cong 0$), and we deduce that

$$k = \frac{\pi}{q} \frac{2v_C T_R \sin\theta_A}{\lambda}.$$

This equation shows that the clutter signals are cancelled by a two-pulse MTI canceller for each value of α, if the conditions that $\Delta\varphi \cong 0$ and α is small are met. Figure 10.3 illustrates the effect of immobilization of the phase center. However, the constraints mentioned above limit the efficiency of this process, and more sophisticated processing techniques have to be used.

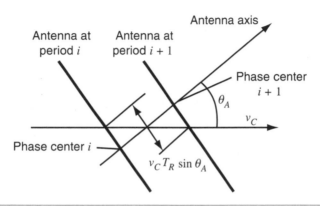

FIGURE 10.3 DISPLACED PHASE CENTER ANTENNA

10.3.2.2 SPACE-TIME ADAPTIVE PROCESSING (STAP)

The limitations of DPCA can be avoided by using Space-Time Adaptive Processing (STAP), the theory of which is described in Chapter 6. DPCA is a special case of STAP, with $M = 2$ and $N = 2$. A description of the normal case ($M \neq 2$ and $N \neq 2$) can be found in the work of Klemm (1992, 1994).

Figure 10.4 illustrates another particular case of STAP. Similar to DPCA, space sampling is obtained from monopulse channels Σ and Δ in azimuth, but a train of M time pulses is used for frequency analysis of the signals. As far as the clutter signal is concerned, any frequency value f_D can be associated with direction α (Equation 10.4) if there is no velocity ambiguity for these ground echoes.

The processing involves the construction of *synthetic diagram* $\vec{\Sigma}' = \vec{\Sigma} - w\vec{\Delta}$ for each frequency (direction), where w is an adaptive coefficient calculated to enable clutter cancellation in the direction of interest. In fact

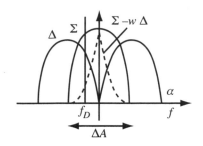

FIGURE 10.4 SPACE-TIME PROCESSING

this coefficient creates a *monopulse notch* in this direction (Figure 10.4). For a moving target located in the same direction, the term v_T causes it to appear in a different filter. For this Doppler filter, the *synthetic monopulse notch* created corresponds to another direction so that the target is not eliminated and can be detected.

REMARK

Unlike the general STAP described in Chapter 6, DPCA (or STAP based on Δ and Σ channels) acts only on main main clutter. Ground echoes located in the side or far lobes have to be rejected by sufficient angular selectivity.

11

MARITIME TARGET DETECTION AND TRACKING

11.1 MARITIME SURVEILLANCE RADARS

Maritime surveillance and maritime patrol use airborne radars designed for the detection, tracking, and classification of sea targets. Their main functions are

- to detect surface vessels
- to detect the exposed periscopes of submersed submarines
- to control a tactical surface configuration
- to designate targets to weapon systems
- to contribute to search and rescue operations
- to classify detected targets

We can achieve all these functions using two waveforms associated with different kinds of processing:

- a waveform with a low pulse-repetition frequency and medium-range resolution to detect surface vessels up to the horizon
- a waveform with a low pulse-repetition frequency and high-range resolution to detect small targets at medium range, in a rough sea

Some radars specialized in the detection of small, fast-moving targets (patrol boats) use a Doppler mode with low-range resolution (75 to 150 meters). In this case, the principle is similar to the one used for the detection and tracking of mobile ground targets (Chapter 10).

Multifunction fire control radars on combat aircraft have an air-to-sea function using low pulse-repetition frequency modes and low- or medium-range resolution (30 to 150 meters). This type of mode matches the requirement, which is to deliver an anti-ship missile from a platform flying at very low altitude. Sea clutter is low at grazing elevation and the radar cross section of likely targets is high (more than 1 000 m^2).

The rest of this chapter deals with the waveforms and processing used in radars specialized in the detection, tracking, and classification of maritime targets.

11.2 SEARCH STRATEGY

11.2.1 POSITIONING OF THE RADAR WITH RESPECT TO WIND DIRECTION

The sea surface backscattering coefficient depends on wind speed, observation incidence angle, wave height, transmission frequency, polarization, and wind direction with respect to radar line of sight.

Location downwind of the target is the most unfavorable position. Upwind positioning, on the other hand, improves performance by approximately one sea state. The crosswind position is an intermediate case. So the best solution is to position the radar upwind of the search area whenever possible.

11.2.2 PLATFORM ALTITUDE

Propagation over the surface of the sea is characterized by a *transition grazing angle*, both for sea clutter, as in the *Katzin* model (Katzin 1957), and for the target (see Chapter 4, Section 4.2.1). At grazing angles greater than the transition grazing angle, that is, at short ranges, propagation obeys an R^{-4} function (or R^{-3} for sea clutter). For grazing angles below the transition grazing angle, that is, for long ranges, propagation obeys an R^{-8} function (or R^{-7} for sea clutter). A transition grazing angle occurs when the axis between the radar and the target passes beneath the last lobe of the RCS pattern due to the image effect above the sea surface. The higher the target, the smaller the transition grazing angle.

The target and sea clutter transition ranges are, respectively,

$$R_{tt} = 4\frac{hH}{\lambda} \text{ and } R_{tc}=k_c\frac{H_{01}H}{\lambda}$$

in which

- H is the platform altitude
- h is the height of the target scatterers above the sea surface
- H_{01} is the height exceeded by 10% of the waves
- $k_c = 2$ for the Mediterranean Sea (short swell) and $k_c = 3$ for the Atlantic Ocean (long swell)

We can divide the range domain into several parts (see Figure 11.1):

- short ranges, that is, ranges less than the sea clutter transition range; targets can be detected
- range interval around the sea clutter transition range; the signal received from the target is below the threshold and targets cannot be detected
- range interval between the two transition ranges; targets can be detected again
- ranges greater or equal to target transition range; targets are no longer detected

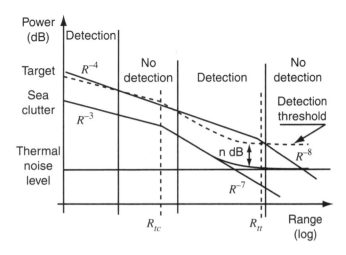

FIGURE 11.1 MARITIME TARGET DETECTION

This phenomenon enables us to determine a strategy for maritime target detection. Three parameters must be known:

- RCS of the target
- height of the target
- height of the waves

The range at which the target can be detected is given by the power budget of the radar in free space. We then choose a platform altitude such that

1. the radar horizon is beyond the chosen detection range:

$$R_{hor} \approx k(\sqrt{h} + \sqrt{H}) > R_{det}s$$

with $k = 1.23$ if h in feet and R_{hor} in nautical miles

2. the sea clutter transition range is less than the detection range:

$$R_{tc} = k_c \frac{H_{01}H}{\lambda} < R_{det}$$

3. the target transition range is greater than the detection range:

$$R_{tt} = 4\frac{hH}{\lambda} > R_{det}$$

EXAMPLE

Table 11.1 shows two examples: detection of a large surface vessel and detection of a smaller target. In both cases, the common characteristics are

- X-band radar: λ = 3 cm
- wave height: H_{01} = 1.25 m, Mediterranean Sea

TABLE 11.1. EXAMPLE OF TWO MARITIME TARGET-DETECTION MISSIONS

Mission n°	Example 1	Example 2
Target to be detected	Frigate	Periscope
Target height	10 m	1 m
Target RCS	1 000 m^2	1 m^2
Radar range in free space	112 NM	20 NM
Condition n°1: $R_{hor} \approx k(\sqrt{h} + \sqrt{H}) > R_{det}$	$H > 7\,280$ ft	$H > 209$ ft
Condition n°2: $R_{tc} = k_{rc}\dfrac{H_{01}H}{\lambda} < R_{det}$	$H < 8\,174$ ft	$H < 1\,460$ ft
Condition n°3: $R_{tt} = 4\dfrac{hH}{\lambda} > R_{det}$	$H > 484$ ft	$H > 911$ ft

For the first mission, the airplane or helicopter altitude must be between 7 500 and 8 000 ft. For the second mission, it must be between 1 100 and 1 300 ft. It is not possible to carry out the missions simultaneously. The platform altitude must be chosen with respect to the characteristics of the target to be detected, the sea state, and radar mode performance.

11.3 SURFACE VESSEL DETECTION

This section describes the radar mode (antenna scan, waveform, and associated processing) for mission 1 described in the previous example.

11.3.1 PULSE-REPETITION FREQUENCY

A waveform range that is unambiguous (low PRF) is chosen, as the surveillance area is large (up to several hundred kilometers) and the number of targets can be high. For this reason it is very difficult to use range ambiguity solvers.

The target must be detectable up to the horizon, which means that the pulse repetition frequency cannot exceed 500 Hz.

11.3.2 RESOLUTION

Detection is based on a contrast criterion. The detection probability is maximum when the power backscattered by a vessel in a resolution cell is greater than the power reflected from the adjoining cells, which contain only the sea returns (or clutter). As the sea clutter power in a resolution cell is proportional to its area, efforts will be made to reduce it. The search for high resolution is, however, limited; when the surface area of the ship can no longer be contained in a single cell, RCS is spread over several cells. In such cases it is no use improving range resolution; contrast enhancement in relation to sea clutter is very small.

As the angle of presentation of the ship with respect to the line of sight can take any value, it is not enough to simply choose a resolution cell equal to the length of the largest vessel to be detected. A good trade-off is to take a value between 30 and 40 m, which ensures the greatest probability of observing a ship in one or two resolution cells, while limiting sea clutter power in the neighboring cells.

11.3.3 POLARIZATION

Polarization is used to minimize sea clutter. If the sea is calm (wind below force 3 on the Beaufort scale), horizontal polarization is preferable, but when the wind force increases, horizontal and vertical polarizations are equally effective. If there is heavy precipitation, it may be useful to use circular polarization in order to reduce rain returns, but doing so reduces the level of the signal received from a maritime target by 4 dB on average.

11.3.4 TRANSMISSION FREQUENCIES

On fixed-frequency transmission, the sea clutter observed in a resolution cell fluctuates but remains partly coherent from one pulse to the next. Pulse-to-pulse post-detection integration does not give the same gain on signal-to-sea clutter ratio as on signal-to-thermal noise ratio. *Frequency agility* is used to achieve this gain. It is then possible to calculate the detection performance using the *Swerling II Model* (see Chapter 3).

11.3.5 PROCESSING

We can divide modern radar processing into the following stages:

- pulse compression (see Chapter 8), using an analog device (*Surface Acoustic Wave filter*, SAW)
- envelope detection

- signal sampling and digital conversion
- non-coherent post-detection integration
- false-alarm regulation (CFAR)
- target extraction
- Track-While-Scan, with automatic track initialization
- synthetic display

Usually the receiver is equipped with a Sensitivity Time Control (STC) device in order to reduce the dynamic range produced by the radar range. Note that there is no Doppler processing.

Antenna scan rate in azimuth is between 6 and 12 rotations per minute.

11.4 DETECTION OF SMALL TARGETS (PERISCOPES)

This section describes the waveform applicable to mission 2 (see Section 11.2), detection of small targets in a rough sea. Some elements remain unchanged: a low PRF waveform, polarization depending on the sea state (see Section 11.3.3), and pulse-to-pulse frequency agility.

11.4.1 PROCESSING

The main difference between standard detection and small-target detection lies in the statistical properties of sea clutter. When the sea surface is observed with a high resolution, sea clutter power fluctuates in accordance with a function known as *K function* (Ward 1990). Sea clutter spikes appear locally, like small point targets. The *K function* is a composite model that involves a *Rayleigh*-type component representing the clutter background and an χ^2 function-type component representing the sea clutter spikes. Frequency agility on transmission helps to reduce the level of the spikes but is not sufficient to eliminate them. This can be achieved if we allow for the fact that the clutter spikes remain visible for less than five seconds, while generally a target can be seen for a longer time, ten seconds for example.

The radar observes each target once per second, for ten seconds. The antenna scan rate must therefore be greater than 60 rotations per minute.

The processing steps are as follows (Figure 11.2):

- pulse compression (analog processing)
- envelope detection
- signal sampling and digital conversion
- non-coherent post-detection integration of the signal received during one path across the target

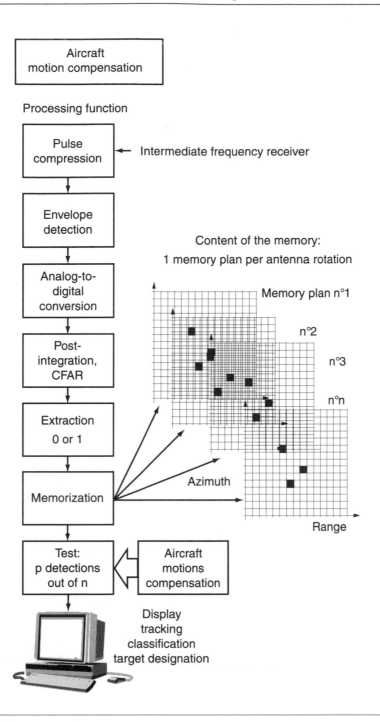

FIGURE 11.2 PROCESSING BLOCK DIAGRAM

- false-alarm regulation (CFAR)
- target extraction

- storage: "0" if the detection threshold is not reached, "1" if it has been exceeded; a memory plane composed of "0" and "1" is filled at each antenna rotation
- scan-to-scan post-detection integration using an M out of N criterion: point-to-point sums of N successive memory planes. When the sum is greater than M, it indicates that a target is present at that point
- Track-While-Scan with automatic track initialization
- synthetic display

Scan-to-scan post-detection integration requires compensation of aircraft motion for ten seconds.

The system works on the double-threshold extractor principle (see Chapter 8). If the detection threshold is exceeded at least M times in N rotations in a range/azimuth cell, then the target can be displayed. Detection performance is linked to performance after post-detection integration in the beam by a binomial function:

$$P_{fa} = C_N^M P_{fa\,0}^M, \qquad P_D = \sum_{i=M}^{N} C_n^i P_{D\,0}^i (1 - P_{D\,0})^{N-i}.$$

11.4.2 RESOLUTION

As in the previous mission, resolution must be adapted to target dimensions. For periscopes, the target is considered a point. However, various effects compromise the achievement of very high resolution:

- The wake of a moving target makes up a considerable part of its radar cross section. In this case, it is useful to be able to observe the target and a part of its wake in the same resolution cell
- Processing will be simplified if we reduce the target range migrations caused by the target's inherent speed during the ten seconds of observation
- Finally, it is useless to achieve a resolution better than the maximum superposition accuracy of the memory planes over ten seconds, which is itself established by the accuracy of the parameters supplied by the navigation system

It follows that resolution of a few meters (2 or 3 m, for example) is sufficient.

11.4.3 PULSE-REPETITION FREQUENCY

For the same reasons as for the previous radar mode, the waveform is range-unambiguous. The aim is to detect a target with a very small radar cross section in a rough sea. Under these conditions, radar range is at best a few tens of kilometers. Pulse-repetition frequency can be higher without ambiguity risks, generally between 800 and 2 000 Hz.

11.5 MARITIME TARGET TRACKING

11.5.1 PURPOSE OF THE TRACKING FUNCTION

The purpose of target tracking is to establish and update a tactical surface situation. This can be achieved simultaneously with detection, using data provided at each antenna rotation (see Chapter 9). The antenna is not continuously pointed toward the target; this is a Track-While-Scan (TWS).

TWS provides the following information:

- target position
- target speed
- a quality criterion for position and velocity data

11.5.2 TRACKING INITIALIZATION

When an object is detected by the radar, a *plot* is created and displayed on a screen. The *plot* becomes a *target* as soon as the tracking process is initialized. A *window* is created, enabling anticipation of the target position at the next antenna rotation.

It is possible to initialize tracking manually by a radar operator who designates the useful target on the screen with a cursor. Initialization can also be automatic. It is then necessary to indicate screen areas where initialization is authorized in order to avoid picking up spurious targets, such as the coastline.

11.5.3 ALGORITHM DESIGN

Figure 11.3 illustrates the various stages of the algorithm.

11.5.3.1 PLOT-TARGET ASSOCIATION

There are three possibilities for plot-target association:

- No plot is detected in the window; the radar maintains the prediction, considering that the predicted position has effectively been measured.
- A plot is detected in the window; its position is measured.
- Several plots are detected in the window; a proximity criterion with the predicted position is applied to select one of the plots, whose position is measured.

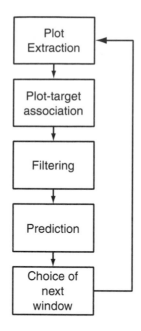

FIGURE 11.3 TWS ALGORITHM

11.5.3.2 FILTERING

The filtering function calculates a position that has been estimated using the predicted position, the measured position, and measurement noise variance:

$$X_{estimated} = X_{predicted} + \alpha \left(X_{measured} - X_{predicted} \right).$$

The coefficient α is a function of the measurement noise variance. Its limit values are

- $\alpha = 0$, measurement too noisy to be considered
- $\alpha = 1$, ideal measurement without any noise

The two main techniques used to calculate the coefficient are briefly described below.

KALMAN FILTER

This is a self-adaptive filter that minimizes the prediction-error variance and generally provides the best results. For each target and at each antenna rotation, it requires the inversion of a matrix whose dimension is equal to the number of the estimated parameters: range, bearing, radial velocity, cross-range velocity, etc. Considerable computing power is required.

α, β FILTER

This is a simplified version of the *Kalman Filter* in which coefficients are not calculated in real time but are previously tabulated. They are selected in accordance with the difference noted between predicted and measured values:

- the α coefficient for position parameters
- the β coefficient for velocity parameters

It requires less computing power than a Kalman Filter, because there is no matrix inversion and performances are more or less equivalent.

11.5.3.3 PREDICTION

The target search window at the next rotation is calculated on the basis of the predicted positions and velocities, as well as on possible target evolution. This requires a physical model of target movements.

11.6 MARITIME TARGET CLASSIFICATION

The targets present in the surveillance area must be classified. Radar can perform this task very efficiently, enabling long-distance classification. This avoids having to reroute an aircraft in order to identify targets of minor interest. Three techniques are used, with very different performances:

- radar cross section measurement
- range profile
- imaging

11.6.1 RADAR CROSS SECTION MEASUREMENT

Radar can provide the operator with an estimate of the RCS of the detected vessel. This measurement is very inaccurate, as it is degraded by the image effect. Moreover, the vessel can be observed in the direction of the minimum of its RCS pattern. Thus a target with a large RCS can occasionally be attributed a small RCS. A small target is rarely associated with a large RCS. This parameter is very simple to access.

11.6.2 RANGE PROFILE

A very high-resolution waveform can be used to measure the ship's RCS in several consecutive range gates. As the ship heading is given by tracking, it is possible to evaluate the length of the vessel.

Each range gate receives a certain amount of power that depends on the superstructure RCS of each section of the ship observed. The profile that is obtained gives an idea of the size of the vessel but is not always in direct relation to its silhouette; a trihedral such as the breakwater on the prow of a vessel has a very high RCS and causes a power peak (see Figure 11.4). On the contrary, the bridge may have a small RCS depending on the observation angle. Although this technique is slightly more complicated than the previous one, it can easily be implemented because it works with available radar waveforms and circuits. This is the reason why many modern radars are equipped with this system.

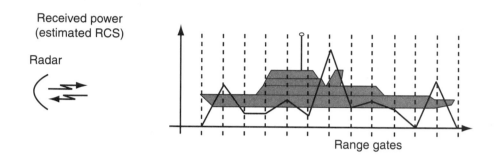

FIGURE 11.4 RANGE PROFILE

11.6.3 IMAGING

SAR or *ISAR* type imaging is the most complex and most successful process. For a detailed description, refer to Chapter 17.

12

ELECTROMAGNETIC POLLUTION

12.1 INTRODUCTION

Radar, which is itself a highly complex system, is generally part of a larger system including many other passive or active electronic devices. For this reason, it is vital that radar components be totally compatible, without any internal interference. In addition, radar must not interfere with other equipment on the platform. Naturally the same holds true for other types of equipment, especially active equipment such as electronic countermeasures (ECM), communications and identification equipment, etc.

Even when its own compatibility is guaranteed, a platform can be subject to electromagnetic interference from the "outside world." This can include

- lightning
- electrical and magnetic fields
- nuclear electromagnetic pulses (EMP)
- unintentional interactions between identical systems (e.g., patrol flights) or between "friendly" or "enemy" systems
- deliberate jamming by an opposing system
- "spurious" reflections from waves transmitted by the system and caused by certain types of clutter, such as clouds, rain, hail, chaff, altitude returns, etc.

12.2 ELECTROMAGNETIC COMPATIBILITY

Great care must be taken right from the start of design of the equipment and its subassemblies to ensure that they are protected from external interference, and also to ensure they are not a source of aggression for other equipment or subsystems (standards: GAM-T-19, MIL-HDBK-237, MIL-E-6051).

Precise determination of the fields, voltages, and currents of the electronic circuits would mean solving Maxwell's equations. However, this is too complex and would only yield approximate results, which would not indicate the relative importance of the different means by which the interfering energy penetrated the system. In practice, optimizing electromagnetic compatibility is a matter of balancing the "resistance" to the different forms of penetration. Field coupling is carried out by diffusion through the casing, or diffraction through the openings. Injected current and voltage coupling can be achieved by using either conduction or crosstalk.

To summarize, the tricky problem of compatibility is best handled using simulation and parametric studies based on existing tools (e.g., finite elements method) and measurements. Experience naturally makes a valuable contribution.

Figure 12.1 shows the possible paths taken by the interfering signals. What is shown is for an electronic unit, but it can equally be applied to components, boards, and modules, as well as to a group of units (i.e., a piece of complex equipment made up of several units). One difficulty is evaluating the level of interference applied to each subassembly and the aggression caused by this same subassembly.

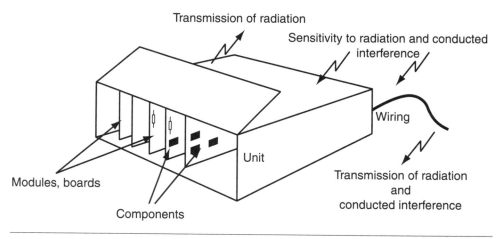

FIGURE 12.1 MODES OF INTERFERENCE COUPLING

When the platform provides only minimal protection (casing, filters, etc.) for electronic equipment, anticipated levels of interference should be taken as follows:

- source external to the platform:
    - nuclear electromagnetic pulse: $E = 50$ kV/m; $H = 130$ A/m
    - lightning: I peak $= 200$ kA, on average: 25 kA
    - electrostatic discharge: $E = 25$ kV/m
    - (these may interfere but do not cause damage)
- source internal to the platform:
    - electromagnetic compatibility: $E = 200$ V/m

12.3 INTERFERENCE FROM OTHER RADAR COMPONENTS

A modern radar is a complex system of subassemblies, each with different characteristics. As a result, in the same small area one finds analog and digital circuits, powerful generators and ultrasensitive receivers, sources with high spectral purity, spurious sources with very wide spectra, etc.

Figure 12.2 is a block diagram of a modern airborne radar with coherent transmission and reception and digital processing. We shall use this scheme to examine the main causes of internal radar interference.

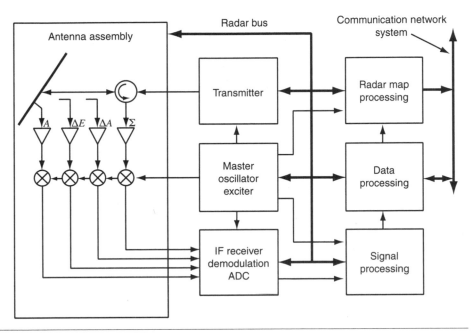

FIGURE 12.2 BLOCK DIAGRAM FOR COHERENT RADAR WITH DIGITAL PROCESSING

12.3.1 FREQUENCY SOURCE (MASTER OSCILLATOR EXCITER)

The frequency source uses an extremely stable quartz oscillator with high spectral purity to generate and supply all the coherent reference signals that the different elements of the radar need in order to function. Each of these signals must have high spectral purity (noise and spurious lines at very low levels). As a result, careful attention must be paid to the frequency plan and the implementation. The frequency oscillator must be protected from outside influences using casing and filters. Moreover, sensitivity to mechanical and acoustic vibrations means such vibration must be absorbed by means of suspension devices and absorbers.

These vibrations are produced by the platform and, in the case of the radar, by the antenna servomechanisms and air conditioning circuits (hydraulic pumps, ventilators, etc.).

The frequency oscillator, like the other units, is coupled to the low-voltage power supply. These power supplies, which transform power generated onboard into DC voltages using rectifiers, filters, and regulators, do not have zero output impedance at all frequencies and therefore introduce coupling. Moreover, these voltage supplies, which have AC residues (several millivolts), pass through cables, plugs, and pins that create additional couplings. There are several ways of dealing with this inconvenience, while still retaining rectifiers, preregulation, and centralized filters in the radar:

- Install regulation circuits at the "board" or module level
- Install regulation circuits at the unit or assembly level; these circuits should be specific to analogue and digital circuits
- Use the previous solution while retaining regulation circuits for very sensitive or interference-generating modules

The final choice will depend on the degree of complexity of the equipment, the performance required, available technology, and, of course, cost.

12.3.2 TRANSMITTER

The transmitter is a microwave amplifier consisting of one or two amplification stages. It receives signals at the transmission frequency from the frequency oscillator. It must then form and amplify these waves without degrading their purity. The transmitter, which often uses over half the total power consumed by the radar (e.g., 5 kVA), transmits high levels of microwave power to the antenna (e.g., 1 kW on average). This makes it a major source of interference for the receiver circuits. As a result, the circuits must be decoupled and given effective protection.

The transmitter is generally housed in a separate unit along with its air-conditioning circuit, its high- and low-voltage power supplies, its protection circuits, tests, etc.

12.3.3 ANTENNA ASSEMBLY

The most important elements of the antenna assembly are the antenna and its servomechanisms, the power link, with the transmitter fitted with rotating joints and a circulator; and microwave receiver circuits with, for each channel, low-noise, broad dynamic-range preamplifiers combined with protection circuits and mixers that supply signals from the four channels at intermediate frequency.

With regard to internal radar interference, the servomechanisms generate vibrations and current surges. The antenna, whose voltage-to-standing wave ratio is always greater than one, reflects part of the energy towards the feed. Finally, leaks between channels can disturb operation. Eliminating these problems involves a combination of various solutions, which we cannot examine here. We can, however, quote, for example, the use of isolators, shielded coaxial cables and plugs, etc.

To give you a more concrete idea of the order of magnitude, here are some typical power values in dBW:

- peak power transmitted: 40 dBW with $T = 1$ μs
- maximum admissible peak power at microwave receiver entrance without causing saturation: –70 dBW
- peak power detectable in low-PRF modes (non-Doppler): –130 dBW with $B = 1$ MHz and S/N $= 10$ dB
- peak power detectable in high-PRF modes with velocity filters: –160 dBW
- power of signals supplied by the frequency oscillator: –20 dBW

Thus, the ratio between the "interference" transmission signal and its detectable level at the microwave receiver input can reach 170 dB, and even 200 dB! Similarly, the ratio of the signals supplied by the frequency oscillator to the detectable input level is over 110. These figures give some idea of the difficulty involved in creating a transmitter-receiver unit with electronic compatibility. In practice, even though the receiver is well protected, it must be blocked during the transmission pulse and cannot be used for detection.

12.3.4 INTERMEDIATE FREQUENCY RECEIVER

The intermediate frequency receiver does not cause interference. However, it should be protected from other components, digital processing, and digital links in particular. This kind of receiver, whose bandwidth matches that of the transmitted signal, is in fact a filter that rejects unwanted frequencies. In the most advanced radars, it features double frequency change to eliminate "image" frequencies (see Section 12.5). The signals are phase- and amplitude-demodulated on output from each of the receiver channels, then digitized using analog-to-digital converters (ADC). The ADCs are installed in the receiver and can be a major source of interference if specific precautions are not taken (casings, specific power supply, etc.).

12.3.5 DIGITAL PROCESSING

Digital processing (signal, data, radar map) generates strong interference phenomena. These systems consist of hundreds of integrated circuits functioning synchronously and producing considerable current surges (several hundred amps) on state transition. These digital circuits, designed to be used in association with each other, have a high built-in immunity to noise and spurious signals (e.g., 0.3 V). It is therefore best to group the logic circuits and provide them with a specific power supply. However, the associated connections and the communication bus radiate energy whose wide spectrum (several hundred MHz) is not stationary. It in fact depends on distribution of the states of each bit transmitted. Because "velocity," density, and the number of circuits continue to increase, it is highly recommended that you use optical technology for the bus and the connections between the boards and the modules (fiber optics, holography, etc.), as it is nonradiating and insensitive to electromagnetic radiation.

12.4 INTER-EQUIPMENT INTERFERENCE ON THE PLATFORM

Integrating several systems onto the same platform means taking account of the compatibilities between the different items of equipment and, in particular, "active" transmitting devices with antennas. This is particularly the case for radar, electronic countermeasures, and communication and identification equipment.

This equipment covers an impressive frequency range, from a few MHz to 100 GHz. The main means of ensuring compatibility between the different systems are

- decoupling the antenna systems
- decoupling frequency
- managing the operating phases of the different equipment items
- adding protective elements to certain equipment

12.4.1 DECOUPLING THE ANTENNA SYSTEMS

In theory, the radar antenna is the most directive antenna. Its diagram has very low side and far lobes. Moreover, it generally uses a narrow band (a few percent). IFF and altimetric radar antennas are also directive, but less so. Often the IFF antenna dipoles are fitted onto the radar antenna, in which case decoupling cannot be very high (40 to 50 dB). Communication equipment and electronic countermeasures antenna systems are in theory omnidirectional, and therefore have low gain. Given the obstruction caused by the platform structures and the distance between antenna systems, decoupling in the radar frequency band can reach 60 to 100 dB. In the case of jammers fitted with electronic scanning antennas with the aim of focusing energy onto the target to be jammed, radar-jammer decoupling can be improved. In summary, most of the decoupling between antenna systems is achieved at radar level due to the qualities of its antenna, its location (e.g., in the platform nose), and certain masks.

12.4.2 FREQUENCY DECOUPLING

In principle, different equipment operating in different frequency bands should not interfere with each other, particularly if saturation is avoided. In practice, the situation is quite the reverse, as the spectral purity of most of this equipment is insufficient and its harmonics may be located in the radar-useful bandwidth. Thus the harmonics 8 to 11 of the IFF interrogator (1 030 MHz, 1 kW peak) are located in the radar X-band. For example, with 0.1% harmonics and 50 dB decoupling, the level at the receiver input is approximately –50 dBW, which is far above the maximum allowable level. In addition, radar jammers, installed on the same platform as the radar, sometimes transmit in the same bandwidth as the radar or in a neighboring bandwidth. Similarly, when the radar is transmitting, the radar detector may suffer from interference. These interactions should therefore be managed at system level.

12.4.3 OPERATION MANAGEMENT

Operation of the different items of equipment must be managed at weapons-system level. In fact, system configuration is modified depending on the mission, the functions, and the modes. To ensure compatibility, priorities have to be defined, temporal decoupling have to be introduced, etc.

Final design of the whole system is achieved by simulation and measurements performed in an anechoic chamber.

12.5 UNINTENTIONAL INTERACTIONS

Even though compatibility on the platform has been established, unintentional interactions can still considerably disturb the operation of some equipment, the radar in particular. Interaction with other friendly or enemy systems may occur either inside or outside the radar bandwidth.

12.5.1 INTERACTIONS OUTSIDE THE RADAR BANDWIDTH

Interactions caused by frequencies outside the radar bandwidth can be avoided if the radar is designed to take these interactions into account. Figure 12.3, a detailed illustration of the radar reception chain shown in Figure 12.2, shows the main ways to ensure compatibility and protection against interference.

The antenna has a dual effect. First, it is a band filter that provides initial protection. However, some antennas use a wide bandwidth and produce little attenuation outside the radar bandwidth. Secondly, the directivity of

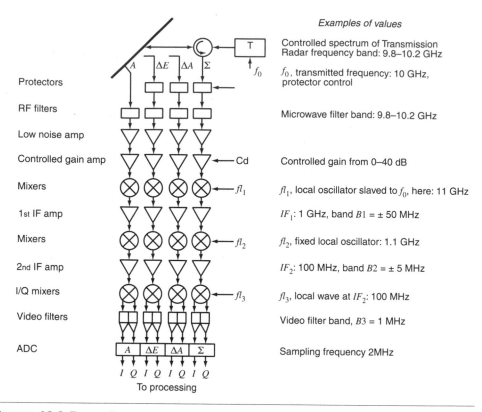

Examples of values

Controlled spectrum of Transmission
Radar frequency band: 9.8–10.2 GHz

f_0, transmitted frequency: 10 GHz,
protector control

Microwave filter band: 9.8–10.2 GHz

Controlled gain from 0–40 dB

fl_1, local oscillator slaved to f_0, here: 11 GHz

IF_1: 1 GHz, band $B1 = \pm 50$ MHz

fl_2, fixed local oscillator: 1.1 GHz

IF_2: 100 MHz, band $B2 = \pm 5$ MHz

fl_3, local wave at IF_2: 100 MHz

Video filter band, $B3 = 1$ MHz

Sampling frequency 2MHz

Protectors

RF filters

Low noise amp

Controlled gain amp

Mixers

1st IF amp

Mixers

2nd IF amp

I/Q mixers

Video filters

ADC

FIGURE 12.3 BLOCK DIAGRAM OF THE RECEPTION CHAIN

its main and the quality of its side and diffuse lobes play an important role in radar function.

Diode protectors except for TR can be time controlled. They are used during radar transmission or to ensure a certain compatibility at platform level. In theory, they have no effect on interactions.

The microwave and band pass filters in the radar bandwidth are highly effective but cause losses of 0.5 to 1 dB.

Controlled-gain amplifiers are used to ensure compatibility or to compensate in part for the R^4 function for low-PRF modes.

12.5.2 INTERACTIONS INSIDE THE RADAR BANDWIDTH

The receiver chain acts as a narrow pass-band filter (in the example shown in Figure 12.3, $B = \pm 1$ MHz to 10 GHz). The gains, bandwidths, and phase rotations of the four channels should be as identical as possible. The amplifier bandwidths should not be under 10%. The video filters match the bandwidth to the form of the transmitted wave or, more often, to the sampling period.

In order to ensure rejection of the "image" frequency band outside the radar bandwidth (for the example shown in Figure 12.3, the bandwidth is between 11.8 and 12.2 GHz), the central frequency of the first intermediate frequency $IF1$ must be clearly higher than the radar bandwidth. Interactions inside the radar bandwidth are further reduced if

- at constant power, the interfering transmission is shifted with respect to f_0
- the interfering wave has a narrow spectrum
- there is major decoupling of the antennas (radiation patterns, range)
- multiplexing inside the radar does little to enrich spectrums on reception

The remaining interactions can be eliminated, or considerably reduced, using the auxiliary channel for off-axis jamming source and by using processing optimized to the waveform emitted by the radar.

For example, taking the values used in Section 12.3.3 and only taking into account the power balance, the range at which perturbation occurs is obtained by applying the formula

$$P_r = P_t \frac{KG_1 G_2 \lambda^2}{(4\pi)^2 R^2},$$

where

$$
\begin{aligned}
K &= 1 \text{ (same frequencies)}, \\
G_1 &= G_2 = 30 \text{ dB, and} \\
R &= 25\ 000\ 000 \text{ km (!!!) for main beam-to-main beam interference;}
\end{aligned}
$$

and $K = 1$, $G_1 = G_2 = -10$ dB (far lobes/far lobes), and $R = 2\ 500$ km for off-boresight configurations.

In conclusion, simultaneous operation of several radars requires frequency management and the optimization of waveforms and processing.

PART III
GROUND MAPPING AND IMAGERY

Two Axis ESCAN Fighter Radar (RBE2 Radar of the Rafale)

13

GROUND MAPPING

13.1 INTRODUCTION

The great majority of airborne radars have an air-to-ground function composed of several modes that invariably use a waveform without range ambiguity (low PRF). This chapter deals with the ground-mapping mode, either with the real antenna beam or with some beam sharpening processing, but using non-coherent radars (Doppler beam sharpening and SAR techniques, as well as spaceborne applications, are addressed in Chapter 15).

Ground mapping consists of providing a ground map with the best possible resolution and accuracy in order to

- enable location with respect to a geographical feature (navigation, position up-date)
- localize a target
- establish a map when optical systems become inoperative

From the map of the ground, the contour mapping modes that will be examined in Chapter 19 ensure selection of ground echoes located above the "clearance planes," thus enabling combat aircraft to carry out low-level flights or blind penetration.

13.2 PRINCIPAL PARAMETERS

13.2.1 AIRCRAFT MOTION

Consider a radar installed on board an aircraft whose velocity is v and that has roll, pitch, or yaw motion. A conformal map of the ground configuration will be obtained by compensating for the platform movements and stabilizing the radar antenna relative to the platform. The antenna is fitted with a roll-stabilizing system (mechanical or electronic), followed by a servo-loop in azimuth "carrying" an elevation servo-mechanism (Figure 13.1). When the elevation gimbal carries the azimuth, the performances of all the air-to-ground modes are degraded; the complex demonstration is not given in this work. Antenna stabilization hardware

FIGURE 13.1 ANTENNA STABILIZATION SYSTEM: AZIMUTH CARRYING ELEVATION (FROM THALES DOCUMENT)

attempts to keep the antenna pointing at the target while a map is being made. As some maps take tens of seconds to produce, the stabilization systems must be able to hold the antenna stable over this period.

13.2.2 BEAM SHAPE

The antenna radiation pattern depends on the antenna dimensions with respect to wavelength (see Chapter 1). Use of the real beam without sharpening produces angular resolution, which improves as the beam narrows. It is sometimes useful to increase the beamwidth in elevation while keeping a narrow beam in azimuth (Figure 13.2). In some radars, the antenna beam in elevation obeys a cosecant-squared function, enabling uniform ground illumination regardless of range, that is, with energy proportional to the square of distance R.

Antenna gain at elevation θ_E is

$$G\left(\theta_E\right) = G_0 \csc^2\left(\theta_E\right) = G_0\left(\frac{R}{h}\right)^2$$

Signal-to-noise ratio, with the cosecant-squared function antenna, for a target at range R and elevation θ_E is

$$\mathrm{SNR} = \frac{P_k\,G^2\left(\theta_E\right)\lambda^2}{\left(4\,\pi\right)^3\,F\,KT_0\,R^4\,L}\,G_t = \frac{P_k\,G_0^2\,\lambda^2}{\left(4\,\pi\right)^3\,F\,KT_0\,h^4\,L}\,G_t$$

SNR is independent of the range; the magnitude of the signal returned by a target is constant whatever the range.

In Figure 13.2, the platform is flying over a flat earth and there is no roll angle. In this case, elevation, grazing angle, and antenna elevation angle are identical.

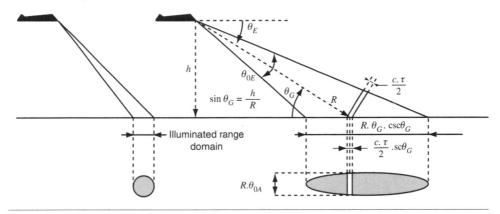

FIGURE 13.2 GROUND ILLUMINATION BY THE BEAM

13.2.3 SIGNAL DYNAMICS ADAPTATION: STC AND LOG RECEIVER

With a narrow beam in both elevation and azimuth (pencil beam), in which the energy radiated is more or less constant (at 3 dB) and regardless of antenna direction, the echo level depends on the following:

- nature of the ground (average backscatter coefficient)
- RCS of point echoes
- range (with R^{-4} for point targets)
- range resolution cell of the radar r
- grazing angle θ_G or the corrected grazing angle for a non-plane ground surface

Depending on the combination of all these parameters, the dynamic range of the ground return signals can be higher than 70 dB (see Section 14.2.7), while that of the display device is only about 20 dB. As a result, dynamic range needs to be compressed without reducing signal contrast too much, as signal contrast is useful in ensuring the quality of ground mapping. One solution is to introduce an attenuator variable in "radar" time, and thus in range on the Σ channel of the microwave receiver, with attenuation being maximum at short ranges (*Sensitivity Time Control*, STC). In practice, the attenuation function, which is theoretically with R^{-4}, should be a function of flight configurations, angle of depression, etc. Such an attenuator operates over a range of about 40 dB.

A second solution, complementary to the first one, is to use a specific receiver at intermediate frequency with gain following a logarithmic function. If the receiver is linear, after detection and analog-to-digital conversion (ADC), use digital transcoding with programmable compression (at the operator's disposal).

As an example, Figure 13.3 shows the same scene, but with two different dynamic compression functions.

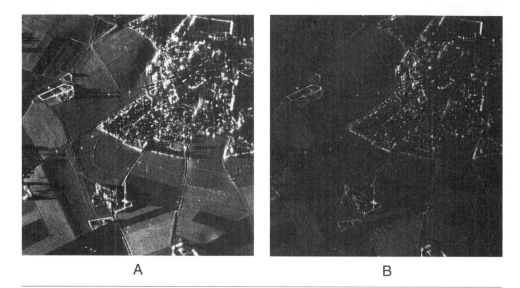

A   B

FIGURE 13.3 (A) DYNAMIC COMPRESSION FUNCTION MATCHED TO THE OBSERVATION OF LOW RCS TARGETS. HIGH-VOLTAGE CABLES CAN BE DISTINGUISHED AT THE BOTTOM LEFT, ON THE DARK FIELD; (B) DYNAMIC COMPRESSION FUNCTION MATCHED TO THE OBSERVATION OF STRONG SCATTERERS. THE CABLES DO NOT APPEAR, BUT THE PYLONS CAN BE MORE EASILY DETECTED. (FROM THALES DOCUMENT)

13.2.4 ANGULAR RESOLUTION

Angular resolution is the minimum angle between two echoes necessary for their discrimination. It improves as aperture diminishes or if the beam is sharpened.

If ground mapping is carried out using the real antenna beam—that is, without beam sharpening—the representation of a point target by a plot is close to the beam aperture in azimuth, as is angular resolution r_A (using *Definition 2* in Section 14.2). In PPI, plot dimensions are proportional to detection range. This is also the case for cross-range resolution r_C ($r_C = R\, r_A$). If post-detection integration is used for a fraction of beam illumination time T_e, time delays in detection occur for both scanning directions. These time delays must be compensated for before any plot representation appears on the display.

EXAMPLES OF VALUES

$PRF = 2\ 000$ Hz

$T = 0.5\ \mu s$

1 000 range gates

$c\ \tau\ /2 = 75$ m

$r_A = \theta_{0A} = 3.5° = 61$ mR

$\omega\ (t) = 100°/s$

$T_e = 35$ ms

at $R = 20$ km,

$r_C = r_A.\ R = 1\ 222$ m

It appears that a real beam cannot achieve extremely high resolution.

13.3 GROUND MAPPING WITH MONOPULSE SHARPENING

As has just been explained, ground mapping without beam sharpening does not produce satisfactory performances in cross-range resolution, in particular at long ranges. If the radar antenna has a monopulse channel in azimuth, the beam can be sharpened but without any actual improvement of resolution. At a given range, if there is only one echo in the beam, this beam will be sharpened. If several echoes can be detected, only the beam with the highest-level echo will be sharpened, the others being masked. When there are several echoes of approximately the same level, sharpening will be achieved on the centroid of these echoes.

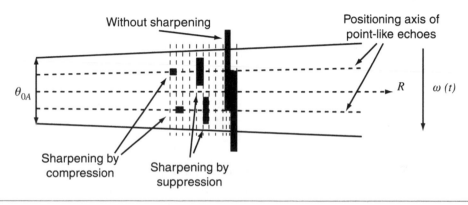

FIGURE 13.4 GROUND MAPPING WITH AND WITHOUT MONOPULSE SHARPENING

Monopulse sharpening is based on knowledge of the off-boresight angle in the beam of the detected echo, δ_A, for each range gate and at each interpulse period. The instantaneous measurement of monopulse angular difference is used:

$$\delta_A = \frac{\Delta \cdot \Sigma}{|\Sigma|^2} = \frac{\Delta}{\Sigma}$$

13.3.1 SHARPENING BY SUPPRESSION

Sharpening by suppression consists of eliminating the presentation of echoes that are beyond a given off-boresight angle in the beam. Thus if δ_A > threshold, the echo is eliminated, while the threshold can be defined as a function of the range (see Figure 13.4).

13.3.2 SHARPENING BY COMPRESSION

Sharpening by compression produces better performance than sharpening by suppression. It consists of positioning the detected echo "in its proper place," at each measurement and throughout illumination time.

For each antenna position in the scan and for each range gate, the position of an echo detected in the beam is measured using the monopulse principle. The beamwidth is shared in about ten smaller "sub-beams." The echo detected is associated to one of them and added to the signal magnitudes already present in the same sub-beam.

The output is a fixed and point-like echo. With this kind of sharpening, the ground map can appear too "plotted," in which case it may be useful to use compression only partially and associate this type of sharpening with sharpening by suppression.

With these techniques, compression of ten can be reached. In the example of Section 13.2.4, this gives an echo width of 122 m, which is not so different than the 75 m range resolution. This means that the map will look more homogeneous, and it may appear as if it would be obtained through Doppler beam sharpening techniques.

In any case, this is only a display technique. As it does not use the Doppler effect for echo positioning, the positioning accuracy in azimuth still depends on antenna boresighting and monopulse measurement. As a result, a root-mean-square error of 5 mR (i.e., 8% relative to the 61 mR antenna beamwidth), representing 100 m at 20 km, is a very good value, where Doppler techniques can be more than ten times better.

14

RADAR IMAGERY

14.1 IMAGING RADAR APPLICATIONS

Imaging function is essentially used in military applications for intelligence, surveillance, reconnaissance, and navigation.

There are numerous possibilities for civilian applications, but the majority of them are still the subject of scientific discussion. Some typical examples of applications are

- vegetation monitoring: identification and estimation of crop growth, surveillance of forests, evaluation of desertification
- ocean surveillance: detection of waves and hydrocarbon pollution
- hydrology: estimation of ground moisture, maps of rivers, expanses of water and flooded areas
- ice surveillance: displacement measurement of ice floes, navigation channel surveillance, iceberg detection
- cartography: production of maps and three-dimensional digital terrain models, measurement of landslides and erosion phenomena
- geology: identification of geological structures using low-frequency waves to pass through plant cover

Imaging radars are not the only sensors capable of providing this kind of information. Their main competitors are

- optical systems operating in the visible or infrared bands
- other types of radar: scatterometers and altimeters
- radiometers (passive electromagnetic systems operating in the radar frequency bands)
- active optical systems (laser radar)

A combination of these different sensors can be used to meet a special requirement. The advantages of imaging radars are numerous: they provide round-the-clock and all-weather capability and are very long-range and high-resolution radars, similar to that of optical systems. The appearance of the image is different than that of optical images, which in certain cases may be considered a drawback.

14.2 IMAGE QUALITY

The image quality required determines the specification for imaging radar design. Image quality is quantified by seven main criteria, each of them involving several parameters:

- resolution
- geometrical linearity
- signal-to-noise ratio
- radiometric resolution
- radiometric linearity
- contrast
- dynamic range

14.2.1 RESOLUTION

Resolution is the ability of a radar to distinguish between two closely spaced point targets (see Section 6.6.2). Figure 14.1 shows the magnitude of two point target signals, which present the same radar cross section, at the receiver output. It illustrates the different possible definitions of resolution.

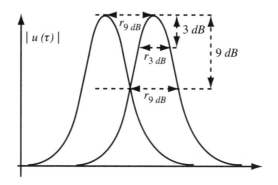

FIGURE 14.1 DEFINITIONS OF RESOLUTION

DEFINITION 1

Resolution is the signal peak width of a single point echo measured at 3 dB (or at 6 dB, or at 9 dB) below its maximum value (Figure 14.1). This definition is the simplest one to verify by measurement, since it deals with one target only.

DEFINITION 2

Resolution is the minimum interval required between two point targets with the same RCS in order to observe a trough between the two peaks on output from the matched filter. It corresponds to the peak width of a single point echo measured at 6 dB.

Consider two identical target signal returns of magnitude a, with a position shift equal to the 6 dB peak width, and a random phase shift, φ, between them. The magnitude of the trough between both targets is

$$\left|u(0)_{a,\,\varphi}\right| = \left|\frac{a}{2} + \frac{a}{2}e^{j\varphi}\right| = a\sqrt{\frac{\cos(\varphi) + \cos^2(\varphi)}{2}}\,.$$

The maximum value of this function is a and is reached only for $\varphi = 0$. This means that for any value of phase shift between both target signal returns, $\left|u(0)_{a,\varphi}\right|$ is equal or lower than a; there will be a trough between both peaks if their position shift is greater than the 6 dB peak width.

This definition has a limited operational interest because, in most cases, the trough is not deep enough to be observable. This leads to the third definition.

DEFINITION 3

Resolution is the minimum interval required between two point targets with the same RCS in order to give a trough of more than 3 dB between the two peaks on output from the matched filter. This definition gives a value about two times as great as the resolution at 3 dB of Definition 1.

To establish a link between the two definitions, consider again the case where the two signals at the matched filter output are added in phase ($\varphi = 0$). If the intersection of the two peaks is 9 dB below their maximum value, then the trough between the two peaks is 3 dB. Under any other configurations of relative phase between signals from the two targets, the trough is greater than 3 dB. Therefore, resolution as given by the second definition corresponds to the resolution at 9 dB of the first definition (9 dB width of the peak at the output of the matched filter).

14.2.1.1 TEMPORAL RESOLUTION AND FREQUENCY BANDWIDTH

Part I explained that the range resolution is inversely proportional to the bandwidth of the signal received. This section proposes a method to verify this result and then to extend it to other parameters, such as the *Doppler frequency* or the *Doppler slope* (derivative of the Doppler frequency).

For signals that are identical but shifted in time, radar resolution is directly linked to the width of the correlation peak (output of the matched filter). An approximate value of resolution—that is, the width of the correlation peak at 3 dB from its maximum value—can be calculated using an order-2 Taylor expansion of the autocorrelation function, around the correlation peak. The autocorrelation function is therefore expressed as an inverse Fourier transform of the signal spectrum. Consequently, as shown in Cook (1967),

$$\tau \approx \frac{\sqrt{2}}{2\pi} \left(\frac{\int_{-\infty}^{+\infty} |U(f)|^2 \, df}{\int_{-\infty}^{+\infty} f^2 |U(f)|^2 \, df} \right)^{1/2} = \left(\int_{-\infty}^{+\infty} f^2 |U(f)|^2 \, df \right)^{-1/2},$$

where τ is expressible as the inverse of a bandwidth. The notion of equivalent bandwidth can be introduced:

$$B_e = 2\sqrt{\pi} \left(\frac{\int_{-\infty}^{+\infty} f^2 |U(f)|^2 \, df}{\int_{-\infty}^{+\infty} |U(f)|^2 \, df} \right)^{1/2} = \left(\int_{-\infty}^{+\infty} f^2 |U(f)|^2 \, df \right)^{1/2}$$

The resolution, which depends on B_e, is given by

$$\tau \approx \sqrt{\frac{2}{\pi}} \frac{1}{B_e} = \frac{0.8}{B_e}.$$

Resolution is equal to the inverse of the equivalent bandwidth of the received signal (to within one factor, close to one).

14.2.1.2 FREQUENCY RESOLUTION AND DURATION OF THE ANALYZED SIGNAL

It is useful to calculate the resolution of the matched processing for signals with a Doppler frequency shift. This is assumed as being constant during the signal but different for each other signal received. Using the same kind of calculation performed in the previous section, we obtain the following expression for frequency resolution:

$$v_3 \approx \frac{\sqrt{2}}{2\pi} \left(\frac{\int_{T_e} |u(t)|^2 \, dt}{\int_{T_e} t^2 |u(t)|^2 \, dt} \right)^{1/2} = \frac{\sqrt{2}}{2\pi} \left(\int_{T_e} t^2 |u(t)|^2 \, dt \right)^{-1/2},$$

where v_3 is expressible as a time inverse. The notion of signal-equivalent duration can be introduced, as for bandwidth:

$$\tau_e = 2\sqrt{\pi} \left(\int_{T_e} t^2 |u(t)|^2 \, dt \right)^{1/2}$$

Depending on τ_e, resolution is

$$v_3 \approx \sqrt{\frac{2}{\pi}\frac{1}{\tau_e}} = \frac{0.8}{\tau_e}.$$

Doppler frequency resolution is equal to the inverse of the equivalent duration of the signal (to within one factor, close to one). The Doppler frequency is a powerful parameter for discriminating targets.

14.2.1.3 RELATION BETWEEN DOPPLER FREQUENCY DERIVATIVE RESOLUTION AND THE SQUARE OF THE DURATION OF THE ANALYZED SIGNAL

This third parameter is less currently used. Here, modulation of the received signal is made more complicated by assuming that the Doppler frequency shift of the received signal is not constant throughout its duration. This is the case for the focused synthetic antenna and for some ISAR configurations. In the latter case, the derivative of the Doppler frequency, also called the *Doppler slope*, may be a discriminative parameter. It then becomes interesting to calculate the discriminating power of the matched processing for signals with a Doppler frequency that varies linearly, that is, whose derivative is constant but different for each received signal.

The received signal has a frequency shift, f_D, at the initial instant and a variation of the frequency shift with a Doppler slope $\dot{f}_D$:

$$s(t) = Au(t - t_0)e^{j(2\pi f_D t + \pi \dot{f}_D t^2)}$$

The transfer function of the receiver matched to frequency shift Δf and gradient $\Delta \dot{f}$ is

$$h(t) = u*(-t)e^{-j(2\pi \Delta f t + \pi \Delta \dot{f} t^2)}.$$

The magnitude of the signal output from the matched receiver is described by

$$y(\Delta t, \Delta f, \Delta \dot{f}) = \left| \int_{T_e} u(t - t_0)u(t - \Delta t)e^{j2\pi \left((f_D - \Delta f + \Delta \dot{f} \Delta t)t + (\dot{f}_D - \Delta \dot{f})\frac{t^2}{2} \right)} dt \right|^2.$$

If the target is placed at the time origin, that is by substituting t with $t - t_0$, and if $\tau = \Delta t - t_0$, $v = \Delta f - f_D$, $\dot{v} = \Delta \dot{f} - \dot{f}_D$, a particular form of the generalized ambiguity function is obtained:

$$\left| x(\tau, v, \dot{v}, \Delta \dot{f}) \right|^2 = \left| \int_{T_e} u(t)u*(t - \tau)e^{-j2\pi \left((v - \Delta \dot{f} \Delta t)t + \dot{v}\frac{t^2}{2} \right)} dt \right|^2
$$

By using a second-degree Taylor expansion, dependent on $\dot{v}$ around $(\tau, v, \dot{v}, \Delta f) = (0, 0, 0, 0)$, and a calculation similar to the calculation already performed for time and frequency, it is possible to express the resolution, $\dot{v}_3$, in accordance with frequency derivative:

$$\dot{v}_3 \approx \frac{\sqrt{2}}{\pi} \left(\frac{\int_{T_e} |u(t)|^2 \, dt}{\int_{T_e} t^4 |u(t)|^2 \, dt} \right)^{1/2} = \left(\int_{T_e} t^4 |u(t)|^2 \, dt \right)^{-1/2},$$

where $\dot{v}_3$ is equivalent to a squared time inverse. Resolution with regard to the frequency derivative is linked to the inverse of the square of the signal duration. For a signal of the rectangular type with time T_e, resolution is given by $\dot{v}_3 \approx 4/T_e^2$. High resolution is achieved only for a long observation time. For most systems, the Doppler slope is a poor discriminating factor. Conversely, it is of great help for ISAR imaging of a vessel, when the target is illuminated during several seconds (Section 17.3.2).

14.2.2 GEOMETRICAL LINEARITY

Geometrical linearity consists of three successive levels of increasingly stringent requirements for the radar and the system into which it is integrated:

* accuracy in localizing a point relative to another on the same image
* accuracy in positioning a point on an image in relation to the radar platform
* accuracy in localizing a point on an image in a system of geographical (i.e., inertial) coordinates

Localization accuracy is developed in Section 16.8 in the case of the synthetic aperture radar.

14.2.3 SIGNAL-TO-NOISE RATIO

For most applications, such as detection, it is useful to maximize the signal-to-noise ratio, which also has an influence on radiometric resolution. Section 16.7 describes the signal-to-noise ratio for an imaging radar.

14.2.4 RADIOMETRIC RESOLUTION

Radiometric resolution is the ability to discriminate between two diffuse targets with similar backscatter coefficients. It is a measurement of the speckle of the map, which is due to three types of spurious signals:

* additive noise, whose power is independent of the power of the received signal, such as thermal noise

- multiplicative noise, whose power is proportional to the power of the received signal, such as natural fluctuation of targets, and phase noise from the frequency source and the transmitter. If the area illuminated by the radar is a homogeneous surface, the backscattered signal has a fluctuation that obeys an exponential function. This phenomenon is entirely independent of the radar power budget. It is caused by amplitude and phase recombining of the signal reflected by the multitude of elementary scatterers that form the observed surface
- noise that cannot be classified in the two previous categories, such as quantization and encoding noise

Radiometric resolution can be quantified in decibels using the following parameters:

$$\rho = 10 \log \left(1 + \frac{\sigma_p}{\sigma_0}\right),$$

in which σ_0 is the average value of the target backscatter coefficient and σ_p is the standard deviation of the fluctuation.

Radiometric resolution can be improved by using the non-coherent sum of several identical images (or looks) of the same site independently of each other. (The multilook process is described at the end of the next chapter.) This method is similar to post-detection integration as described in the general theory of radar, with the following assumptions:

- quadratic detector: the value given is proportional to signal power
- non-coherent post-detection integration, on totally decorrelated samples
- Swerling 2 target: quickly fluctuating, in accordance with an exponential function

From detection theory (Berkowitz 1965), it is possible to determine radiometric resolution as a function of the number of averaged looks and the signal-to-noise ratio of each look. The probability density of the signal output from post-detection integration is given by a Rice function:

$$f(Q) = \frac{1}{(S+N)^2(n-1)!} Q^{n-1} e^{-\frac{Q}{S+N}} .$$

Its average value and its variance are, respectively,

$$m_1 = n(S+N) \quad \text{and} \quad v = n(S+N)^2$$

Thus, radiometric resolution takes the form

$$\rho = 10\log\left(1 + \frac{\sqrt{n}(S+N)}{nS}\right) = 10\log\left(1 + \frac{(1+S/N)}{\sqrt{n}S/N}\right).$$

Ideal radiometric resolution (zero dB) is achieved for an infinite number of averaged looks. An infinite signal-to-noise ratio gives only 3 dB radiometric resolution. So the number of looks rather than the signal-to-noise ratio should be maximized in order to improve radiometric resolution.

14.2.5 RADIOMETRIC LINEARITY

In order to measure radar cross section (σ) or backscattering coefficients (σ_0), the radar must be calibrated. Noise level and, generally, spurious signal level (far lobes, ambiguities, encoding noise) must be known in order to perform this measurement.

14.2.6 CONTRAST

The radar image consists of two parts:

- a useful image that concentrates the major part of the energy
- a spurious image that reduces contrast

The spurious image depends on the ambiguity function of the waveform and the receiver. It is a 2-D surface, but with two principal directions in which spurious signals are located: range and cross-range. The spurious image is produced by two phenomena: ambiguity peaks, which create shifted, attenuated, and distorted images with respect to the principal image on the one hand; and the pedestal (side and far lobes), which creates a spurious image, on the other. These effects are quantified by several parameters, which apply to range and cross-range:

- signal-to-ambiguity ratio (S/A), defined both for point and diffuse targets
- peak-to-side-lobes ratio (*PSLR*, Figure 14.2). The *PSLR* is a power ratio between the main peak and the side lobes located in an interval of ten times the peak width
- secondary-side-lobes ratio (*SSLR*, Figure 14.3). The *SSLR* is a power ratio between the main peak and the side lobes located in an interval between ten times and 20 times the peak width
- integrated side-lobes ratio (*ISLR*, Figure 14.4). The *ISLR* is an energy ratio. The *ISLR* in range is given by

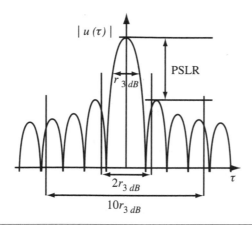

FIGURE 14.2 DEFINITION OF THE PSLR

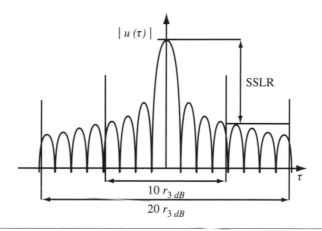

FIGURE 14.3 DEFINITION OF THE SSLR

$$ISLR = \frac{\text{energy in the main peak}}{\text{energy in the side lobes}}$$

$$= \frac{\int_{1/B}^{-1/B} |u(\tau)|^2 d\tau}{\int_{-T}^{-1/B} |u(\tau)|^2 d\tau + \int_{-1/B}^{T} |u(\tau)|^2 d\tau} \quad ,$$

where $|u(\tau)|$ is the impulse response, that is, the signal at the output of the matched filter. T is the pulse length. The length of the impulse response is $2T$. In practice, the far side lobes are very low and might be impossible to measure. That is why there is a second definition of *ISLR* in which the integration is limited to an interval of 20 times the peak width: the *ISLR Measured* (Figure 14.5)

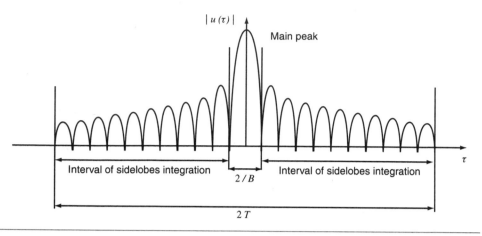

FIGURE 14.4 ISLR INTERVAL OF INTEGRATION

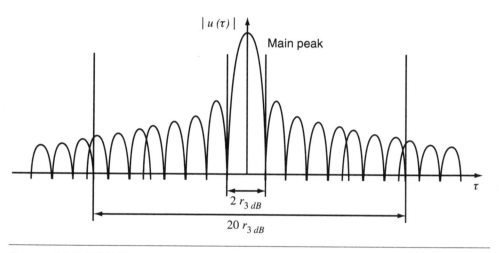

FIGURE 14.5 ISLR MEASURED: INTERVAL OF INTEGRATION

14.2.7 DYNAMIC RANGE

The dynamic range of an imaging radar can be measured at various points of the receiving chain (see Section 1.4):

- At the receiver input (Figure 14.6), it is the ratio between the maximum signal it can receive and the thermal noise. The maximum received signal is the one for which saturation noise and the effects of the nonlinearities remain under a given threshold. This threshold is chosen to avoid degrading the image contrast so much that no- or low-return areas and shadows can no longer be distinguished

- On the image itself, at the signal processing output, it is the ratio between the maximum-point target return and the noise

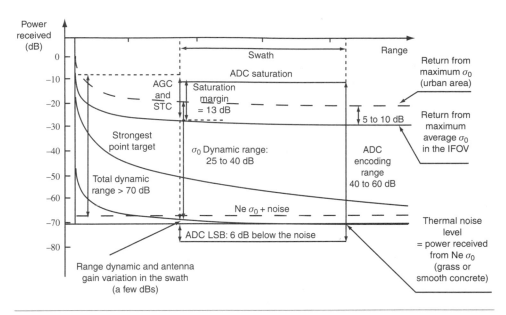

FIGURE 14.6 DYNAMIC RANGE IN THE RECEIVER OF AN X-BAND SAR

The signal received is an average of the signal returned by all surfaces located in the instantaneous field of view (IFOV), defined by the transmitted pulse length and the azimuth beamwidth. Note that due to this averaging effect, the maximum signal power received corresponds to a mean backscatter coefficient that is lower than the maximum backscatter coefficient encountered in the observed area.

The minimum detectable background reflectivity corresponds usually to the thermal noise level. It is known as the "noise equivalent σ_0" ($Ne\sigma_0$).

For an X-band SAR, $Ne\sigma_0$ is set between –25 dBm^2/m^2 and –35 dBm^2/m^2. It is the backscatter coefficient of a smooth surface. As reflectivity increases with the grazing angle, $Ne\sigma_0$ depends on the geometry of the observation. In particular, it is higher for a spaceborne radar than for an airborne one, due to the difference in the incidence angle.

In the receiver, the amplifier's gain is set so that the largest signals pass without noticeable saturation. The main sources of non-linearities and saturation are the analog-to-digital converters. Due to the Gaussian statistics of the signal received, the saturation level of the ADCs has to be set far above the level of the maximum received signal (13 dB in the example of Figure 14.6).

14.2.7.1 NOISE INTRODUCED BY ADCS

The noise produced by ADCs has been treated in the literature by several authors. Max (1960) has calculated the minimization of signal distortion by a saturating ADC with both uniform and nonuniform step sizes. Gray (1971) presents a more detailed noise analysis brought by ADCs with uniform steps that identify both quantization and saturation effects. Sappl (1986) considers a Gaussian random variable with a density probability function, $p(x)$, given by

$$p(x) = \frac{1}{\sigma\sqrt{2\pi}} e^{-\frac{x^2}{2\sigma^2}},$$

where σ stands for the standard deviation of the input signal for the I and Q channels.

The signal-to-quantization noise ratio ($SQNR$) is then given by

$$SQNR = \frac{\sigma^2}{\int\limits_{-\infty}^{+\infty} (x - q(x))^2 p(x) dx},$$

where q is the quantization function.

For an ADC with uniform step size Q, with $2M$ quantization steps, one gets

$$SQNR = \left\{ 1 - \sqrt{\frac{2}{\pi}} \frac{Q}{\sigma} \sum_{m=-M+1}^{M-1} e^{-\left(\frac{mQ}{\sigma\sqrt{2}}\right)^2} + \frac{Q^2}{\sigma^2} \left(\left(M - \frac{1}{2} \right)^2 - \sum_{m=-M+1}^{M-1} m \ \mathrm{erf}\left(\frac{mQ}{\sigma\sqrt{2}} \right) \right) \right\}^{-1}.$$

Figure 14.7 yields the value of the $SQNR$ for different numbers of quantization bits as a function of σ/V_{sat}, where the saturation voltage is given by $V_{sat} = MQ$.

This set of curves enables to draw the following conclusions:

- For values of σ/V_{sat} inferior to the optimal value, the $SQNR$ expressed in dB is a linear function of σ/V_{sat}. In this region, the saturation is negligible. The offset between two adjacent curves is about 6 dB, as in the case of the classical quantization of sinusoids
- Beyond the optimal value of σ/V_{sat}, saturation effects degrade the $SQNR$ very abruptly

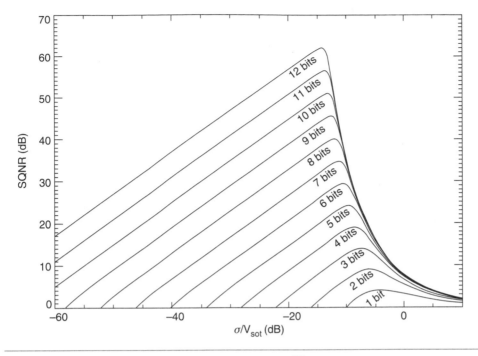

FIGURE 14.7 CHARTS OF SNR AS A FUNCTION OF σ/V_{sat} FOR DIFFERENT NUMBERS OF BITS

That is why, as said earlier, the receiver gains are set so that no saturation occurs for the largest expected signals.

Table 14.1 summarizes the parameters associated with the optimal values of SQNR for different numbers of bits.

TABLE 14.1. OPTIMAL VALUES OF Q/σ, σ/V_{sat}, AND SQNR AS A FUNCTION OF THE NUMBER OF BITS

Number of Bits	Q/σ opt.	σ/V_sat opt. (dB)	SQNR opt. (dB)
1	1.596 E+0	−4.06	4.40
2	9.957 E−1	−5.98	9.25
3	5.860 E−1	−7.40	14.27
4	3.352 E−1	−8.57	19.38
5	1.881 E−1	−9.57	24.57
6	1.041 E−1	−10.45	29.83
7	5.687 E−2	−11.22	35.17
8	3.076 E−2	−11.90	40.57
9	1.650 E−2	−12.51	46.03
10	8.785 E−3	−13.06	51.55
11	4.650 E−3	−13.55	57.11
12	2.448 E−3	−14.00	62.71

14.2.7.2 TOTAL AND INSTANTANEOUS INPUT SIGNAL DYNAMIC RANGE

The received SAR signal shows a considerable variation in the signal amplitude. The largest part of this variation is due to the different σ_0 inside the observed area, and the range domain.

DYNAMIC RANGE AT THE INPUT OF THE RECEIVER

In Figure 14.6, the main contributors to the total dynamic range at the input of the receiver are

- the range dynamic: A range of 1 to 100 km, with a range equation function in R^{-3}, corresponds to $30 \log (100/1) = 60$ dB. The illumination by the antenna beam is reducing this value. If the maximum gain is oriented in the direction of the maximum range, then the shortest range is illuminated by the side lobes of the antenna diagram. The attenuation is higher than 20 dB
- the backscatter coefficient variation, for an X-band airborne radar, is between 5 dBm^2/m^2 (strong urban area backscatter) and -35 dBm^2/m^2 (smooth concrete at low grazing angle). Note that for pulse compression radars, the transmitted pulse is long. The *IFOV* is quite wide and the averaging effective: the maximum signal returned corresponds to a backscatter coefficient 5 to 10 dB lower than the maximum backscatter present in the observed area

The total dynamic range is higher than 70 dB.

DYNAMIC RANGE IN THE SWATH

The signal is now considered only in an interval of time corresponding to the observed swath. In this interval, a variable gain control such as an Automatic Gain Control (AGC) or a Sensitivity Time Control (STC) is adjusting the signal power to the level acceptable by the ADCs. The range effect is far lower than at the input of the receiver. The main contributors to the dynamic range are here:

- the backscatter coefficient variation, still of the order of 20 to 35 dB
- the variation of the antenna gain from the middle of the swath to the edges, that contributes between 2 dB up to 6 dB (two-way), depending on the system parameters. The difference in distance to the ground entails a signal attenuation from the near edge to the far edge of the swath. It may vary from zero to 8 dB

The encoding range of the ADCs must present some margins: the saturation is far above the maximum signal received (13 dB in Figure 14.6) to minimize the saturation noise. The LSB is below the noise to minimize the quantification noise (6 dB in Figure 14.6). The total encoding range is between 40 and 60 dB. This corresponds to ADCs with 8 bits (7 bits + sign) to 11 bits (10 bits + sign).

INSTANTANEOUS DYNAMIC RANGE

The receiver gain can even be adapted to the average signal level received during a shorter period of time. The relevant parameter is then the *instantaneous input signal dynamic range.*

Even when two areas with extreme backscatter coefficients are adjacent, the instantaneous signal dynamic range is considerably smaller because the rise or fall of the signal amplitude is spread over the entire pulse length and the antenna azimuth beamwidth. If we look at a fraction of the pulse length, the instantaneous dynamic range is much smaller than the total dynamic range due to the averaging effect. It is also interesting to note that the signal level variation due to the antenna elevation beamwidth and the difference in distance to the ground within this fraction of the pulse length is generally a few tenths of a decibel and therefore negligible.

If we consider our example of two adjacent areas in range with extreme σ_0, and if we look, let us say, at 1/20 of the pulse length, the instantaneous signal amplitude variation within this fraction of time is reduced to about 5 dB. We shall see in the next section that we can use the fact that the instantaneous input signal dynamic range is much smaller than the total input signal dynamic range to efficiently compress the raw data issued from the ADCs.

To conclude this section about the input signal dynamic range, let us take up the question of the influence of point targets to the dynamic range. To understand why the individual point targets do not contribute significantly to the dynamic range, it is useful to compute the order of magnitude of the *IFOV*. Generally the pulse length is several tens of microseconds long, and the projected azimuth beamwidth on the ground is several hundreds of meters. Therefore the area covered by the *IFOV* is, most of the time, greater than 2.10^6 m^2. Even with σ_0 of moderate level, let us say -25 dBm^2/m^2, this represents an equivalent radar reflectivity larger than 6 000 m^2. That is why only very strong scatterers may emerge from the clutter at the ADC level.

This fact is not in discrepancy with the characteristic of SAR images, which show very large dynamic range after processing (between 60 dB and 80 dB). Even if point targets are buried within the overall return of the *IFOV*, the quantization by ADCs does not remove weak or strong scatterer information from the composite signal. The processing gain of pulse compression and azimuth compression preserves low- and high-level signals. Coherent signals from small and strong scatterers benefit from this processing gain, whereas thermal noise and background clutter do not. We just have to ensure that the number of bits used by the processor is large enough to accommodate the final output image dynamic range.

14.2.7.3 BLOCK FLOATING POINT QUANTIZER

As mentioned earlier, the total signal dynamic range is as high as 50 dB, whereas the instantaneous signal dynamic range measured on a very short period of time is at most a few dBs. This leads to the concept of the so-called "Block Floating Point Quantizer" (BFPQ) (Joo 1985) where an ADC with a large number of bits (let us say eight in our case) digitizes the received signal and passes it to a compression unit.

Dynamic compression is required in most SAR systems, mainly for storage purposes or for transmission of the raw data to the ground. BFPQ is a technique commonly used to compress the raw data, in a ratio of 2 to 1. It corresponds to the state of the art. Note that the compression ratio for SAR raw data is very low compared to the ratio achieved in other fields.

Here the received stream of data is divided up into small blocks of, for example, 64 samples and the standard deviation of this block of data is computed. The next step in processing the data is basically a scaling operation (or gain adjustment) followed by a mapping of the data into an ideal ADC with fewer bits (let us say a 4-bit ADC to preserve a 20 dB SQNR). The gain setting of each individual block of data is always transmitted along with the 4-bit data in order to recover the original data after reception on the ground.

With this technique, the 20 dB optimal SQNR of the 4-bit ideal ADC is extended over the range of the actual receiver ADC for which the SQNR is superior to 20 dB. The choice of the block size is important to ensure that the instantaneous dynamic range is compatible with the number of bits used to code the data by the BFPQ. The points to take into account to tackle this problem are as follows:

- The block should contain a sufficient number of samples so that the Gaussian hypothesis is met; this implies a rough number of samples between 32 to 128
- The block should be small enough so that the signal amplitude variations due to the antenna elevation beamwidth and the difference in distance to the ground within the block are negligible
- The block should be small compared to the pulse length to benefit from the averaging of the input scene reflectivity

14.3 SPECIAL TECHNIQUES FOR RANGE RESOLUTION

Many processes are used to obtain high range resolution. Pulse compression using correlation, as described in Chapter 8, is the most common. Three other methods are presented in the following sections:

Deramp, Stepped Frequency, and *Synthetic Bandwidth.* The last two methods are based on frequency jumps and may involve chirp pulses.

14.3.1 DERAMP

In the deramp process, the principle of continuous radar with a linearly frequency-modulated wave is applied to a pulsed radar. Advantage is taken of the fact that the correlation between the received signal and the reference function can be achieved via demodulation by the reference signal followed by a Fourier transform.

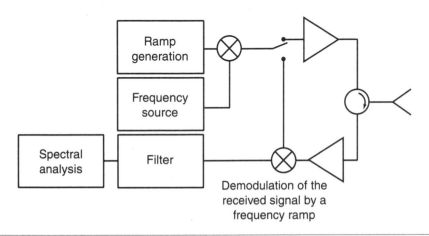

FIGURE 14.8 BLOCK DIAGRAM OF A RADAR USING DERAMP

14.3.1.1 BASIC PRINCIPLES

The transmitted signal is a linearly frequency-modulated pulse. As a first step, the received signal is demodulated by the transmission signal. An echo located at a given range is then represented by a signal whose frequency is constant and proportional to the range. The useful frequency spectrum (i.e., the range domain observed, or the swath) can be selected by means of a filter placed after demodulation. Processing continues with spectral analysis, which separates the various received frequencies.

14.3.1.2 PERFORMANCE CHARACTERISTICS

The received signal is identical to that of a pulse-compression radar:

$$u(t - t_0) = \text{Rect}_T(t - t_0)\, e^{\, j\, \pi\, \frac{B}{T}(t - t_0)^2}$$

The demodulation signal is triggered at instant τ_0, at the beginning of the swath:

$$h(t) = e^{\, -j\, \pi\, \frac{B}{T}(t - \tau_0)^2}$$

After demodulation, the next operation is a Fourier transform performed in a time window of duration T_A (signal analysis time, see Figure 14.9). This window is set such that all the echoes to be displayed in the swath actually send back a signal throughout its duration:

$$S(f) = \int_{T_A} u(t - t_0) h(t) e^{-j 2 \pi f t} dt$$

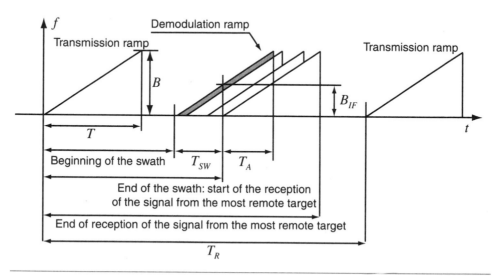

FIGURE 14.9 DERAMP TIMING

Consider $f = (\tau_0 - \Delta t) B / T$. The signal at the receiver output can be written as

$$c(\Delta t) = T_A e^{j \pi \frac{B}{T} (t_0^2 - \tau_0^2)} \frac{\sin\left(\pi B \frac{T_A}{T} (t_0 - \Delta t) \right)}{\pi B \frac{T_A}{T} (t_0 - \Delta t)}.$$

When this expression is compared with the output signal from the true matched filter, we get

$$c(\Delta t) = T \frac{\sin\left(\pi B (t_0 - \Delta t) \left(1 + \frac{(t_0 - \Delta t)}{T} \right) \right)}{\pi B (t_0 - \Delta t)}.$$

Two main differences are noted:

- This time, the magnitude corresponds exactly to a sinc function, but its width is greater by a ratio of T/T_A. It follows that resolution is degraded in the same ratio. This result is simply due to the fact that the signal band effectively used for processing is no longer B but BT_A/T

- A phase term remains. It is a function of target range, in accordance with a quadratic function. It will have to be corrected if synthetic antenna type processing (with range migration) or interferometry is subsequently performed

14.3.1.3 SIZING CONSTRAINTS

A trade-off must be found between resolution and swath in order to use the deramp mode. Given that

$$T = T_{sw} + T_A,$$

the resolution is

$$r = \frac{k}{B(T_A/T)} = \frac{kT}{B(T - T_{sw})},$$

where k is a coefficient close to 1 that expresses the influence of the weighting window used to reduce the level of side lobes.

The bandwidth of the filter that selects the swath is proportional to swath duration:

$$B_{IF} = B\frac{T_{sw}}{T}$$

The sampling frequency of the analog-to-digital converters located after the filter is determined by B_{FI}, which is smaller than B.

EXAMPLE

radar resolution:	$r = 1$ m
swath to be displayed:	$R_{sw} = 1\ 500$ m
duration of transmitted pulses:	$T = 50\ \mu s$
duration of analysis window:	$T_A = T - 2\ R_{sw}/c = 40\ \mu s$
bandwidth to be analyzed:	$B\ T_A/T = k\ c/2\ r = 180$ MHz (with $k = 1.2$)
transmitted bandwidth:	$B = 225$ MHz
filter bandwidth:	$B_{IF} = B\ (1 - T_A/T) = 45$ MHz

In this case, a radar with a typical resolution of 4 m (reception circuit bandwidth of the order of 45 MHz) can reach, on a small area, a resolution four times better. It must, however, be able to transmit a frequency ramp whose bandwidth is larger than the bandwidth that would strictly be required to reach 1 m resolution (225 MHz instead of 180 MHz).

Swath can be extended by oversweeping the demodulation ramp.

14.3.1.4 Applications

The *deramp* technique produces radars with very high resolution, transmitting pulses with a very broad bandwidth while limiting the receiver bandwidth on intermediate frequency to a much lower value. On the other hand, swath is reduced. One application is very high-resolution imaging of small ground surfaces. In this case, this process of high range resolution can be combined with the synthetic antenna spotlight process described in the next chapter. Another application is altimetry, where the range domain observed at a given moment is effectively very small.

14.3.2 Stepped Frequency

Stepped frequency enables radars with a narrow instantaneous bandwidth, but with a transmission frequency agility, to achieve high range resolution.

14.3.2.1 Basic Principles

The transmitted waveform is composed of pulses of duration T and bandwidth B (Figure 14.10). They are spaced at intervals of ΔT for time and Δf for frequency. The pattern is the result of N successive pulses.

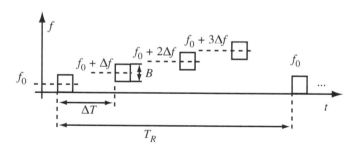

FIGURE 14.10 Stepped Frequency Waveform

Pulses are received by a matched receiver. The output signals from this receiver are sampled and converted to digital. The next step in processing is a discrete Fourier Transform on N samples. The samples belong to the same range gate and are acquired successively for N transmission frequencies.

14.3.2.2 Performance Characteristics

The transmitted signal is

$$u_e(t) = \sum_{n=0}^{n=N-1} \Re[u(t-n\Delta T)e^{j2\pi(f_0+n\Delta f)(t-n\Delta T)}].$$

The received signal can be written (in the case of a fixed target, zero Doppler frequency, and no range migration) as

$$s_r(t) = \sum_{n=0}^{n=N-1} \Re[u(t - n\Delta T - t_0)e^{j2\pi(f_0 + n\Delta f)(t - n\Delta T - t_0)}].$$

After demodulation by the carrier frequency of each pulse, the signal can be written as

$$s(t) = \sum_{n=0}^{n=N-1} u(t - n\Delta T - t_0)e^{-j2\pi n\Delta f t_0}e^{-j2\pi f_0 t_0}.$$

In Figure 14.11, triangles represent the response of the pulse-matched receiver filter to three echoes located at various ranges.

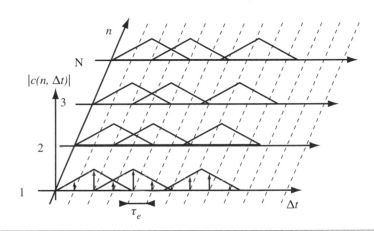

FIGURE 14.11 OUTPUT OF THE PULSE-MATCHED RECEIVER

The receiver matched to the waveform carries out the following operation:

$$c(\Delta t) = \int_{T_R} \sum_{n=0}^{n=N-1} u(t - n\Delta T - t_0)e^{-j2\pi n\Delta f t_0} \sum_{m=0}^{m=M-1} u^*(t - m\Delta T - \Delta t)e^{j2\pi m\Delta f\Delta t}\, dt$$

$$|c(\Delta t)| = \left| \int_T u(t - t_0)u^*(t - \Delta t)dt \frac{\sin(\pi N\Delta f(t_0 - \Delta t))}{\sin(\pi \Delta f(t_0 - \Delta t))} \right|$$

The response of the pulse-matched receiver is multiplied by a function similar to the sinc function ($\sin x / x$), whose width is: $\tau_{3dB} = 1 / N\, \Delta f$. It is composed of a periodic term, the period being $1 / \Delta f$. Consequently, resolution is determined by the spectrum width of the transmitted waveform, and ambiguity is given by the interval between two successive frequencies.

14.3.2.3 SIZING CONSTRAINTS

Figure 14.12 shows the signal on output from the pulse-matched receiver. Pulses of width $1/B$ are sampled with period τ_e. To comply with the sampling theorem, $\tau_e < 1/B$ (assuming that sampling is on two channels). The signal reflected by a target is received successively at each of the N transmission frequencies.

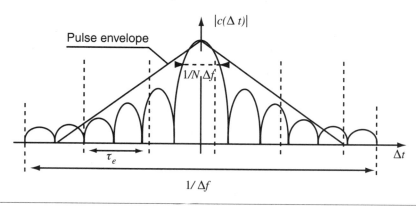

FIGURE 14.12 SIZING CONSTRAINTS

For a given frequency, we observe that a target is present in several successive range gates ($m = 4$ in Figure 14.12), with a different power. To avoid range ambiguities, the Fourier transform analysis must cover at least the m range gates:

$$\frac{1}{\Delta f} > m \tau_e \gg \frac{1}{B}$$

Practically, the constraint is at least

$$\frac{1}{\Delta f} > \frac{1}{2B}.$$

Section 14.3.3 describes the concept of a different processing, *synthetic bandwidth*, which suppresses this sampling ambiguity effect. The resulting constraint is then only

$$\frac{1}{\Delta f} > \frac{1}{B}.$$

It means that the number of pulses required by the synthetic bandwidth principle is less than half the number required by stepped frequency to obtain the same result.

14.3.2.4 APPLICATIONS

The advantage of the process is that the bandwidth of transmitted pulses and the instantaneous bandwidth of the receiver are limited to a value that is much lower than the bandwidth of the resolution aimed for. On the other hand, assuming that N successive frequencies are used, the time required for the coverage of a given swath with a given resolution is N times as great as the time taken for a pulse-compression radar. Inversely, if the required constraint is to maintain a minimum pulse-repetition frequency, as in SAR, the dimension of the swath will be reduced as a ratio of N.

It should be noted that due to the longer time required to sequentially transmit and receive all the frequencies, a radar operating in this way may have limited performance compared to a radar capable of transmitting the whole required bandwidth in a single pulse.

14.3.3 SYNTHETIC BANDWIDTH

Synthetic bandwidth is a particular form of stepped frequency. It makes optimal use of the transmitted bandwidth, leading to shorter sequences of pulses. The computation load, however, is heavier.

14.3.3.1 BASIC PRINCIPLES

As with stepped frequency, the synthetic bandwidth waveform is built on the repetition at the required *PRF* of a sequence made of N identical elementary pulses. Each one is modulated by a different carrier frequency. The difference with a classical stepped frequency waveform is that the overlap between the spectra of the elementary pulses is negligible, while a classical sinc (sin x / x) impulse response after pulse compression is maintained (Queen 1995). Figure 14.13 gives an example of such a waveform with $N = 3$.

The design of the waveform implies that the N carrier frequencies are separated by B. Let us denote k, the rank of a particular pulse within the sequence, and i, its index, such that

$$
k = \begin{cases} i - \dfrac{N-1}{2}, & N \text{ odd} \\[2mm] i - \dfrac{N}{2} + 1, & N \text{ even} \end{cases} \quad, 0 \le i < N
$$

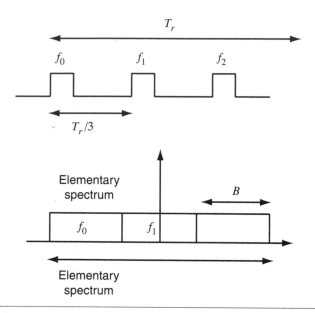

FIGURE 14.13 SYNTHETIC BANDWIDTH WAVEFORM

and

$$f_i = \begin{cases} f_0 + kB, & N \text{ odd} \\ f_0 - B/2 + kB, & N \text{ even} \end{cases} = f_0 + \Delta f_i$$

generates the list of carrier frequencies used by the sequence, where f_0 corresponds to the central frequency of the sequence.

14.3.3.2 PULSE-COMPRESSION ALGORITHM

Figure 14.14 describes the pulse-compression algorithm. The N signals associated with each elementary pulse are first multiplied by a complex exponential term to shift their spectrum to their actual location within the synthetic bandwidth. Each one is then Fourier transformed, usually through an FFT. In the next step, the results are replicated N times so that the equivalent sampling frequency is compatible with the synthetic bandwidth.

Following the spectrum translation, the current pulse spectrum is at its correct position within the synthetic spectrum. The useful frequency bins are selected, while the others are set to zero. Once this has been achieved for each one of the N channels, the synthetic spectrum is created by summing the individual results. To perform the pulse compression, the synthetic spectrum is then multiplied by the spectrum of the matched filter, which is possibly weighted to give lower side lobes.

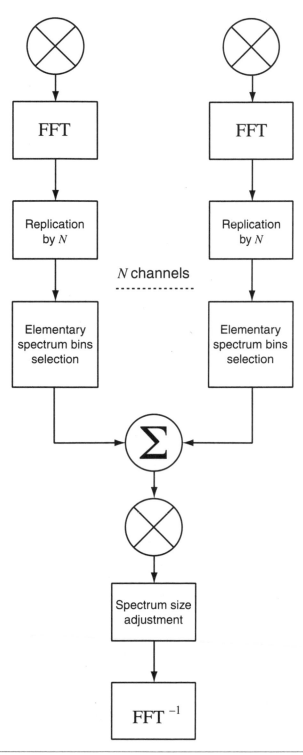

FIGURE 14.14 BLOCK DIAGRAM OF SYNTHETIC BANDWIDTH PULSE COMPRESSION

The last stage consists in coming back to the time domain by taking the inverse Fourier transform of the signal. Seeing that the FFT algorithm is only efficient for a signal-length multiple of powers of 2, 3, and 5, it is often desirable to adjust the size of the synthetic spectrum by zero-padding the high-frequency bins.

To generate the matched filter, two strategies may be envisaged. The first one begins with the analytic computation of the temporal samples of the elementary pulse. These samples then undergo the same operations of synthetic spectrum generation as those that have just been described. The transfer function of the matched filter is eventually obtained by taking the complex conjugate of the result.

It is more operational to use actual samples of the N elementary pulses instead of their analytic expression. Calibration is performed by recording a given number of transmitted sequences before starting the imaging period. By collecting transmitted signals directly at the receiver output (after proper attenuation), it is possible to take into account the transfer function of the waveform generator, the transmitter, the receiver, and possibly the antenna, depending on the calibration hardware.

14.3.3.3 APPLICATION

The main advantages of this waveform are as follows:

- The waveform generator and the receiver are matched to an elementary bandwidth that is N times as small as the synthetic bandwidth used to get the range resolution. Thus the associated hardware is simpler
- As the bandwidth is transmitted by parts, an electronic scanning antenna is less dispersive because weights can be adapted to the center frequency of each pulse
- The radar is better protected against jamming and saturation because of the frequency agility and the smaller receiver bandwidth
- The main disadvantage of the synthetic bandwidth is the reduction by N of the swath width for a given pulse-repetition frequency of the block

15

SYNTHETIC APERTURE RADAR

15.1 DESIGN PRINCIPLE

Synthetic Aperture is a method used to improve radar resolution in azimuth or, more precisely, in the direction of the velocity vector of the platform. This resolution can be compared with that which could be obtained by a very large physical antenna.

The principle can be described from several points of view:

- The radar specialist would describe it as *Doppler processing*. The Doppler frequency is used to discriminate targets and to position them in azimuth
- The antenna specialist would establish a link with *a linear radiating array*. Synthetic Aperture takes its name from this process because it is based on array antenna synthesis
- The physics expert would note that in this kind of processing *the receiver is matched to the signal received*. The output from processing is the autocorrelation function of the input signal

This chapter emphasizes the optimal aspect of Synthetic Aperture processing and explains why it is now used in all airborne and spaceborne imaging radar applications.

The Synthetic Aperture effect can be obtained by displacement of either the platform or the target. In this case, it is called Inverse Synthetic Aperture (Chapter 17).

A SHORT HISTORY

The SAR principle was established in 1951 in the US at the Goodyear Company by Carl Wiley (Wiley 1985). Applications to airborne radars followed immediately. In Europe the first studies were undertaken in France by the CSF company in 1960, and an in-flight experiment was conducted in 1964. The *Raphaël TH* system, used by the French Air Force since the end of the '80s, has remained the unique operational military system of this kind conceived in Europe until now.

15.1.1 SYNTHETIC APERTURE RADAR: A TYPE OF DOPPLER PROCESSING

The radar moves along a rectilinear trajectory, at velocity v and at a constant altitude, above a plane and horizontal surface. The first target direction (M_1) makes an angle, θ_A, with the velocity vector (Figure 15.1). A second target (M_2) is located at the same range as the first one, making angle $\theta_A + \delta\theta_A$.

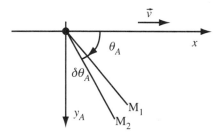

FIGURE 15.1 THE OBSERVATION GEOMETRY IN THE VELOCITY VECTOR AND TARGET PLAN: DISCRIMINATION OF TWO TARGETS LOCATED IN THE ANTENNA LOBE AT THE SAME RANGE

The two targets are located in the antenna main lobe; it is impossible to distinguish one from the other with the antenna beamwidth in azimuth. Relative to the radar, each target has an approach velocity that is the projection of the platform velocity along the target radar axis:

$$v_{r_1} = v\cos\theta_A, \quad v_{r_2} = v\cos\left(\theta_A + \delta\theta_A\right)$$

There is a Doppler frequency for each target velocity (Chapter 3):

$$f_{D_1} = \frac{2v_{r_1}}{\lambda} = \frac{2v}{\lambda}\cos\theta_A, \quad f_{D_2} = \frac{2v_{r_2}}{\lambda} = \frac{2v}{\lambda}\cos\left(\theta_A + \delta\theta_A\right)$$

Leaving aside ambiguity phenomena, measuring the Doppler frequency means measuring target direction, independent of antenna position. Assuming that $\delta\theta_A$ is small compared to θ_A, the difference between these two Doppler frequencies is

$$f_{D_1} - f_{D_2} = \delta f_D = \frac{2v}{\lambda}\left(\cos\theta_A - \cos\left(\theta_A + \delta\theta_A\right)\right) \approx \frac{2v}{\lambda}\sin\theta_A \cdot \delta\theta_A \ .$$

This difference is proportional to the angular separation between the targets.

RESOLUTION IN RELATION TO THE ILLUMINATION TIME OF THE TARGET

The two Doppler frequencies can be found by simple spectral analysis using a Fourier transform. If the analysis takes illumination time T_e, the frequency resolution is $0.886 / T_e$ (if no previous weighting factor has been applied to the signal). Therefore, the smallest Doppler frequency difference that can be established is given by

$$\delta f_D = r_{f_D} = \frac{0,886}{T_e} = \frac{k_r}{T_e} \approx \frac{1}{T_e},$$

where k_r is a coefficient close to one that depends on the quality of the processing performed on the received signal. The corresponding angular resolution is

$$r_{\theta_A} = \frac{k_r \lambda}{2 v T_e \sin\theta_A}.$$

The cross-range resolution, projected onto the ground at distance R_0 is

$$r_c = R_0 r_{\theta_A} = \frac{k_r \lambda R_0}{2 v T_e \sin\theta_A}.$$

This formula shows that in order to achieve high resolution, it is necessary

- to cover the greatest distance possible during target illumination time (product $v\, T_e$). This term should increase in proportion to observation distance in order to maintain constant cross-range resolution
- to point the radar perpendicular to the direction of flight. Optimum resolution is obtained when $\sin\theta_A$ is maximum, that is, for a sight angle of 90° in relation to platform velocity. This is the reason why many SAR are also side-looking antenna radars: the line of sight is perpendicular to the direction of flight

When the platform is an aircraft at low altitude, that is, when the height above the ground is small in comparison to the range, the depression angle approaches zero. The term r_c can be identified as the azimuth resolution.

15.1.2 FOCUSED AND UNFOCUSED SYNTHETIC APERTURE

15.1.2.1 GEOMETRY AND ANGLES

Figure 15.2 describes the geometry of observation.

In this part of the book, the azimuth angle, θ_A is in the plane containing the velocity vector and the target. It is the angle between the velocity vector, $\vec{v}$, and the line-of-sight vector, $\vec{u}$ (unit vector of the radar-to-target axis).

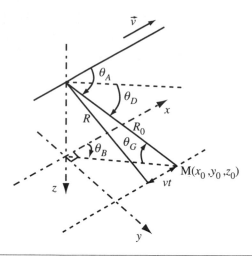

FIGURE 15.2 OBSERVATION GEOMETRY

θ_B is the bearing angle. It is the projection of the azimuth angle on the horizontal plane. θ_B and θ_A are identical when the depression angle is equal to zero.

θ_D is the depression angle. This angle is in the vertical plane. It is the angle between the horizontal plane at the location of the platform and the line of sight.

θ_G is the grazing angle. This angle is in the vertical plane. It is the angle between the horizontal plane at the location of the target and the line of sight. In the case of a flat Earth assumption, it is equal to the depression angle.

In the aircraft-related terrestrial coordinate system (x_h, y_h, z_0) of Section 9.2, the line-of-sight vector coordinates are

$$\vec{u}\begin{pmatrix} \cos\theta_B\ \cos\theta_D \\ \sin\theta_B\ \cos\theta_D \\ \sin\theta_D \end{pmatrix} = \begin{pmatrix} \cos\theta_A \\ \sin\theta_B\ \cos\theta_D \\ \sin\theta_D \end{pmatrix}.$$

15.1.2.2 UNFOCUSED SYNTHETIC APERTURE

In Section 15.1.1, it is assumed that the Doppler frequency of the signals reflected by the targets is constant throughout illumination time. The geometry of the problem remains unchanged: target bearing is not modified and it appears as though the targets are located at an infinite distance from the radar.

Therefore the processing described performed by a simple spectral analysis is matched to targets located at infinity. It is called Unfocused Aperture Synthesis. It is assumed that illumination time is short, and this limits bearing resolution to medium values (around some 10 m, for typical airborne applications).

This kind of processing was used during initial trials of airborne SAR, due to its simplicity.

15.1.2.3 FOCUSED SYNTHETIC APERTURE

Focused Synthetic Aperture processing is more elaborate and takes into account the variation of target angular position with radar velocity during illumination. The received signal can be expressed in relation to the transmitted signal and target range:

$$s(t) = u(t - t_0) e^{-j\frac{4\pi R(t)}{\lambda}}$$

In this section, it is assumed that the waveform is continuous during illumination. The consequences of pulsed waveform will be described later.

At any given moment, the radar-target range is a function of the initial distance and the time. Applying Pythagoras' theorem, it follows that

$$R^2(t) = (x_0 - vt)^2 + y_0^2 + z_0^2$$
$$= R_0^2 - 2R_0 \, vt \cos\theta_A + v^2 t^2$$

and

$$R(t) = R_0 \sqrt{1 - \frac{2\, vt \cos\theta_A}{R_0} + \frac{v^2 t^2}{R_0^2}} \, .$$

In the usual applications, the distance between radar and target is much greater than the length of the synthetic antenna:

$$\forall t \in [-T_e/2, \, T_e/2], \, R_0 \gg |vt| \, .$$

R(t) can consequently be expanded by using

$$(1 + \varepsilon)^{1/2} = 1 + \frac{1}{2}\varepsilon - \frac{1}{8}\varepsilon^2 + ...$$

Limiting the expansion to the second order,

$$R(t) = R_0 - vt\cos\theta_A + \frac{v^2 t^2}{2R_0}\sin^2\theta_A .$$

The phase difference corresponding to wave propagation along the two-way path between radar and target is

$$\varphi(t) = \frac{4\pi R(t)}{\lambda} = -\frac{4\pi R_0}{\lambda} + \frac{4\pi}{\lambda}v\cos\theta_A t - \frac{2\pi v^2}{\lambda R_0}\sin^2\theta_A t^2 .$$

By differentiating this phase difference, we obtain the signal Doppler frequency to within a factor of 2π. It represents the sum of a constant term and a linear term (Figure 15.3):

$$f_D(t) = \frac{2v}{\lambda}\cos\theta_A - \frac{2v^2}{\lambda R_0}\sin^2\theta_A t$$

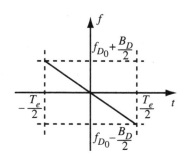

FIGURE 15.3 DOPPLER FREQUENCY VALIDATION DURING ILLUMINATION TIME

The constant term is the central Doppler frequency, f_{D_0}. It is independent of range. It is the only one taken into consideration for an unfocused synthetic antenna. Thus this technique involves first-order expansion of the received signal, while focused aperture is concerned with the higher-order terms.

The linear term corresponds to linear frequency modulation of the received signal, whose slope depends on range. The modulation bandwidth is

$$B_D = \frac{2v^2}{\lambda R_0}\sin^2\theta_A T_e$$

After being demodulated by the RF carrier frequency, the received signal is

$$s(t) = \text{Rect}_{T_e}(t)\, e^{j\left(-\frac{4\pi R_0}{\lambda} + 2\pi f_{D_0} t - \pi \frac{B_D}{T_e} t^2\right)}.$$

Its duration is T_e, and it is linearly frequency modulated on bandwidth B_D around central Doppler frequency f_{D_0}. In practice, it is necessary to take into account the signal modulation by the antenna lobe, which introduces amplitude weighting.

FOCUSED SYNTHETIC APERTURE RESOLUTION

The chirp represented in Figure 15.4 is characteristic of a signal whose phase is a quadratic function of time. Matched-filter processing carries out the following operation (by placing time origin at the minimum-distance crossing point, R_0):

$$|C(\Delta t)| = \left| \int s(t) s^*(t - \Delta t)\, dt \right| = \left| \int_{-T_e/2}^{T_e/2} \text{Rect}_{T_e}(t - \Delta t) e^{-j2\pi \frac{B_D \Delta t}{T_e} t}\, dt \right|$$

$$|c(\Delta t)| = \left| \int_{-\frac{T_e}{2} + \Delta t}^{+\frac{T_e}{2}} e^{-j2\pi \frac{B_D \Delta t}{T_e} t}\, dt \right|$$

$$|c(\Delta t)| = T_e \left| \frac{\sin\left(\pi B_D \Delta t \left(1 - \frac{\Delta t}{T_e} \right) \right)}{\pi B_D \Delta t} \right|$$

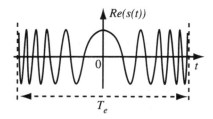

FIGURE 15.4 RECEIVED SIGNAL (REAL PART); HYPOTHESIS $f_{D_0} = 0$

We calculate only the magnitude of the signal output from processing, which is sufficient for most cases. For some other types of processing, it may be necessary to calculate the signal phase as well. This is the case, for instance, for polarimetric and interferometric processing, which are based on phase differences, from pixel to pixel, between several images.

For the small values of δt, that is, near the correlation peak, the function $|c(\Delta t)|$ can be approximated by a sinc function. Its width at 3 dB determines the radar resolution:

$$\Delta t_{3\ \text{dB}} = \frac{0.886}{B_D} = \frac{k_r}{B_D}$$

The matched receiver output signal (Figure 15.5) presents side lobes 13.3 dB below the correlation peak. The difference is insufficient, as it is clearly smaller than the dynamic range of the received echoes. In order to achieve a lower level of side lobes, the filter pulse response is multiplied by an amplitude weighting (for example, Hamming or Taylor windows). The matched receiver output signal is then

$$|C(\Delta t)| = \left| \int_{-T_e/2}^{T_e/2} \rho(t - \Delta t) e^{-j\pi \frac{B_D \Delta t}{T_e} t} dt \right|.$$

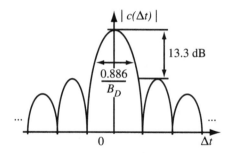

FIGURE 15.5 SIGNAL OUTPUT FROM FOCUSED SYNTHETIC APERTURE PROCESSING

A substantial decrease in side lobes leads to a 20% or 40% widening of the correlation peak, depending on the type of window used.

Two identical targets (M_1 and M_2) located on a straight line parallel to the flight path (Figure 15.6) reflect the same signals toward the radar, but they are shifted in time. This shift corresponds to the time required by the platform to cover the distance δx between the two targets:

$$\Delta t = \frac{\delta x}{v}$$

The resolutions along the flight path axis and along the cross-range axis are, respectively,

$$r_x = v \Delta t_{3\ \text{dB}} = \frac{k_r v}{B_D},$$

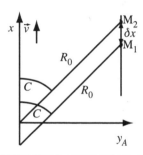

FIGURE 15.6 OBSERVATION OF TWO TARGETS SEPARATED BY THE DISTANCE D X

and

$$r_c = \frac{k_r v}{B_D} \sin \theta_A \approx \frac{v}{B_D} \sin \theta_A .$$

When expressing the Doppler bandwidth in relation to illumination time, we find the same formula as for the unfocused synthetic aperture:

$$r_c = \frac{k_r \lambda R_0}{2 v T_e \sin \theta_A}$$

RESOLUTION IN RELATION TO SYNTHETIC APERTURE LENGTH

The equation $l_e = v T_e \sin \theta_A$ is the projection of the flight path during target illumination along the axis normal to the line of sight (see Figure 15.7). This is the *Synthetic Aperture length*. Angular resolution and cross-range resolution projected onto the ground as a function of l_e can be written as

$$r_A = \frac{k_r \lambda}{2 l_e}$$

and

$$r_c = \frac{k_r \lambda R_0}{2 l_e} \approx \frac{\lambda R_0}{2 l_e} .$$

This expression should be compared to the formula for the angular aperture of the pattern of a real antenna whose length is l:

$$\theta = \frac{k \lambda}{l}$$

The SAR resolution is twice as good as the resolution obtained by means of a real antenna of the same length.

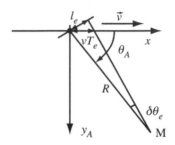

FIGURE 15.7 SYNTHETIC APERTURE LENGTH

15.1.2.4 RESOLUTION IN RELATION TO VARIATION OF THE ANGLE OF VIEW OF THE TARGET

Depending on the variation, $\delta\varphi_e$, of the angle of view of the target during illumination (see Figure 15.7), resolution in relation to the direction normal to range is given by

$$r_c = \frac{k_r \lambda}{2\,\delta\varphi_e} \approx \frac{\lambda}{2\,\delta\varphi_e}.$$

Although, for convenience, we discuss this relationship after having discussed three other expressions of resolution, it is undoubtedly the most fundamental one. It involves only simple and directly measurable physical values. The different expressions of resolution are summarized in Table 15.1.

TABLE 15.1. VARIOUS EXPRESSIONS OF RESOLUTION WITH FOCUSED SAR

Input Parameter	Illumination Time T_e	Doppler Bandwidth B_D	Variation of Angle of View of Target $\delta\varphi_e$	Synthetic Aperture Length l_e
Cross-Range Resolution, r_c	$\dfrac{\lambda R_0}{2vT_e \sin\theta_A}$	$\dfrac{v}{B_D}\sin\theta_A$	$\dfrac{\lambda}{2\,\delta\varphi_e}$	$\dfrac{\lambda R_0}{2l_e}$

15.1.2.5 UNFOCUSED SAR LIMIT

Processing of unfocused synthetic aperture is optimum when the illumination time is very short. If the illumination time is sufficiently long, the received signal is linearly frequency modulated. In this case, the unfocused processing output is the Fourier transform of a linearly frequency-modulated signal. Its envelope tends toward a rectangle whose width, equal to the modulation bandwidth, increases with illumination time.

Contrary to the indications of the expression of resolution, the resolution deteriorates when the illumination time exceeds a certain limit. Figure 15.8 shows changes in the resolution obtained via unfocused SAR processing as a function of illumination time. This resolution cannot be better than r_0, which corresponds to the intersection point of the curves characterizing the two domains. It is calculated for an optimum illumination time, T_{e0}, such that

$$\frac{1}{T_{e0}} = \frac{2v^2\sin^2\theta_A}{\lambda R_0}T_{e0}$$

and

$$T_{e0} = \frac{1}{v\sin\theta_A}\sqrt{\frac{\lambda R_0}{2}}.$$

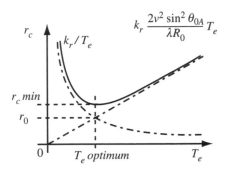

FIGURE 15.8 RESOLUTION ACHIEVED BY UNFOCUSED SAR PROCESSING IN RELATION TO TARGET ILLUMINATION TIME

The resolution consequently cannot be better than

$$r_0 = \frac{k_r}{\sin\theta_A}\sqrt{\frac{\lambda R_0}{2}}.$$

The Doppler frequency difference at the end of the illumination produces a shift in phase that increases with the square of illumination time:

$$\delta\varphi = \pi\frac{B_D}{T_e}\frac{T_e^2}{4}$$

When $T_e = T_{e0}$, the phase shift is $\pi/4$. For an unfocused SAR, the best resolution is achieved when the phase error due to linear variation of the Doppler frequency reaches $\pi/4$. This frequently used criterion is called the Rayleigh criterion.

15.1.3 A REMARKABLE CONFIGURATION: THE SIDE-LOOKING ANTENNA RADAR

SAR OR SLAR?

Consider a radar fitted with a fixed antenna directed at 90° to the flight path (see Figure 15.9). In the case of airborne radar, this configuration is called SLAR *(Side-looking Airborne Radar)*. It should be noted that a radar can be SLAR without being a SAR, and vice-versa:

- The first expression (SLAR) refers only to the radar antenna direction, independent of signal processing.
- The second (SAR) indicates that synthetic aperture processing is performed on the received signal even if it is not a side-looking radar.

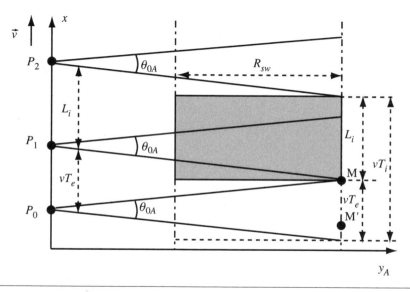

FIGURE 15.9 SIDE-LOOKING RADAR: GEOMETRY OF THE IMAGING PROCEDURE IN THE VELOCITY AND LINE-OF-SIGHT PLANE OF THE RADAR; DEFINITION OF IMAGE DIMENSIONS

OBSERVATION GEOMETRY

The SLAR delivers an image of a strip of ground that lengthens as the aircraft advances. Its dimension along the range axis is called the *swath* R_{sw} (in the plane containing the radar line of sight; see Figure 15.10). It is limited by the ambiguous range of the radar. The perpendicular dimension of the image L_i, parallel to the flight path is a function of the distance covered by the platform during time T_i while the radar is operating: $L_i = vT_i - vT_e$. Processing is matched to illumination time T_e. The points in the antenna lobe at the initial moment (for example, the point M') are illuminated for a time less than T_e and do not appear in the radar image. Figure 15.19 shows the image obtained in gray.

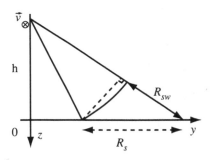

FIGURE 15.10 SIDE-LOOKING RADAR: IMAGING GEOMETRY IN THE VERTICAL PLANE;
DEFINITION OF SWATH IN GROUND RANGE Rs AND RADAR RANGE Rsw

This imaging technique for a continuous strip of ground is known as stripmap or swath mode. Unlike the acronym SLAR, we can apply this term to either airborne or spaceborne radar systems.

SIDE-LOOKING FOCUSED SAR RESOLUTION

Target illumination time depends on antenna aperture, platform velocity, and the distance between the radar and the target. The illumination time is voluntarily limited to the time the target needs to pass through the 3 dB beamwidth. The illumination time and resolution are given respectively by

$$T_e = \frac{R_0 \theta_{0A}}{v}$$

and

$$r_c = \frac{k_r \lambda R_0}{2 v T_e} = \frac{k_r \lambda}{2 \theta_{0A}}.$$

The aperture of the main lobe of the antenna pattern at 3 dB can be expressed as a function of its length,

$$\theta_{0A} = \frac{k_{0A} \lambda}{l},$$

in which k_{0A} is a coefficient close to one depending on amplitude and phase distribution over the antenna. This results in

$$r_c = k_c \frac{l}{2} \approx \frac{l}{2}.$$

This expression represents the basic characteristic of a side-looking focused synthetic antenna radar. Resolution is independent of

- *wavelength*: the radar can transmit on any frequency; frequency selection is determined by criteria different than cross-range resolution

- *platform velocity*: the radar can operate either onboard a helicopter flying at 20 m/s or a satellite moving at 7500 m/s
- *range*: very high resolution can be achieved even at ranges of hundreds of kilometers

This characteristic is, however, paradoxical:

- With a real antenna, the smaller the antenna, the poorer is the natural resolution
- With SAR, it is exactly the opposite: the smaller the real radar antenna, the higher the SAR resolution

Under such conditions, the temptation is to use an extremely small antenna. However, this is not always possible, due to ambiguity and power budget reasons that will be discussed in the sections and chapters that follow.

EXAMPLES

A radar with an antenna of length 2 m produces a resolution of 1 m along the flight path.

A radar with a 10 m antenna produces a resolution of 5 m.

SIDE-LOOKING FOCUSED SAR LIMITS

Unfortunately, SAR is subject to a number of spurious phenomena: aircraft motion, instability of the frequency source, and amplitude modulation of the signal received by the antenna lobe. These phenomena cause a slight widening of the correlation peak at the processing output and make it impossible to reach the theoretical limit resolution. In general, resolution is degraded by 10% to 20% in relation to $l/2$.

However, this rule holds for fixed antennas only. On the other hand, if the antenna can be steered in azimuth, the illumination time of a given area can be increased and resolution improved. This is not, however, continuous imaging of the strip-map type: the area illuminated by the antenna is no longer exactly side-looking and accumulates a certain delay. Observation must be interrupted when the antenna reaches the end of its travel (see Section 15.4.2).

15.1.4 ULTIMATE SAR RESOLUTION

A certain number of assumptions and approximations have been made in order to obtain the result presented in Table 15.1. One of them is that illumination time is sufficiently short to ensure generation of the matched filter by using an expansion limited to the second order of the phase of the received signal. This amounts to considering that there is little change in the angle of view:

$$\delta\varphi_e \ll 1$$

But what if the illumination time is very long, even tending towards infinity? How can resolution be defined in this case?

In Figure 15.11, the variation of the angle of view is

$$\delta\varphi_e = \varphi_1 + \varphi_2.$$

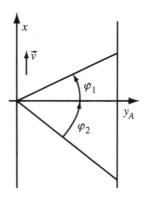

FIGURE 15.11 OBSERVATION WITH LARGE VARIATION OF THE TARGET ANGLE OF VIEW

During illumination, the Doppler frequency of the received signal varies between

$$f_{D_1} = \frac{2v}{\lambda}\sin\varphi_1$$

and

$$f_{D_2} = -\frac{2v}{\lambda}\sin\varphi_2.$$

Although the signal is not linearly frequency modulated, it can be demonstrated that its equivalent bandwidth remains close to

$$B_e = 2\sqrt{\pi}\left(\frac{\int_{-\infty}^{+\infty} f^2 |S(f)|^2\, df}{\int_{-\infty}^{+\infty} |S(f)|^2\, df}\right)^{\!\frac{1}{2}} \approx f_{D_1} - f_{D_2}.$$

The width of the correlation peak at 3 dB is then given by (see Chapter 14)

$$\Delta t_{3\ dB} \approx \sqrt{\frac{2}{\pi B_e}\frac{1}{}} \approx \frac{\sqrt{2}}{\sqrt{\pi(f_{D_1} - f_{D_2})}}.$$

The resolution along the flight path is

$$r_x = v \Delta t_{3\,\text{dB}},$$

which gives (Raney 1992)

$$r_x \approx \sqrt{\frac{2}{\pi}} \frac{\lambda}{2(\sin\varphi_1 + \sin\varphi_2)}.$$

LIMIT FOR SHORT ILLUMINATION TIME

For illumination limited to a small angle of variation, the expression from Table 15.1 still holds true:

$$\sin\varphi_1 + \sin\varphi_2 = \varphi_1 + \varphi_2 = \delta\varphi_e \Rightarrow r_x \approx \sqrt{\frac{2}{\pi}} \frac{\lambda}{2\delta\varphi_e} = \frac{k_r \lambda}{2\delta\varphi_e}$$

LIMIT FOR LONG ILLUMINATION TIME

The longest illumination time is obtained when φ_1 and φ_2 tend toward $\pi/2$. As a result, resolution tends toward its ultimate value:

$$\lim_{(\theta_1,\theta_2)\to(\pi/2,\pi/2)} r_x \approx \sqrt{\frac{2}{\pi}} \frac{\lambda}{4} \approx \frac{\lambda}{5}$$

The ultimate value of SAR resolution is therefore a fraction of the wavelength. For a radar operating in X-band (3 cm wavelength), the theoretical ultimate resolution is less than 1 cm. This level of performance can be approximated provided that measurement conditions are perfectly controlled, as in a laboratory for instance.

15.2 SAR AMBIGUITIES

A pulse waveform of low repetition frequency is generally used in SAR. The time and Doppler ambiguities are related to the value of the repetition period.

15.2.1 RANGE AMBIGUITY

The ambiguous range is $R_a = \dfrac{cT_R}{2}$.

The swath must not be interrupted by blind areas, and there must be no ambiguity between its points. On the other hand, as is shown for space-based radars, it is not absolutely necessary for the swath to be located in the first period of reception (see Figure 15.12).

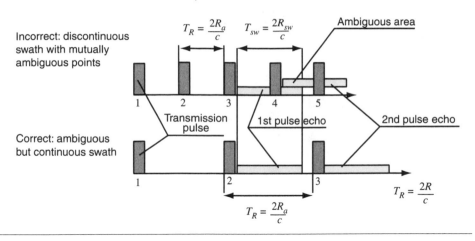

FIGURE 15.12 RANGE AMBIGUITY AND SWATH POSITION IN THE INTERPULSE PERIOD

15.2.2 CROSS-RANGE AMBIGUITY

Doppler frequency ambiguity creates angular ambiguity. The spectrum of the received signal (see Figure 15.13) is composed of lines. It is centered on the transmission frequency and shifted by the target Doppler frequency: $f_0 + f_{D_0}$. Its envelope is the transmission pulse spectrum envelope. The repetition-frequency value separates each line. An enlargement of the central part is shown around the carrier frequency that is considered a reference. The Doppler frequency axis is then the abscissa. The Doppler frequency of the targets in the antenna main lobe is contained inside an interval, of width B_{D_a}, around the Doppler frequency along the antenna axis, with

$$B_{D_a} = \frac{2v}{\lambda} \sin \theta_A \cdot \theta_{0A} = k_{0A} \frac{2v}{l} \sin \theta_A.$$

A target located at angle θ_A and with Doppler frequency f_{D_0}, is ambiguous with all targets whose Doppler frequency equals f_{D_0}, modulo PRF. The PRF must be greater than the instantaneous Doppler bandwidth, B_{D_a}, to prevent the presence of two ambiguous targets simultaneously in the antenna lobe:

$$PRF = k_D B_{Da} = k_D \frac{2v}{\lambda} \sin \theta_A \cdot \theta_{0A}$$

where k_D, usually greater than or equal to 1.2, is chosen in relation to the level of ambiguous signal tolerated. The angular difference between two ambiguous targets is such that the difference between their Doppler frequencies equals the PRF (see Figure 15.14):

$$\delta \varphi_a = \frac{\lambda PRF}{2v \sin \theta_A}$$

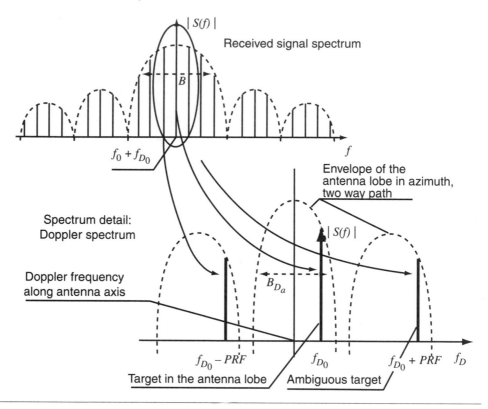

FIGURE 15.13 CROSS-RANGE AMBIGUITY: RELATIONSHIPS BETWEEN PRF AND DOPPLER BANDWIDTH

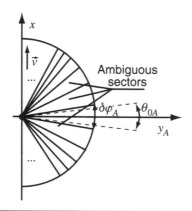

FIGURE 15.14 ANGULAR AMBIGUITY AND CONSTRAINT ON ANTENNA LOBE APERTURE

In the case of a focused synthetic antenna, the variation of Doppler frequency during illumination is allowed for during processing. It depends on range and direction of each target, but it is insufficient to constitute a discriminating criterion. Figure 15.15 shows the effect of processing on signals reflected by ambiguous targets. They undergo mismatching,

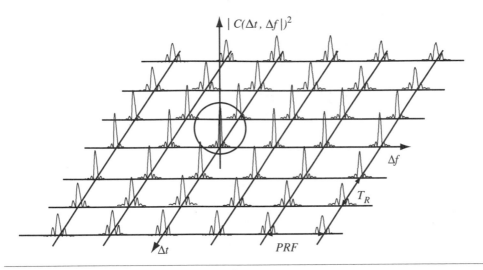

FIGURE 15.15 IMPULSE RESPONSE OF A FOCUSED SAR RECEIVER

depending on their range and angular shift. This produces a loss in level, peak broadening, and increase of the side lobes.

15.3 SPACEBORNE SAR

Installing a SAR onboard a satellite introduces a new set of constraints linked to the geometry of the mapping process:

- The altitude required by flight into orbit places targets at long range, and a range-ambiguous waveform is therefore used. Moreover, it is necessary to take the curvature of the Earth into account
- The Earth's rotation modifies the Doppler frequency of the targets
- The angle of incidence, which is around 80° for airborne radars (that is a 10° grazing angle), is between 15° and 70° for radars in terrestrial orbit
- A detailed study of spaceborne SAR is given in the work of Curlander and McDonough entitled Synthetic Aperture Radar (Curlander 1991)

SHORT HISTORY OF SPACEBORNE RADARS

On August 21, 1965, a *Titan II* rocket launched the *Gemini 5* spacecraft and the small radar satellite *REP*, which was going to be used in the first space rendezvous experiment. Later, other *Gemini* spacecraft and, finally, *Apollo* program lunar modules were fitted with docking radars.

Following this initial use, the application of radar in space changed and has since been concentrated on observation of the Earth and the other planets. The first experimental use of SAR in Earth orbit was in 1978. On July 26 that year, NASA launched the satellite *SEASAT*. Among other instruments, it carried an L-band SAR and a wind scatterometer. After only three months of operation, the radar sent back signals of such high quality and quantity that processing and analysis continued for more than ten years. During the SIR-A (Shuttle Imaging Radar, November 12, 1981) and SIR-B (October 5, 1984) experiments, a radar derived from the SEASAT radar operated onboard the Shuttle for two, five, and eight days, respectively. Finally, it must be pointed out that the US has put a number of military radar satellites of the Lacrosse category into Earth orbit, for surveillance and Intelligence missions. The first was probably launched by the Shuttle Atlantis on December 13, 1988.

The first spaceborne Soviet SAR, Kosmos 1870, probably operated for two years, starting on July 25, 1987. Transmission was made in S-band as in the case of the radar Almaz, which succeeded it on March 31, 1991.

In the same year, the first European remote sensing satellite, ERS-1 (Earth Resources Satellite), was placed into orbit by an Ariane IV, on July 16. The ERS-1 payload comprised a C-band SAR, a wind scatterometer, and an altimeter. ERS-2, largely identical to ERS-1, was injected into its sun-synchronous polar orbit April 20, 1995. Both satellites have been operating simultaneously, for interferometric applications experiments. ENVISAT-1, carrying among several instruments an active array antenna C-band SAR, is foreseen to be launched in June 2001.

The Canadian RADARSAT-1 has been operating since November 1995. It carries a C-band SAR fitted with an electronic steered array. This 1-D elevation scanning provides the radar with a wide instantaneous accessibility, compatible with new SAR modes like SCANSAR.

In Japan the NASDA agency satellite JERS-1 was placed in orbit with an L-band radar very similar to the SEASAT radar, on February 11, 1992.

Finally, the SIR-C and X-SAR radars, which are not only capable of operating on several simultaneous transmission frequencies (L-, C-, and X-band) but also have polarimetric capacity (for the first two bands), performed their first mission onboard a Shuttle in March 1994. The SRTM (Shuttle Radar Topography Mission) project flights took place between 11 and 23 February 2000. The C-band and the X-band radars, improved with interferometric capabilities, acquired data over approximately 80% of the Earth's surface in order to compute elevation models.

15.3.1 SIDE-LOOKING FOCUSED SAR RESOLUTION

Assuming a circular orbit, the satellite trajectory is an arc of a circle. In order to reproduce the same conditions as for flat Earth geometry, the velocity used is the ground projection of satellite velocity (see Figure 15.16):

$$r_c = \frac{k_r\, v_g}{B_D} = \frac{k_r\, v}{B_D} \frac{R_T}{R_T + H}$$

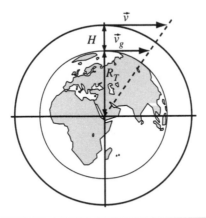

FIGURE 15.16 SATELLITE VELOCITY AND ITS GROUND PROJECTION

The velocity of the satellite is related to the orbit altitude (Kepler functions, where g_0 is the gravitation constant):

$$v = \sqrt{g_0}\, \frac{R_T}{\sqrt{R_T + H}}$$

On the other hand, the Doppler frequency of the targets is the same as that used for flat Earth surface geometry. Consequently, the Doppler bandwidth of the signal reflected by a target passing through the beam is

$$B_D = \frac{2v}{\lambda} \theta_{0A} = k_{0A}\, \frac{2v}{l}.$$

This gives

$$r_c = k_c \frac{R_T}{R_T + H} \frac{l}{2}.$$

In comparison with the flat Earth surface geometry, a small gain in resolution is achieved, resulting from the additional variation of the angle of view due to the slightly curved trajectory of the radar in orbit.

15.3.2 A Range-ambiguous Waveform

The antenna length determines both the width of the Doppler spectrum of the ground returns, B_{D_a}, and the resolution that can be obtained by means of strip-map processing. In the Table 15.2, the relationships between the various parameters are summarized and illustrated.

As target range is greater than the altitude, it appears clearly that the radar is highly ambiguous in range.

TABLE 15.2. RELATIONSHIP BETWEEN PARAMETERS OF SAR IN EARTH ORBIT

Parameter	Expression and Example
Flight altitude	$H = 800$ km
Earth's radius	$R_T = 6\,400$ km
Gravitation constant	$g_0 = 9.81 \text{m/s}^2$
Antenna length	$l = 10$ m
Various coefficients	$k_D = 1.2;\ k_c = k_{0A} = 1$
Satellite velocity	$v = \sqrt{g_0}\,\dfrac{R_T}{\sqrt{R_T + H}} = 7470\,\text{m/s}$
Minimum PRF	$PRF = k_D B_{D_a} = k_D k_{0A}\dfrac{2v}{l} = 1793\,\text{Hz}$
Resolution in stripmap mode	$r_c = k_c \dfrac{R_T}{R_T + H}\dfrac{l}{2} = 4.4\text{m}$
Ambiguous range	$R_a = \dfrac{c}{2PRF} = \dfrac{cl}{4v k_D k_{0A}} = 83.7\text{km}$

It should be noted that on the one hand, all these parameters are independent of frequency, and on the other, that altitude has little influence. Figures 15.17 and 15.18 show that several ambiguities intercept the ground. The antenna pattern in elevation is used to discriminate the available swath from the ambiguous swath. The aperture of the antenna main lobe in elevation must be sufficiently small to enable illumination of the usable swath only:

$$\theta_{0E} < \frac{R_{sw}}{R\,\tan i} < \frac{R_a}{R\,\tan i}$$

Switching: to avoid saturation, the receiver is protected during transmission by a switch or a limiting device. The switching time is the period of time between these two statuses: *Transmitting* and *Receiving*. The value depends on the technology of the switch or the limiter.

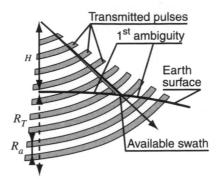

FIGURE 15.17 RANGE AMBIGUITIES

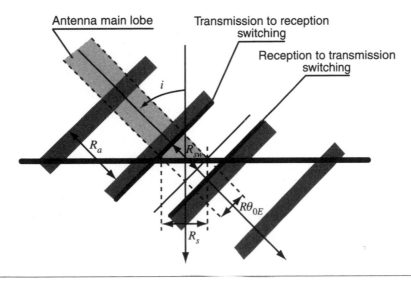

FIGURE 15.18 APERTURE OF ANTENNA LOBE IN ELEVATION (ENLARGEMENT OF
FIGURE 15.17; HYPOTHESIS: EARTH SURFACE LOCALLY FLAT)

15.3.3 ANTENNA SURFACE AREA

MINIMAL SURFACE AREA

Antenna height is affected by the constraint on lobe aperture in elevation,
and ambiguous range can be replaced by its expression as a function of
antenna length:

$$\theta_{0E} = k_{0E}\, \frac{\lambda}{h} \quad \Rightarrow \quad h > k_{0E}\, k_D\, k_{0A} \frac{4\,v\,\lambda\,R\,\mathrm{tg}\,i}{c\,l}$$

Target range is written in relation to altitude and angle of incidence, when resolving the triangle whose apexes are the satellite, the center of the Earth, and the target:

$$(R_T + H)^2 = R_T^2 + R^2 + 2RR_T \cos i \Rightarrow R \approx \frac{H}{\cos i} + \frac{H^2}{2R_T \cos i}$$

We finally obtain a constraint on the surface area of the antenna:

$$lh > k_{0E} k_D k_{0A} \frac{4v\lambda H \sin i}{c \cos^2 i} \left(1 + \frac{H}{2R_T}\right)$$

The surface area of the antenna is proportional to

- the altitude of the satellite
- the transmitted wavelength

INFLUENCE OF ANTENNA PROPORTIONS ON PERFORMANCE

The antenna area is defined by the choice of orbit, the transmission frequency, and the observation incidence domain. It should be noted that it is *independent of resolution*.

The only possibility remaining for the radar designer is to use the ratio between antenna length and height that is adjusted to suit the satellite mission. There are two extreme cases:

- An antenna that is short in length and great in height (said to be in the *vertical* position as its longest dimension is upwards) will enable high resolution with reduced swath in the strip-map mode.
- An antenna that is long but not great in height (said to be in the *horizontal* position) will enable medium resolution with large swath.

THE CONSTRAINT ON THE ELEVATION ANTENNA PATTERN

As shown in Figure 15.19, the elevation antenna pattern that illuminates the required swath must have constant gain and present very low side lobes in the direction of ambiguous swaths. In order to obtain such a pattern, the antenna dimensions must be larger than in the case of a more "conventional" main lobe, the aperture being the same. The power distribution over the antenna area can be a $(\sin x/x)^2$ function, thus forming a nearly square main lobe. Another method of slightly enlarging the lobe is to add a quadratic phase function. In this case, the k_{0E} coefficient is between 1.5 and 2, depending on the rejection level of the desired ambiguities.

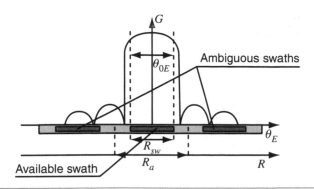

FIGURE 15.19 AMBIGUOUS SWATH REJECTION USING THE ANTENNA PATTERN IN
ELEVATION

EXAMPLE

On the basis of the parameters described in Section 15.3.2, Table 15.3
shows the variation of antenna area in relation to wavelength and angle of
incidence (the coefficient k_{0E} equals 1.5).

TABLE 15.3. EXAMPLES OF ANTENNA AREAS FOR A SAR IN EARTH ORBIT AT 800 KM
ALTITUDE

	L-band Radar $f_0 = 1\,\text{GHz}\,;\quad \lambda = 30\,\text{cm}$	X-band Radar $f_0 = 10\,\text{GHz}\,;\quad \lambda = 3\,\text{cm}$
Incidence 20°	18 m²	2 m²
Incidence 60°	158 m²	16 m²

Antennas can be of very significant dimensions. As the available volume
inside a launcher nose cone is limited, complex mechanisms are required
to keep it folded during launch. As soon as the satellite is in orbit, the
antenna is deployed. Under such conditions, the deployment and the
flatness of the deployed antenna are critical stages in the design of a space-
based radar.

15.3.4 DOPPLER FREQUENCY AND YAW STEERING

EARTH ROTATION EFFECT

The relative motion of the antenna phase center with respect to a point
located on the ground results from a combination of the satellite motion
(velocity $\vec{v}$) and the rotating motion of the Earth (vector $\vec{v}_T$, see
Figure 15.20). Given the high value of this second component, which
introduces displacement (or migration) of targets along the range axis
during illumination, it is useful to maintain antenna line of sight normal to
the relative velocity; the Doppler frequency is maintained at equal to zero

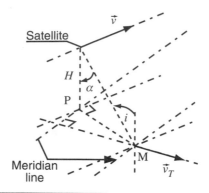

FIGURE 15.20 EFFECTS OF EARTH ROTATION: RELATIVE MOTION OF SATELLITE AND GROUND POINT

along the antenna axis. This kind of control requires either *yaw steering* of the satellite or an antenna with electronic scanning in the azimuth plane.

CENTRAL DOPPLER FREQUENCY

This section deals with the expression of the Doppler frequency of a ground-based point M located along the axis of the radar line of sight. The calculation basis is the simple case of a circular orbit around the Earth, which is supposed to be perfectly spherical (see Figure 15.21).

The satellite position in the orbit plane is determined by angle W. This angle is zero when the satellite is located on the orbital node, that is, at the intersection of the equatorial and the orbital planes. γ is the longitude difference between the node of the orbit and the instantaneous position of the satellite. ψ is the satellite latitude. In can be demonstrated that the Doppler frequency of a ground-based point is given by

$$f_D = -\varepsilon_v \frac{2v}{\lambda} \sin\alpha \, \sin Y \left[1 + \frac{\omega_t}{\omega_s} \left(\cos\psi \, \sin i \cot Y - \cos i \right) \right],$$

where

- ε_v indicates the viewing position: $\varepsilon_v = -1$ for a left-hand sight, $\varepsilon_v = +1$ for a right hand sight
- α is the angle formed by the line of sight and the local horizontal line (depression angle, Figure 15.20)
- ω_t is the angular velocity of the Earth
- ω_s is the angular velocity of the satellite around the Earth
- i is the orbit inclination
- Y is the angle formed by the trajectory and the line of sight in the yaw plane

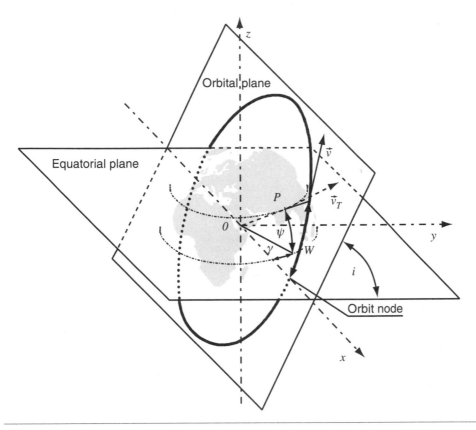

FIGURE 15.21 DEFINITION OF THE ANGLES

In order to maintain a zero Doppler frequency along the line of sight, the yaw angle must satisfy the following equation:

$$\tan Y = -\frac{\sin i \cos \psi}{(\omega_s / \omega_t) - \cos i}$$

This angle depends on the altitude of the satellite, and as a result, either the beam or the satellite must be steered continuously. On the other hand, the yaw angle does not depend on the angle of incidence. Yaw steering is therefore effective over the entire swath.

15.4 SAR OPERATING MODES

Previous sections dealt with the side-looking mode. Other characteristic modes are now discussed.

15.4.1 DOPPLER BEAM SHARPENING, WITH ROTATING ANTENNA

In airborne radars, DBS is used to obtain an image of the ground around the aircraft. The antenna has a rotating motion in bearing with an angular velocity ω (see Figure 15.22). There is a very short illumination time; this is an unfocused *SAR* mode:

$$T_e = \frac{\theta_{0B}}{\omega}$$

The cross-range resolution is

$$r_c = \frac{k_r \lambda R \omega}{2v\theta_{0B}\sin\theta_A}.$$

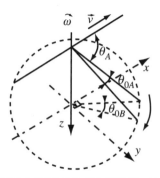

FIGURE 15.22 DOPPLER BEAM SHARPENING: ANTENNA ROTATION AROUND THE VERTICAL AXIS

By modulating the rotation speed to maintain the ratio $\omega/\sin\theta_A$ as a constant, it is possible to obtain constant resolution, independently of bearing. At all events this speed cannot be kept constant when $\theta_A = 0$, that is, when the antenna crosses the vertical plane containing the velocity vector of the aircraft. Under such conditions, the resolution is the real antenna resolution. In practice, DBS can be used from both sides of the aircraft, for bearings from 15° to 165°. If the platform is flying at low altitude, the azimuth angle, θ_A, is equivalent to the bearing angle, θ_B.

EXAMPLE

Let us consider typical parameters:

$v = 300\,\text{m}/\text{s}$

$\omega = 30°/\text{s}$

$$\theta_{0B} = 2°$$

$$\lambda = 3\,\text{cm}$$

$$R = 50\,\text{km}$$

$$\theta_A = 45°$$

$$\theta_D = 5°$$

$$k_r = 1 : r_a \approx 50\text{m}$$

The illumination time is equal to 100 ms. This kind of mode produces a map with medium resolution, suitable for infrastructure detection for instance.

15.4.2 SPOTLIGHT SAR

The spotlight (or spotbeam) mode can be used for airborne or space-based radars. It provides cross-range resolution better than $l/2$. The radar antenna is kept aimed at the target whose image is required (see Figure 15.23). Illumination time and, therefore, Doppler bandwidth are increased until the desired resolution is obtained. The display is not continuous and the image is only available at the end of illumination.

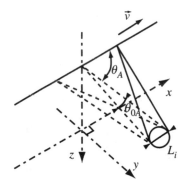

FIGURE 15.23 SPOTLIGHT SAR: ANTENNA IS TRACKING A PATCH OF GROUND

EXAMPLE

A resolution of 0.5 m can be achieved with the same parameters used for the previous mode. Illumination time will have to be at least

$$T_e = \frac{k_r\,\lambda\,R}{2\,v\,r_c\,\sin\theta_A} = 7\,\text{s}.$$

The antenna look angle, equal to the variation of the angle of view of the target is

$$\delta\varphi_e = \frac{k_r\lambda}{2r_c} = 1.7°.$$

The cross-range dimension of the image L_i is limited by the aperture of the antenna beam in azimuth: $L_i = R\,\theta_{0A} = 1700$ m.

15.4.3 SCANSAR

Scansar can be used by a radar operating at high altitude and in particular by spaceborne radars in order to obtain a swath wider than the ambiguous range, at the cost of degraded resolution. Image display is continuous. The natural illumination is divided into n segments. Each segment is assigned to the observation of a different swath. The selected swaths are adjacent. The number of segments is adjusted so that the entire swath is obtained; the natural swath is multiplied by $(n-1)$. On the other hand, the illumination time and the Doppler bandwidth of each target are divided by n. This difference of one unit is necessary to ensure the continuity of the imaging process, as shown in Figure 15.24.

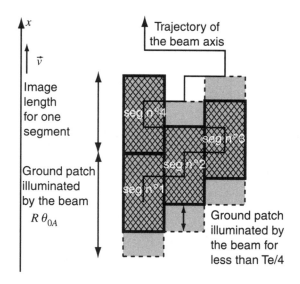

FIGURE 15.24 EXAMPLE OF SCANSAR: ILLUMINATION DIVIDED INTO FOUR SEGMENTS; THREE ADJACENT SWATHS; RESOLUTION DIVIDED BY FOUR

To use this technique, the antenna must be switched rapidly in elevation, which can only be achieved with an electronic scanning antenna.

15.4.4 SQUINT OR OFF-BORESIGHT MODE

The *squint* mode can be compared to *strip-map*, except that the antenna beam is not directly at right angles to the velocity vector (see Figure 15.25). A continuous display is obtained. The illumination time is given by

$$T_e = \frac{R_0\, \theta_{0A}}{v \sin \theta_{0A}}.$$

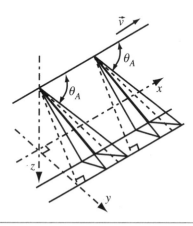

FIGURE 15.25 THE SQUINT MODE: GENERATION OF A CONTINUOUS DISPLAY IMAGE WITH AN OFF-BORESIGHT ANTENNA

By introducing this expression into the resolution formula, we obtain the same result as in the side-looking case:

$$r_c = k_c\, \frac{l}{2} \approx \frac{l}{2}$$

The choice of antenna pointing direction has no effect on resolution provided that the line of sight is away from the platform velocity axis. The side-looking mode remains the best solution because it minimizes illumination time. Moreover, it is the easiest method from the processing point of view, as it causes the smallest variation in target range during illumination (migration).

15.4.5 MULTILOOK MODE

All SAR modes can be converted to another version, called *multilook*, where several images of the same site are formed from observations made at various angles of view or at different frequencies (see Figure 15.26). These images, statistically independent because of target fluctuation with respect to angle of view or frequency, are then summed in power, resulting in a non-coherent post-detection integration effect that reduces speckle.

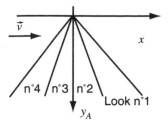

FIGURE 15.26 MULTILOOK MODE

The difference between this mode and the basic mode lies in the processing described in Chapter 16. Multilook is widely used to improve radiometric resolution; you can apply it to both focused and unfocused SAR (in particular to DBS).

15.4.6 OTHER MODES

The *SAR* modes presented in this chapter are among the most commonly used. It is not an exhaustive list, and they can be developed or combined in order to create new modes.

16

SYNTHETIC APERTURE RADAR SPECIFIC ASPECTS

16.1 MIGRATIONS

One of the most troublesome phenomena of SAR images is range migration. Throughout the illumination time, the range of an echo varies in accordance with the quadratic function calculated in Section 15.1. However, regular sampling is carried out on return signals by the radar receiver. The range of an echo changes from one interpulse period to the next, and the signal is not always received in the same range gate. To perform Synthetic Aperture processing, the signal samples need to be sought in the correct gates and migrations must be compensated for. Migrations can be of two types, linear or parabolic:

- Linear migrations are directly proportional to the average Doppler frequency of the signal. The range and number of gates covered are given respectively by (see Figure 16.1)

$$R = \frac{\lambda f_{D_0} T_e}{2} \quad \text{and} \quad N_{RL} = \frac{\lambda f_{D_0} T_e F_s}{c} \,.$$

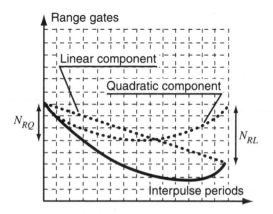

FIGURE 16.1 TARGET MIGRATION DURING ILLUMINATION

where F_s is the sampling frequency of the receiver. There is no linear migration for airborne side-looking radars or for spaceborne radars, if the antenna beam is yaw-steered in order to maintain a null average Doppler frequency along its axis

- Parabolic migrations depend on the variation of the Doppler frequency, that is, on the quadratic component of the phase. The range and number of range gates covered are, respectively,

$$R = \frac{\lambda}{16} B_D T_e \quad \text{and} \quad N_{RQ} = \frac{\lambda B_D T_e F_s}{8c}$$

Migrations are proportional to illumination time and, consequently, to range. At short or medium ranges, migrations can be smaller than one range gate. In this case, they do not need to be taken into account (see Section 16.5.1).

16.2 PHASE ERRORS

Usually, the phase function of return signals is not the ideal phase function deduced from the observed geometry. It is interfered with by

- the spurious motion of the platform during image acquisition
- the instabilities of the frequency source and the radar transmitter
- quantization noise and the inaccuracy of parameters used to calculate the reference signal and the corrections applied to the received signal
- the fact that the exact geometry of the terrain observed is not taken into account. It is usually considered a flat, horizontal surface

This explains why the correlation peak at the processing output is distorted and the quality of the image is degraded. Phase errors have consequences for resolution, geometric linearity, contrast, and, to a certain extent, radiometric linearity. It is possible to extract analytic expressions that quantitatively link the phase errors with their effects on the image. Note that all effects described here for the azimuth apply also to the range in the pulse compression processing; in the following expression, the illumination time, T_e, should be replaced by the pulse duration, T.

After being demodulated by the carrier frequency, the received signal corresponds to the one described in Chapter 15 supplemented by a spurious phase $\varphi(t)$:

$$s(t) = \text{Rect}_{T_e}(t) e^{j\left(-\frac{4\pi R_0}{\lambda} + 2\pi f_{D_0} t - \pi \frac{B_D}{T_e} t^2 + \varphi(t)\right)}$$

The processing, matched to the ideal signal and limited to calculation of the magnitude as presented in Chapter 15, performs the following operation:

$$|C(\Delta t)| = \left| \int_{-T_e/2}^{T_e/2} \mathrm{Rect}_{T_e}(t - \Delta t) e^{-j2\pi \frac{B_D \Delta t}{T_e} t + \varphi(t)} \, dt \right|$$

Assuming that $\varphi(t)$ is a sample of a periodic random process with period T_i (image acquisition time) and whose mean value is null, it is then possible to break down $\varphi(t)$ into a Fourier series,

$$\forall t \in \left[-\left(\frac{T_i}{2}, \frac{T_i}{2} \right) \right], \ \varphi(t) = a_0 + \sum_{n=1}^{\infty} \left(a_n \cos 2\pi \frac{n}{T_i} t + b_n \sin 2\pi \frac{n}{T_i} t \right),$$

where a_n and b_n are random processes with zero mean. Sine and cosine notation is preferable to complex exponential notation in this case, as the sine and cosine terms produce different effects on the image. The constant phase term has no influence on the magnitude of the signal output from processing and can consequently be ignored:

$$\varphi(t) = \sum_{n=1}^{\infty} \varphi_n(t),$$

where

$$\varphi_n(t) = a_n \cos 2\pi \frac{n}{T_i} t + b_n \sin 2\pi \frac{n}{T_i} t$$

The random variables a_n and b_n are linked to the spectrum of φ:

$$\lim_{T_i \to \infty} T_i E(a_n^2) = \lim_{T_i \to \infty} T_i E(b_n^2) = 2 S_\varphi \left(\frac{n}{T_i} \right)$$

In the rest of this section, we show the effect of a term $\varphi_n(t)$, whose spectrum is a pair of lines at frequencies $-n/T_i$ and n/T_i. The effect of the entire series $\varphi(t)$ is then deduced.

16.2.1 EFFECT OF A PERIODIC PHASE ERROR OF FREQUENCY f_n

Consider one of the terms of the Fourier series expansion, whose behavior is studied during illumination time T_e. The time T_e is itself included in the image acquisition time, T_i:

$$\varphi_n(t) = a_n \cos(2\pi f_n t) + b_n \sin(2\pi f_n t),$$

with

$$f_n = n/T_i$$

and

$$t \in \left[t_e - T_e/2, t_e + T_e/2 \right] \subset \left[-T_i/2, T_i/2 \right].$$

If $t = t_e + \theta$, the time origin is centered in the middle of the illumination time, which gives

$$\varphi_n(\theta) = \alpha_n \cos\left(2 \pi f_n \theta\right) + \beta_n \sin\left(2 \pi f_n \theta\right)$$

with

$$\alpha_n = a_n \cos\left(2 \pi f_n t_e\right) + b_n \sin\left(2 \pi f_n t_e\right),$$

and

$$\beta_n = - a_n \sin\left(2 \pi f_n t_e\right) + b_n \cos\left(2 \pi f_n t_e\right).$$

PHASE ERROR PERIOD LONGER THAN ILLUMINATION

A phase error with a period that is long compared to the illumination time (Figure 16.2) can be approximated by a parabolic branch by performing an expansion limited to the second order:

$$\varphi_n(\theta) = \alpha_n + 2 \pi \beta_n f_n \theta - 2 \pi^2 \alpha_n f_n^2 \theta^2, \text{ with } f_n \theta \ll 1$$

$$|C(\Delta t)| = \left| \int_{-T_e/2}^{T_e/2} \mathrm{Rect}_{T_e}(\theta - \Delta t) e^{-j\left(2\pi \frac{B_D \Delta t}{T_e} + 2\pi B_n fn\right)\theta} e^{j2\pi^2 \alpha_n f_n^2 \theta^2} d\theta \right|$$

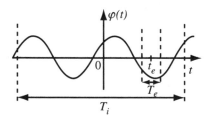

FIGURE 16.2 PHASE ERROR PERIOD LONGER THAN ILLUMINATION

The maximum value of the correlation function is at time shift

$$\Delta t = -\frac{\beta_n f_n T_e}{B_D}.$$

If the criterion of image quality is to restrict the shift to less than half of one angular resolution cell, then the constraint on the amplitude of the phase error is given by

$$\left| \frac{\beta_n \, f_n \, T_e}{B_D} \right| < \frac{1}{2 \, B_D} \quad \Rightarrow \quad |\beta_n| < \frac{1}{2 \, f_n \, T_e}.$$

The quadratic phase term causes a broadening of the correlation peak. No simple and precise analytic expression can be established to relate the width of the correlation peak to the value of this quadratic term. Nevertheless, it can be shown that if this term remains below $\pi/2$ at the end of illumination, then the deterioration of resolution is less than 10%. In this case, the constraint on the phase error amplitude is given by

$$\left| 2 \pi^2 \alpha_n f_n^2 \frac{T_e^2}{4} \right| < \frac{\pi}{2} \Rightarrow |\alpha_n| < \frac{1}{\pi f_n^2 T_e^2}.$$

PHASE ERROR PERIOD SHORTER THAN ILLUMINATION

Now let us consider the same term from the Fourier series, assuming that its period is short in relation to illumination time (see Figure 16.3):

$$\varphi_n(\theta) = \alpha_n \, \cos(2 \, \pi \, f_n \, \theta) + \beta_n \, \sin(2 \, \pi \, f_n \, \theta),$$

with

$$f_n \theta \gg 1, \, \alpha_n \ll 1 \quad \text{and} \quad \beta_n \ll 1$$

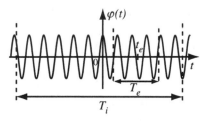

FIGURE 16.3 PHASE ERROR PERIOD SHORTER THAN ILLUMINATION

In the expression for the magnitude of the signal output from processing, the spurious phase term is "small":

$$\forall \theta, \, \varphi_n(\theta) \ll 1$$

The complex exponential of which it is the argument can be broken down under the integral, to the first order. The magnitude of the signal at the processing output becomes

$$|C(\Delta t)| \approx \left| \int_{-\frac{T_e}{2}}^{\frac{T_e}{2}} \mathrm{Rect}_{T_e}(\theta - \Delta t)\left(1 - j\alpha_n \cos(2\pi f_n \theta) - j\beta_n \sin(2\pi f_n \theta)\right)e^{-j2\pi \frac{B_D \Delta t}{T_e}\theta} \, d\theta \right|$$

In the case of no phase error, this gives

$$|C(\Delta t)| \approx \left| C_0(\Delta t) - j\frac{\alpha_n - j\beta_n}{2}C_0\left(\Delta t - \frac{f_n T_e}{B_D}\right) - j\frac{\alpha_n + j\beta_n}{2}C_0\left(\Delta t + \frac{f_n T_e}{B_D}\right) \right| \, ,$$

where $C_0(\Delta t)$ is the processing output.

Because the function $|C_0(\Delta t)|$ is close to zero almost everywhere, except in the vicinity of the correlation peak, which is located at zero, we can say

$$|C(\Delta t)| \approx |C_0(\Delta t)| + \frac{\sqrt{\alpha_n^2 - \beta_n^2}}{2}\left(\left|C_0\left(\Delta t - \frac{f_n T_e}{B_D}\right)\right| + \left|C_0\left(\Delta t + \frac{f_n T_e}{B_D}\right)\right|\right) \, .$$

The ratio between the magnitude of the correlation peak and that of the side lobes is (see Figure 16.4)

$$\frac{\sqrt{\alpha_n^2 + \beta_n^2}}{2} = \frac{\sqrt{a_n^2 + b_n^2}}{2} \, .$$

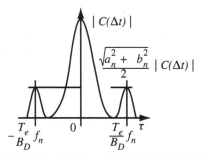

FIGURE 16.4 EFFECT OF PHASE ERROR AT HIGH FREQUENCY

In order to limit the side lobe level (which is added to the lobes of the ideal pulse response of the radar) to −25 dB below the correlation peak, the phase error amplitude must be below 7°.

16.2.2 EFFECT OF A RANDOM ERROR

The phase error function is the result of adding a function composed of low frequencies to a function composed of high frequencies:

$$\varphi(t) = \sum_{n=1}^{n=n_e} \varphi_n(t) + \sum_{n=n_e+1}^{\infty} \varphi_n(t)$$

The limit between the two domains is determined by the frequency $1/T_e$, that is, integer n_e, such that $n_e/T_i = 1/T_e$.

INFLUENCE OF LOW FREQUENCIES ON IMAGE GEOMETRICAL LINEARITY

The frequencies ranging from $1/T_i$ and $1/T_e$ are low frequencies. In each case, the linear term of the phase error shifts the correlation peak:

$$\Delta t = -\frac{T_e}{B_D} \sum_{n=1}^{n=n_e} f_n \beta_n, \text{ with } f_n = \frac{n}{T_i}$$

As the mean value of β_n is zero, that of the shift is zero too:

$$E(\Delta t) = 0$$

With the variables β_n being uncorrelated, the standard deviation of the time-shift of the correlation peak on output from processing for image acquisition time T_i takes the form

$$\sigma_{L_{Ti}} = \frac{T_e}{B_D} \sqrt{\sum_{n=1}^{n=n_e} f_n^2 E(\beta_n^2)} \ .$$

When the image acquisition duration is sufficiently long, we can write

$$\sigma_L \approx \frac{T_e}{B_D} \sqrt{2 \int_{1/T_i}^{1/T_e} f^2 S_\varphi(f) df} \ .$$

The following condition is required to maintain the standard deviation of the shift at less than half a resolution cell:

$$\sqrt{2 \int_{1/T_i}^{1/T_e} f^2 \, S_\varphi(f) \, df} < \frac{1}{2 T_e}$$

INFLUENCE OF LOW FREQUENCIES ON RESOLUTION

Each frequency helps to broaden the correlation peak. At the end of illumination, the quadratic term of phase error φ_Q represents the sum of the errors of each term in the Fourier series,

$$\varphi_Q\left(t_e + \frac{T_e}{2}\right) = -\frac{\pi^2}{2}\, T_e^2 \sum_{n=1}^{n=n_e} f_n^2\, \alpha_n,$$

where the mean value of α_n is zero:

$$E\left(\varphi_Q\left(t_e + \frac{T_e}{2}\right)\right) = 0$$

Given that all the α_n are uncorrelated, the standard deviation of φ_Q for image acquisition time T_i can be written as

$$\sigma_{Q_{Ti}} = \frac{\pi^2}{2} T_e^2 \sqrt{\sum_{n=1}^{n=n_e} f_n^4 E(\alpha_n^2)}.$$

When the illumination time is sufficiently long, we can say

$$\sigma_Q \approx \frac{\pi^2}{2} T_e^2 \sqrt{2\int_{1/T_i}^{1/T_e} f^4 S_\varphi(f)\,df}.$$

The standard deviation of the quadratic phase error remains below $\pi/2$ at the end of illumination provided that

$$\sqrt{2\int_{1/T_i}^{1/T_e} f^4\, S_\varphi(f)\,df} < \frac{1}{\pi\, T_e^2}.$$

INFLUENCE OF HIGH FREQUENCIES ON CONTRAST

Each frequency of φ broken down into a Fourier series above $1/T_e$ creates a pair of side lobes. The part of the φ spectrum above $1/T_e$ creates a "floor" of far lobes, resulting in a reduction of image contrast. The ratio between the power of the correlation peak and the total power of the far lobes is designated by the acronym ISLR (see Section 14.2.6). For an image acquisition with duration T_i, the ISLR is

$$\text{ISLR}_{T_i} = \sum_{n=n_e}^{\infty} E(a_n^2) + E(b_n^2).$$

In the limit, for a sufficiently long image acquisition time, ISLR is

$$\text{ISLR} = \lim_{T_i \to \infty} \text{ISLR}_{T_i} = 2\int_{1/T_e}^{\infty} S_\varphi(f)df.$$

The ISLR is equal to the standard deviation of the phase error, limited to high frequency disturbances, that is, of frequency greater than $1/T_e$:

$$\text{ISLR} = \sigma_{H\varphi}^2$$

16.3 PLATFORM MOTION

For an airborne radar, the platform motion has a determining influence on image quality. Any deviation from an ideal rectilinear flight path results in a phase error that degrades the processing output if there is no compensation.

The effects of platform trajectory errors are comparable to the effects caused by a distortion in a real antenna. Platform movements are divided into three broad categories (see Figure 16.5), each of them corresponding to specific effects on the image:

- *motion along the flight path.* Generally, there is a bias on velocity or a slight variation in velocity during illumination. Variations of these disturbances are very slow, due to the inertia of the aircraft along its longitudinal axis. The frequency of these movements is usually very low and "small" relative to the illumination time inverse, $1/T_e$ (typically less than 0.1 Hz)
- *transverse motion,* linked to lateral and vertical flight control, as well as to atmospheric effects (wind gusts). Their spectrum is limited to low frequencies, generally less than $1/T_e$ (usually between 0.1 Hz and 1 Hz)
- *vibrations,* which can have either aerodynamic (turbulence) or mechanical (engines, propellers) origins, and whose spectrum is above $1/T_e$

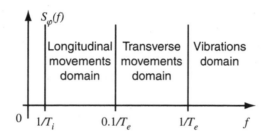

FIGURE 16.5 DIVISION OF THE SPECTRUM OF PLATFORM MOTION INTO THREE DOMAINS

In Figure 16.6, point O follows the ideal trajectory of the antenna phase center. It is located at point P at the moment when it should be at point O. The spurious movement of the platform can be characterized by a position, velocity, or acceleration error along each of the three axes of the coordinate system (O,x,y,z). The spectra of these three types of errors can be deduced from each other using derivation, and they are themselves linked to the power spectrum density of phase error $S_\varphi(f)$.

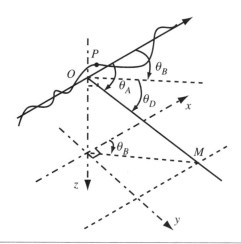

FIGURE 16.6 PLATFORM TRAJECTORY

Figure 16.7 describes the qualitative effect of the platform motion. The quantitative aspect is presented in the following sections.

16.3.1 CALCULATION EXAMPLE: MOTION ALONG PLATFORM FLIGHT AXIS

This section shows calculation of the effect of longitudinal and slow platform motion in detail. In regard to other kinds of motion, we shall present only the hypothesis and results because these can be calculated in the same way as for longitudinal motion using the expressions given in Section 16.2. The position error, $x(t)$, of the antenna phase center is expressed as a Fourier series:

$$x(t) = \sum_{n=1}^{\infty} x_n(t), \text{ with } x_n(t) = a_{xn} \cos 2\pi f_n t + b_{xn} \sin 2\pi f_n t, \text{ and } f_n = n/T_i$$

The derivative of $x_n(t)$ represents the velocity error:

$$\dot{x}_n(t) = a_{\dot{x}n} \cos 2\pi f_n t + b_{\dot{x}n} \sin 2\pi f_n t$$
$$= -2\pi f_n a_{xn} \sin 2\pi f_n t + 2\pi f_n b_{xn} \cos 2\pi f_n t$$

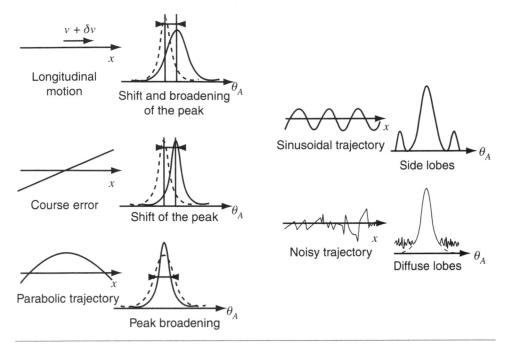

FIGURE 16.7 EFFECT OF PLATFORM MOTION ON THE CORRELATION PEAK ON OUTPUT FROM SAR PROCESSING

Thus a relationship between the Fourier coefficients of $x_n(t)$ and those of $\dot{x}_n(t)$ is obtained. Similar relationships can be established with the Fourier acceleration coefficients $\ddot{x}_n(t)$.

16.3.1.1 SLOW PERIODIC MOTION

Retaining the notations and the hypotheses of Section 16.2, we get

$$x_n(\theta) = \alpha_{xn} + \alpha_{\dot{x}n}\,\theta + \alpha_{\ddot{x}n}\frac{\theta^2}{2},$$

with

$$f_n\theta \ll 1.$$

The target-range error is calculated from the geometry in Figure 16.8: a second-order expansion is performed, retaining only the significant terms. The phase error is deduced. The constant phase term is ignored, and the linear and quadratic phase terms are, respectively,

$$\varphi_{Ln}(\theta) = \frac{4\pi}{\lambda}\left(\alpha_{\dot{x}n}\cos\theta_A - \frac{v}{R_0}\alpha_{xn}\sin^2\theta_A\right)\theta$$

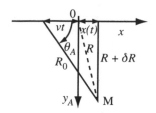

FIGURE 16.8 POSITION ERROR ALONG THE FLIGHT AXIS

and

$$\varphi_{Qn}(t) = \frac{2\,\pi}{\lambda}\left(\alpha_{\ddot{x}n}\,\cos\theta_A - \frac{2\,v}{R_0}\,\alpha_{\dot{x}n}\,\sin^2\theta_A\right)\theta^2.$$

CRITERION OF GEOMETRICAL IMAGE LINEARITY

The criterion of geometrical linearity (echo shift less than half a resolution cell) is met if

$$\left|\alpha_{\ddot{x}n}\cos\theta_A - \frac{v}{R_0}\alpha_{xn}\sin^2\theta_A\right| < \frac{\lambda}{4T_e}.$$

RESOLUTION CRITERION

The resolution degradation criterion (less than 10%) is met if

$$\left|\alpha_{\ddot{x}n}\cos\theta_A - \frac{2v}{R_0}\alpha_{\dot{x}n}\sin^2\theta_A\right| < \frac{\lambda}{T_e^2}.$$

The two previous inequalities are composed of two terms: the $\cos\theta_A$ term corresponds to a displacement along the radar-target axis; the $\sin\theta_A$ term is the variation of the synthetic antenna length.

16.3.1.2 RANDOM MOTION AT LOW FREQUENCY

The spectra of position, velocity, and acceleration errors are linked to the terms of the Fourier series with regard to the following equations:

$$\lim_{T_i\to\infty} T_i\, E\!\left(\alpha_{xn}^2\right) = \lim_{T_i\to\infty} T_i\, E\!\left(\beta_{xn}^2\right) = 2\,S_x(f)$$

$$S_{\dot{x}}(f) = (2\,\pi\,f)^2\, S_x(f)$$

$$S_{\ddot{x}}(f) = (2\,\pi\,f)^4\, S_x(f)$$

CRITERION FOR IMAGE GEOMETRICAL LINEARITY

The shift of an echo for imaging operation time T_i, resulting from the contribution of each term of the Fourier series, can be written as

$$\Delta t = -\frac{2T_e}{\lambda B_D} \sum_{n=1}^{n=n_e} \left(\alpha_{xn} \cos\theta_A - \frac{v}{R_0} \alpha_{xn} \sin^2\theta_A \right) .$$

Its mean value is zero and its standard deviation is

$$\sigma_{L_{Ti}} = \frac{2T_e}{\lambda B_D} \sqrt{\sum_{n=1}^{n=n_e} E(\alpha_{xn}^2)\cos^2\theta_A + \frac{v^2}{R_0^2} E(\alpha_{xn}^2)\sin^4\theta_A } .$$

Over a very long operating time, the standard deviation of the shift becomes

$$\sigma_L \approx \frac{2T_e}{\lambda B_D} \sqrt{2\cos^2\theta_A \int_{1/T_i}^{1/T_e} S_{\dot{x}}(f)\,df + 2\frac{v^2}{R_0^2}\sin^4\theta_A \int_{1/T_i}^{1/T_e} S_x(f)\,df } .$$

This expression can be simplified by using the standard deviation of position and velocity errors σ_{Bx} and $\sigma_{B\dot{x}}$ at low frequency:

$$\sigma_L = \frac{2T_e}{\lambda B_D} \sqrt{\sigma_{B\dot{x}}^2 \cos^2\theta_A + \frac{v^2}{R_0^2}\sigma_{Bx}^2 \sin^4\theta_A }$$

Finally, the criterion of linearity of the image is

$$\sqrt{\sigma_{B\dot{x}}^2 \cos^2\theta_A + \frac{v^2}{R_0^2}\sigma_{Bx}^2 \sin^4\theta_A } < \frac{\lambda}{4T_e} .$$

SIDE-LOOKING RADAR

When $\theta_A = 90°$, the inequality is limited to a very simple constraint on the standard deviation of the position error:

$$\sigma_{Bx} < \frac{\lambda R_0}{4 v T_e} \Leftrightarrow \sigma_{Bx} < \frac{r_c}{2}$$

RESOLUTION CRITERION

The same calculation is used for the resolution criterion, which gives

$$\sqrt{\sigma_{B\dot{x}}^2 \cos^2\theta_A + \frac{4v^2}{R_0^2}\sigma_{Bx}^2 \sin^4\theta_A } < \frac{\lambda}{T_e^2} .$$

SIDE-LOOKING RADAR

$$\sigma_{B\dot{x}} < \frac{\lambda R_0}{2vT_e^2} \Leftrightarrow \sigma_{B\dot{x}} < \frac{r_c}{T_e}$$

EXAMPLE

Consider a side-looking airborne radar in X-band ($\lambda = 3\,\text{cm}$) supplying an image with a cross-range resolution equal to 1 m, with the range at 10 km. The velocity of the radar platform is 150 m/s. The image has a dimension of 3 km along the flight axis. The illumination time and image-acquisition time are

$$T_e = 1\,\text{s} \quad \text{and} \quad T_i = 21\,\text{s}.$$

Considering longitudinal motion only, the standard deviation of the position and velocity errors calculated between 0.05 Hz and 1 Hz should be

$$\sigma_{Bx} < 0.5\text{m} \quad \text{and} \quad \sigma_{B\dot{x}} < 1\text{m/s}.$$

The contribution of lateral and vertical movements can lead to an additional margin on these values to ensure that the image quality criteria are met. Note that these constraints ensure linearity of the image for a flight path of 3 km but do not guarantee its positioning in relation to either the platform or the geographical axes. In this case, the errors at zero frequency (i.e., the bias) needs to be taken into consideration.

16.3.2 CALCULATION OF TRANSVERSE MOTION AND VIBRATION EFFECTS

For first-order terms, spurious transverse motion, that is, motion along the Oy or Oz axis, creates a variation in the distance between the antenna phase center and the target phase center:

$$\delta R(t) = y(t)\sin\theta_B\,\cos\theta_D$$

or

$$\delta R(t) = z(t)\sin\theta_D$$

Generally speaking, the higher-order terms do not have a great influence. The resulting phase error and its spectral power density are given by

$$\varphi(t) = -\frac{4\pi}{\lambda}\delta R(t),$$

and

$$S_\varphi(f) = \left(\frac{4\pi}{\lambda}\sin\theta_B\cos\theta_D\right)^2 S_y(f)$$

or

$$S_\varphi(f) = \left(\frac{4\pi}{\lambda} \sin\theta_D\right)^2 S_z(f).$$

Using the same calculation as in Section 16.3.1, we obtain the conditions that need to be met to satisfy the image quality criteria.

16.3.3 SUMMARY OF PLATFORM MOTION

The expressions given in the following tables refer to motion along each axis considered separately. In practice, it is necessary to apply slightly more severe constraints to the standard deviations because the movements occur simultaneously along three axes.

16.3.3.1 LONGITUDINAL MOTION

GENERAL CASE

TABLE 16.1. LONGITUDINAL MOTION ALLOWED

	Geometrical Linearity Criterion: Echo Shift Less Than Half One Resolution Cell	Resolution Criterion: Degradation Less Than 10%
Periodic longitudinal motion (Ox axis)	$\left\|\alpha_{\ddot{x}n}\cos\theta_A - \dfrac{v}{R_0}\alpha_{xn}\sin^2\theta_A\right\|$ $< \lambda/(4T_e)$	$\left\|\alpha_{\ddot{x}n}\cos\theta_A - \dfrac{2v}{R_0}\alpha_{\dot{x}n}\sin^2\theta_A\right\|$ $< \lambda/(T_e^2)$
Random longitudinal motion	$\sqrt{\sigma_{B\ddot{x}}^2\cos^2\theta_A + \dfrac{v^2}{R_0^2}\sigma_{B\dot{x}}^2\sin^4\theta_A}$ $< \lambda/(4T_e)$	$\sqrt{\sigma_{B\ddot{x}}^2\cos^2\theta_A + \dfrac{4v^2}{R_0^2}\sigma_{B\dot{x}}^2\sin^4\theta_A}$ $< \lambda/T_e^2$

SIDE-LOOKING RADAR

TABLE 16.2. LONGITUDINAL MOTION ALLOWED, SIDE-LOOKING CASE

	Geometrical Linearity Criterion: Echo Shift Less Than Half One Resolution Cell	Resolution Criterion: Degradation Less Than 10%
Periodic longitudinal motion (Ox axis)	$\|\alpha_{xn}\| < \dfrac{r_c}{2}$	$\|\alpha_{\ddot{x}n}\| < \dfrac{r_c}{T_e}$
Random longitudinal motion	$\sigma_{Bx} < \dfrac{r_c}{2}$	$\sigma_{Bx} < \dfrac{r_c}{T_e}$

16.3.3.2 TRANSVERSE MOTION

GENERAL CASE

TABLE 16.3. TRANSVERSE MOTION ALLOWED

	Geometrical Linearity Criterion: Echo Shift Less Than Half One Resolution Cell	Resolution Criterion: Degradation Less Than 10%
Periodic lateral motion (Oy axis)	$\left\|\alpha_{\dot{y}n}\right\| < \left\|\dfrac{\lambda}{4T_e \sin\theta_B \cos\theta_D}\right\|$	$\left\|\alpha_{\ddot{y}n}\right\| < \left\|\dfrac{\lambda}{T_e^2 \sin\theta_B \cos\theta_D}\right\|$
Periodic vertical motion (Oz axis)	$\left\|\alpha_{\dot{z}n}\right\| < \dfrac{\lambda}{4T_e \sin\theta_D}$	$\left\|\alpha_{\ddot{z}n}\right\| < \dfrac{\lambda}{T_e^2 \sin\theta_D}$
Random lateral motion	$\sigma_{B\dot{y}} < \left\|\dfrac{\lambda}{4T_e \sin\theta_B \cos\theta_D}\right\|$	$\sigma_{B\ddot{y}} < \left\|\dfrac{\lambda}{T_e^2 \sin\theta_B \cos\theta_D}\right\|$
Random vertical motion	$\sigma_{B\dot{z}} < \dfrac{\lambda}{4T_e \sin\theta_D}$	$\sigma_{B\ddot{z}} < \dfrac{\lambda}{T_e^2 \sin\theta_D}$

SIDE-LOOKING RADAR

TABLE 16.4. TRANSVERSE MOTION ALLOWED, SIDE-LOOKING CASE

	Geometrical Linearity Criterion: Echo Shift Less Than Half One Resolution Cell	Resolution Criterion: Degradation Less Than 10%
Periodic lateral motion (Oy axis)	$\left\|\alpha_{\dot{y}n}\right\| < \dfrac{\lambda}{4T_e \cos\theta_D}$	$\left\|\alpha_{\ddot{y}n}\right\| < \dfrac{\lambda}{T_e^2 \cos\theta_D}$
Periodic vertical motion (Oz axis)	$\left\|\alpha_{\dot{z}n}\right\| < \dfrac{\lambda}{4T_e \sin\theta_D}$	$\left\|\alpha_{\ddot{z}n}\right\| < \dfrac{\lambda}{T_e^2 \sin\theta_D}$
Random lateral motion	$\sigma_{B\dot{y}} < \dfrac{\lambda}{4T_e \cos\theta_D}$	$\sigma_{B\ddot{y}} < \dfrac{\lambda}{T_e^2 \cos\theta_D}$
Random vertical motion	$\sigma_{B\dot{z}} < \dfrac{\lambda}{4T_e \sin\theta_D}$	$\sigma_{B\ddot{z}} < \dfrac{\lambda}{T_e^2 \sin\theta_D}$

NOTATIONS

The standard deviation for motion at low frequency is denoted by

$$\sigma_{B\frac{d^k u}{dt^k}} = \left(2 \int_{1/T_i}^{1/T_e} \left(2\pi f \right)^{2k} S_u(f)\, df \right)^{1/2}$$

with $k = 0$ or 1 or 2, and $u = x$ or y or z.

EXAMPLE

Let us consider the example presented in Section 16.3.1.2. The elevation angle is chosen to be equal to 10°, which is a common value for an airborne radar. The maximum values for the standard deviation of velocity and acceleration errors are given in Table 16.5:

TABLE 16.5. VELOCITY AND ACCELERATION ERROR VALUES

Random lateral motion	$\sigma_{B\ddot{y}} < 8.10^{-3} (m/s)$	$\sigma_{B\ddot{y}} < 3.10^{-2} m/s^2$ or $\sigma_{B\ddot{y}} < 3.10^{-3} g$
Random vertical motion	$\sigma_{B\ddot{z}} < 4.10^{-2} (m/s)$	$\sigma_{B\ddot{z}} < 0.2 m/s^2$ or $\sigma_{B\ddot{z}} < 2.10^{-2} g$

Note the especially critical aspect of the lateral velocity error.

VIBRATIONS

TABLE 16.6. EFFECTS OF VIBRATIONS

	Side Lobe Level or ISLR Contrast Criterion
Sinusoidal motion with amplitude a along Ox	$\dfrac{2\pi a}{\lambda} \cos \theta_A$
Sinusoidal motion with amplitude a along Oy	$\dfrac{2\pi a}{\lambda} \sin \theta_B \cos \theta_D$
Sinusoidal motion with amplitude a along Oz	$\dfrac{2\pi a}{\lambda} \sin \theta_D$
High-frequency random motion along Ox, Oy, and Oz	$ISLR = \left(\dfrac{4\pi}{\lambda}\right)^2 ((\sigma_{Hx}\cos\theta_A)^2 + (\sigma_{Hy}\sin\theta_B\cos\theta_D)^2 + (\sigma_{Hz}\cos\theta_A)^2)$

EXAMPLE

In the case of the side-looking radar in the previous example, the longitudinal vibrations have no effect on the image. The most critical axis is again Oy. To obtain an integrated level of far lobes smaller than –20 dB, the standard deviation of the displacement along Oy calculated for frequencies above 1 Hz must remain below 0.2 mm.

16.3.4 X-BAND OR L-BAND?

If we wish to minimize the influence of platform motion while maintaining the same quality of image, should we use high or low frequency for transmission?

GEOMETRICAL LINEARITY

A low frequency requires a longer illumination time, which minimizes the standard deviation of low-frequency motion. The criteria of geometric linearity lead to inequalities whose upper term is independent of wavelength, such that

$$\sigma_{B\frac{d^k u}{dt^k}} < g\left(v, R_0, r_c\right).$$

Consequently, image geometrical linearity is better at low transmission frequencies.

RESOLUTION

In the case of resolution, the two terms of the inequalities vary in the same way. The result depends on the nature of the motion spectrum. For spectra with a gradient between f^{-2} and f^{-4} (position error), it can be shown that the shorter the illumination time, the easier it is to respect the inequalities. It is therefore easier to achieve image resolution using a high transmission frequency.

CONTRAST

With regard to the contrast criterion, the evolution of the ISLR also depends on the nature of the motion spectrum. Position error spectra with an f^{-2} slope favor the low transmission frequencies. Conversely, spectra with an f^{-4} slope favor high frequencies.

CONCLUSION

In the absence of accurate data on platform motion spectrum, this motion does not constitute a criterion for selection of transmission frequency.

16.4 SPECTRAL PURITY

16.4.1 MODELING

Figure 16.9 presents a radar block diagram. The signal to be transmitted, which is generated on a carrier at intermediate frequency, is converted to microwave frequency by a local oscillator signal supplied by the frequency source. The same signal enables conversion of the received signal from microwave frequency to intermediate frequency.

The stability of the frequency source is characterized by slow deviations in frequency, spurious lines, and phase noise spectrum $\mathcal{L}(f)$, whose effects are, respectively,

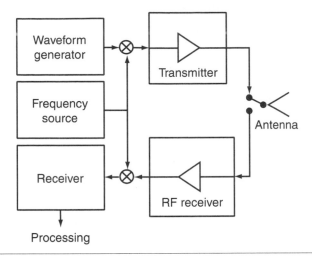

FIGURE 16.9 SIMPLIFIED RADAR BLOCK DIAGRAM

- slow deviations in frequency: displacement and broadening of the correlation peak
- spurious lines: side lobes
- phase noise: far lobes

16.4.2 EFFECTS OF INSTABILITIES

16.4.2.1 SLOW DEVIATIONS IN FREQUENCY

The phase error is modeled in the same way and with the same notations as in Section 16.2:

$$\varphi_n(\theta) = \alpha_n + 2\pi\beta_n f_n \theta - 2\pi^2 \alpha_n f_n^2 \theta^2 \ ,$$

with

$$f_n \theta << 1$$

The conditions required in order to obtain image linearity and resolution are also the same, but the weighting effect of modulation and demodulation is taken into account such that

$$|\beta_n| < \frac{1}{4\pi f_n^2 T_e t_0}$$

and

$$|\alpha_n| < \frac{1}{2\pi^2 f_n^3 T_e^2 t_0} \ ,$$

where t_0 is the two-way propagation time of the signal between radar and target.

For practical reasons, stability may be specified in terms of frequency rather than in terms of phase. Assuming that

$$\delta f(\theta) = \frac{1}{2\pi}\frac{d\varphi}{d\theta} ,$$

$$\delta f_n(\theta) = \alpha_{\delta f_n}\cos(2\pi f_n\theta) + \beta_{\delta f_n}\sin(2\pi f_n\theta) ,$$

and

$$\frac{d\delta \dot{f}_n}{d\theta}(\theta) = \alpha_{\delta \dot{f}_n}\cos(2\pi f_n\theta) + \beta_{\delta \dot{f}_n}\sin(2\pi f_n\theta) ,$$

the conditions on the derivative of the frequency error may be written as

$$\left|\beta_{\delta \dot{f}_n}\right| < \frac{1}{2T_e t_0}$$

and

$$\left|\alpha_{\delta \dot{f}_n}\right| < \frac{1}{\pi f_n T_e^2 t_0} .$$

The coefficients $\beta_{\delta \dot{f}_n}$ and $\alpha_{\delta \dot{f}_n}$ are expressed in Hz/s. They represent the frequency drift.

REMARKS

- The greater the target range, the more severe the constraint
- Stability is required only for the duration of image acquisition T_i, which is generally only a few seconds. This is very short-term stability

16.4.2.2 SPURIOUS LINES

The phase error is expressed as in Section 16.2:

$$\varphi_n(\theta) = \alpha_n \cos(2\pi f_n\theta) + \beta_n \sin(2\pi f_n\theta) ,$$

with $f_n\theta \gg 1$, $\alpha_n \ll 1$ and $\beta_n \ll 1$.

A spurious line of amplitude

$$\frac{\sqrt{\alpha_n^2 + \beta_n^2}}{2}$$

and at frequency f_n, creates a pair of side lobes whose shift in relation to the correlation peak is given by

$$\tau = \frac{T_e}{B_D} f_{fr},$$

with

$$f_{fr} = f_n (\text{modulo } f_r).$$

Their magnitude is (using the notations used in Section 16.2)

$$\left| \sqrt{\alpha_n^2 + \beta_n^2} \ \sin(\pi \, f_n \, t_0) \, u_0 \left(\tau \pm f_{fr} \, \frac{T_e}{B_D} \right) \right|.$$

16.4.2.3 PHASE NOISE

Phase noise produces a line of far lobes whose level is

$$\text{ISLR} = 8 \int_{1/T_e}^{B/2} \sin^2(\pi f t_0) \mathcal{L}(f) df$$

Because the cut-off frequency, $1/T_e$, is small compared to $1/t_0$, the approximation

$$4\sin^2(\pi f t_0)\mathcal{L}(f) \approx 2\mathcal{L}(f)$$

cannot be used for this calculation. The integration bandwidth is limited to the radar receiver bandwidth, and the other frequencies are eliminated by filtering. Note that for a 1 s illumination time, the phase noise spectrum must be integrated starting at 1 Hz, that is, from frequencies that are very close to the carrier. However, the f^2 weighting results in strong attenuation of the effect of these low frequencies.

16.4.3 OTHER SOURCES OF FREQUENCY INSTABILITY

TRANSMITTER

The transmitter is the other main source of frequency instability. The effects of frequency drifts, spurious lines, and phase noise are identical to those due to the instabilities of the frequency source. The main difference is that modulation by spurious signal only occurs during transmission.

There is no weighting by a term depending on echo range. Low frequencies are not therefore attenuated and have a significant effect. Fortunately, for modern TWT transmitters, the low-frequency phase noise is very weak and does not limit the image quality.

SATURATION, NON-LINEARITY

Receiver non-linearities produce effects on the received signal similar to those caused by frequency instabilities. The reception chain must be designed so that receiver saturation occurs when the signal is at the carrier frequency and not at baseband. A part of the intermodulation lines is then naturally rejected outside the usable frequency bandwidth.

16.5 SIGNAL PROCESSING

In Chapter 14, the operation performed by synthetic antenna processing is discussed for a simple case of a continuous signal, without any particular waveform modulation. It is the purpose of this chapter to calculate the received signal considering the pulsed nature of the waveform. The transmitted signal is given by

$$s_0(t) = \sum_{n=-T_e/2T_R}^{T_e/2T_R} u(t - n\,T_R),\, t \in \left[-\frac{T_e}{2}, \frac{T_e}{2} \right].$$

In order to express the return signal received by the radar, radar-target range needs to be known at each moment in time (Figure 16.10). It can be expressed in relation to geometrical parameters, but calculation is more efficient if it is expressed in relation to the Doppler parameters of the target, that is, the central Doppler frequency f_{D_0} (at $t = 0$) and Doppler slope B_D/T_e of the received signal:

$$R(t) = R_0 - vt\cos\theta_A + \frac{v^2 t^2}{2R_0}\sin^2\theta_A = R_0 - \frac{c}{2f_0}\left(f_{D_0}t - \frac{B_D t^2}{T_e 2} \right) = R_0 + \Delta R(t)$$

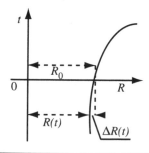

FIGURE 16.10 TARGET RANGE IN RELATION TO TIME

Thus, the received signal can be written (to within one factor) as

$$s(t) = \sum_{n=-T_e/2T_R}^{T_e/2T_R} u\left(t - nT_R - \frac{2R(t)}{c}\right) e^{-j\frac{4\pi D(t)}{\lambda}}, t \in \left[-\left(\frac{T_e}{2}, \frac{T_e}{2}\right)\right].$$

We can approximate that the radar-target range does not change over one pulse duration but only from one interpulse period to the next. The received signal can then be written as

$$s(t) = \sum_{n=-T_e/2T_R}^{T_e/2T_R} u\left(t - n\,T_R - \frac{2\,R(n\,T_R)}{c}\right) e^{-j\frac{4\pi\,R(n\,T_R)}{\lambda}}.$$

We now set

$$t_1 = t - nT_R - \frac{2R_0}{c}$$

and

$$t_2 = nT_R,$$

where t_1 and t_2 are temporal variables. t_1 designates the "fast" time, on which the signal depends within one radar interpulse period. t_2 is the "slow" time on which the signal depends from one interpulse period to the next. In relation to these two variables, the received signal is given by

$$s(t_1, t_2) = u\left(t_1 - \frac{2\Delta R(t_2)}{c}\right) e^{-j\frac{4\pi\Delta R(t_2)}{\lambda}} e^{-j\frac{4\pi R_0}{\lambda}}.$$

The complex exponential term depending on t_2 is the same as that in the expression for the received signal used in Chapter 14, but extra terms have been introduced to take into account the modulation of the waveform.

16.5.1 TRANSFER FUNCTION

The matched receiver performs the following operation:

$$c(\Delta t_1, \Delta t_2) = \int\int_{T_e T_R} s(t_1, t_2) s^*(t_1 - \Delta t_1, \Delta t_2 - \Delta t_2) dt_1 dt_2$$

Using the expression computed above yields

$$c(\Delta t_1, \Delta t_2) = \int\int_{T_e T_R} u\left(t_1 - \frac{2\Delta R(t_2)}{c}\right) u^*\left(t_1 - \Delta t_1 \frac{2\Delta R(t_2 - \Delta t_2)}{c}\right) e^{j\frac{4\pi(\Delta R(t_2 - \Delta t_2) - \Delta R(t_2))}{\lambda}} dt_1 dt_2.$$

There are two ways to carry on with the computation, depending on the amount of range migration.

LOW RANGE MIGRATION

If the geometry of image acquisition is such that target migration is small, then the term t_2 in the modulation function u can be ignored:

$$c(\Delta t_1, \Delta t_2) = \int\limits_{T_e T_R} \int u(t_1)u^*(t_1 - \Delta t_1)e^{j\frac{4\pi(\Delta R(t_2 - \Delta t_2) - \Delta R(t_2))}{\lambda}} dt_1 dt_2$$

The two integrals are now independent:

$$c(\Delta t_1, \Delta t_2) = \int\limits_{T_R} u(t_1)u^*(t_1 - \Delta t_1)dt_1 \int\limits_{T_e} e^{j\frac{4\pi(\Delta R(t_2 - \Delta t_2) - \Delta R(t_2))}{\lambda}} dt_2$$

$$c(\Delta t_1, \Delta t_2) = c_D(\Delta t_1) \cdot c_A(\Delta t_2)$$

Processing is composed of two consecutive stages that can be clearly distinguished:

- The receiver is matched to the received pulses in each interpulse period:

$$c_R(\Delta t_1) = \int\limits_{T_R} u(t_1)u^*(t_1 - \Delta t_1)dt_1$$

- Synthetic Aperture processing is performed from one interpulse period to the next, taking the form of correlation in each range gate:

$$c_A(\Delta t_2) = \int\limits_{T_e} e^{j\frac{4\pi(\Delta R(t_2 - \Delta t_2) - \Delta R(t_2))}{\lambda}} dt_2$$

Certain digital computers in Synthetic Aperture radars perform this kind of correlation processing in the time domain. This operation can also be carried out in the frequency domain by taking the Fourier transform of each of the two terms to be correlated and multiplying one Fourier transform by the conjugate of the other.

HIGH RANGE MIGRATION

When range migration is significant, the problem needs to be considered in its entirety. Calculations are very difficult in the time domain. The programmable computers currently available can perform them more simply in the frequency domain.

Synthetic antenna processing performs the correlation between the received signal and a reference signal in accordance with the two variables t_1 and t_2. This correlation can be performed in the frequency domain by multiplying the Fourier transform of the received signal by the conjugate of the Fourier transform of the reference signal. To simplify the calculations, let us assume that the transmitted pulses are Dirac impulses and the constant phase terms are not considered:

$$s(t_1, t_2) = \delta\left(t_1 - \frac{2\Delta R(t_2)}{c}\right)e^{-j\frac{4\pi\Delta R(t_2)}{\lambda}}$$

The signal is a linear frequency modulation with respect to the dimension t_2. If the product $B_D T_e$ is large, the magnitude of its spectrum is a rectangle centered on the central Doppler frequency, f_{D0}, and its width is equal to the Doppler bandwidth, B_D. The Fourier transform of the received signal can then be written as

$$S(f_1, f_2) = \text{Rect}_{B_D}(f_2 - f_{D_0})e^{j\frac{\pi c R_0}{2v^2\sin^2\theta_A(f_1+f_0)}\left(f_2+\frac{2}{c}v\cos\theta_A(f_1+f_0)\right)^2}.$$

The phase in relation to the Doppler parameters f_{D0} and B_D is given by

$$S(f_1, f_2) = \text{Rect}_{B_D}(f_2 - f_{D_0})e^{j\pi\frac{T_e}{B_D}\left(-\frac{f_0}{(f_1+f_0)^2}f_2^2 + 2f_{D_0}f_2 - \frac{(f_1+f_0)f_{D_0}^2}{f_0}\right)}$$

In this expression, f_0 is the carrier frequency of the transmitted signal, f_1 is the instantaneous frequency of the range-compressed pulses, and f_2 is the instantaneous Doppler frequency.

The processing transfer function is in the 2-D-frequency domain:

$$H(f_1, f_2) = S^*(f_1, f_2)$$

This type of processing is interesting as it solves the problem of migration in the frequency domain by simple phase rotation. It avoids use of interpolation algorithms in the time domain and correction of the errors generated by these algorithms. However, correlation calculation in the frequency domain assumes that the same transfer function is applied to each pixel of the image. This condition is only met for small images; the transfer function along the cross-range axis depends on range. Large images have to be split into small areas in which the Doppler parameters can be considered constant. An algorithm that avoids the variation effect of the Doppler parameters in relation to range is described in the referenced literature (Cumming 1992).

Other algorithms, based on a polar format formalism and close to tomography techniques, enable processing of very high-resolution images and are used in US radars for spotlight mode implementation (Carrara 1995; Queen 1996).

16.5.2 PROCESSING BLOCK DIAGRAM

Figure 16.11 presents a global block diagram of processing. It shows the pulse compression function, the Synthetic Aperture processing function, and various corrections and processing methods applicable once the image has been calculated. Some of the corrections are useful to airborne radars for

- calculation of the Doppler parameters (central Doppler frequency and Doppler slope) based on inertial measurements or analysis of the received signal (autofocus)
- compensation of high-order motion: fast oscillations around the theoretical trajectory, or vibrations, on the basis of inertial measurements or analysis of the received signal

Other corrections can be applied to spaceborne radars: calculation of Doppler parameters based on knowledge of the orbit or analysis of the received signal.

Finally, other corrections can be applied to both airborne or spaceborne radars:

- corrections of differences between amplitude, gain, and phase of the I and Q channels (real part and imaginary part of signal modulation)
- gain calibration (radiometric corrections)
- phase calibration, if the radar has several reception channels (for polarimetric or interferometric applications)
- geometrical corrections (ground projection)
- synthetic Aperture processing in the 2-D-frequency domain can itself be broken down as shown in Figure 16.12. Note that the corrections of range migration are performed simultaneously with correlation

16.5.3 "SINGLE-PASS" PROCESSING

Certain pulse compression radars use analog or digital processes to perform range compression prior to Doppler processing. When a programmable computer is available, it is possible to take advantage of the 2-D Fourier transform calculation to perform both processing operations simultaneously in the frequency domain.

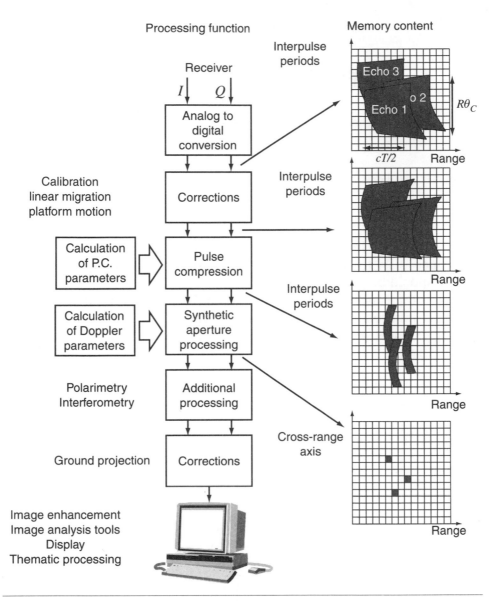

FIGURE 16.11 PROCESSING BLOCK DIAGRAM

If pulse modulation is linear, with a bandwidth B and a duration T, the transfer function that performs both processing operations in a "single pass" can be written as

$$H(f_1, f_2) = e^{j\pi \frac{T_e}{B_D}\left(\frac{f_0}{(f_1 + f_0)}f_2^2 - 2f_{D_0}f_2 + \frac{(f_1 + f_0)f_{D_0}^2}{f_0} \right) + j\frac{\pi T}{B}f_1^2}.$$

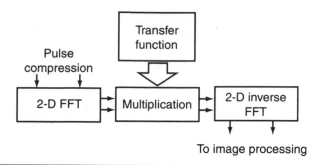

FIGURE 16.12 BLOCK DIAGRAM OF SYNTHETIC APERTURE PROCESSING

16.5.4 MULTILOOK PROCESSING

The quality of the images to be analyzed depends on the end use. It is not always necessary to provide the best geometrical resolution. In some cases, priority should be given to radiometric resolution, which means producing images with reduced speckle. The method commonly used consists of making a non-coherent sum of several independent images of the same site. Independent images can be created in different ways. We shall detail three of them.

RANGE PARALLEL MULTILOOK

In this type of processing, the bandwidth B of the received signal is divided into several separate parts. This provides ground images for different transmission frequencies. If B is divided into n sub-bands, n images can be formed, but with range resolution degraded n times (Figure 16.13).

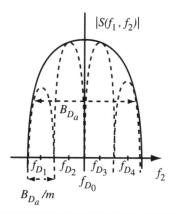

FIGURE 16.13 DOPPLER BANDWIDTH SPLIT INTO M SUB-BANDS PRIOR TO MULTILOOK
           PARALLEL PROCESSING ALONG THE CROSS-RANGE AXIS

CROSS-RANGE PARALLEL MULTILOOK

This processing divides the instantaneous Doppler bandwidth of ground echoes observed by the antenna, B_{D_a}, into several separate parts. This division provides ground images for different aspect angles. If B_{D_a} is divided into m sub-bands, m images can be formed, and cross-range resolution is degraded m times. Figure 16.14 shows an example of the

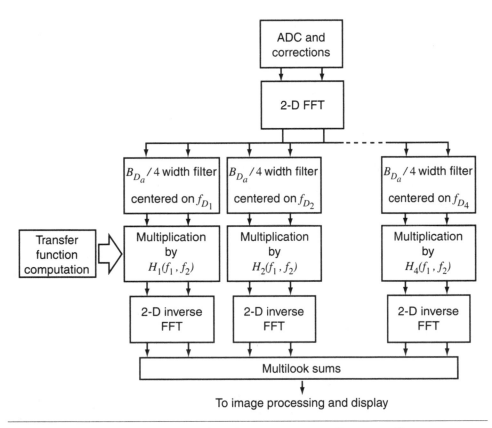

FIGURE 16.14 BLOCK DIAGRAM OF SAR PROCESSING WITH PARALLEL MULTILOOK IN CROSS-RANGE

Doppler spectrum divided into four sub-bands. It is the block diagram of Synthetic Aperture processing with parallel multilook in cross-range generating four looks. The 2-D Fourier transform is followed by a series of four filters whose width is $B_{D_a}/4$. Each of the resulting four channels is multiplied by a 2-D transfer function centered on a different Doppler frequency:

$$\forall i = 1, ..., 4, H_i(f_1, f_2) = e^{j\pi \frac{T_e}{B_D}\left(\frac{f_0}{(f_1+f_0)^2}f_2^2 - 2f_{D_i}f_2 + \frac{(f_1+f_0)f_{D_i}^2}{f_0} \right) + j\frac{\pi T}{B}f_1^2}$$

Processing continues with a 2-D inverse Fourier transform on each of the four channels. After detection (computation of the magnitude of each pixel), the sum of the four resulting images is then made.

The two methods of parallel multilook processing (in-range and cross-range) can be carried out simultaneously. n x m images are then calculated and summed.

The inverse Fourier transform used in the multilook processing methods can be calculated with fewer points than the direct Fourier transform. As the bandwidth of the signal is reduced by filtering, it can be subsampled without any risk of spectrum overlap. This reduces the amount of calculation involved and makes parallel multilook attractive when compared to image calculation at full resolution.

SERIES MULTILOOK

This processing consists of calculating the image at full resolution and applying a filter to it. The most commonly used filter performs a simple sum of several adjacent pixels, along the range or cross-range axis, and calculates their mean value.

The first two methods are called parallel multilook because several images are simultaneously calculated and then added. The last one is named series multilook as only one image calculation is carried out before summing.

16.6 AUTOFOCUS

16.6.1 INTRODUCTION

To yield good image quality, the SAR processing has to rely on precision data describing the position of the antenna phase center. The focusing only requires that this data be relative to a reference position (for instance, at the time origin) during a period at least equal to the illumination time.

Now, as slant range increases and desired resolution improves, the associated augmentation of the illumination time, T_e, generates a very significant constraint on the precision of the antenna phase center position measurement (see Section 16.2).

For an airborne X-band SAR flying at 200 m/s, a 50 km range and a 0.3 m resolution require an illumination time of about 12.5 s in a side-looking configuration. Radial acceleration has to be measured with a precision of $10\,\mu g$, which is far below the capability of a state-of-the-art inertial navigation system (whose typical accelerometer bias of inertial measurement unit is about $50\,\mu g$).

In any case, supposing that we could obtain a perfect inertial device, phase errors originated from propagation effects may also limit our ability to focus high-resolution SAR images. Finally, even for spaceborne SAR, where the trajectory is stable, the accuracy of the orbital parameters is not generally sufficient. That is why every high-resolution SAR processing includes an autofocus stage for phase error estimation using the radar data itself.

In addition to the influence of instantaneous radial acceleration errors on focusing quality, it has also been shown in Blacknell (1989) that these errors introduce expansions and compressions along the azimuth axis in strip-map mode. Thus, supposing that the imaging direction toward the reference trajectory is fixed by the Doppler centroid, f_{d0ref}, every spurious radial acceleration or Doppler rate variation, $\delta f_{d1}(t)$, implies an instantaneous Doppler frequency shift given by

$$\delta f_{d0}(t) = \int_0^t \delta f_{d1}(x)dx \cdot$$

Thus, for a linear flight at constant speed where

$$f_{d0ref} \approx \frac{2V_p}{\lambda}\cos\theta_{Aref} \, ,$$

the Doppler rate shift, $\delta f_{d0}(t)$, entails fluctuations of the imaging direction according to

$$\delta\theta_A = -\frac{\lambda}{2V_p\sin\theta_{Aref}}\delta f_{d0}(t) \, ,$$

that are at the origin of the expansions and compressions in azimuth.

These geometric distortions limit some image processing such as automatic change detection between two images of the same area (White 1991). On the contrary, the compensation of these defects enables the registration of an image pair after some simple geometric operations (translations, rotations).

That is why we look for an autofocus technique that simultaneously corrects focusing and geometric distortions. This constraint discards the method that consists of applying additional SAR processing over defocused areas without modifying the focus position. Conversely, applying a compensation function directly on the raw data or after range compression in order to "bring back" the antenna phase center to the reference trajectory enables us to fulfill both requirements (Oliver 1998).

Before examining the state-of-the-art of autofocus algorithms, we stress the differences between their applications to spotlight and strip-map modes. So far, the majority of the contributions regarding autofocus relate to spotlight mode. For this mode, all the points are focused from the same part of the trajectory because they share the same illumination time. Under this condition, every phase error at the center of the image slowly varies according to azimuth angle. Because the azimuth beamwidth is small, generally the variability of phase errors in azimuth is simply neglected. The autofocus algorithm is then able to benefit from the redundancy of the error information everywhere else in the image. This is all the more true for small swath images where the radial variability may also be neglected.

In strip-map mode, matters are far less simple for autofocus. The difficulties arise from the fact that each point corresponds to a specific illumination period. To evaluate the impact of this feature on the variability of the phase error in azimuth, we consider the following generic error term:

$$\varphi_e(t) = \Phi_0 \sin(2\pi f_e t + \varphi_0)$$

Let C_1 and C_2 be two targets placed Δt apart in azimuth. If we suppose that the phase error is perfectly measured on C_1 and used to process C_2, then the compensation error for C_2 is given by

$$\Delta\varphi_e(t) = \varphi_e(t) - \varphi_e(t - \Delta t) \approx 2\pi f_e \Delta t \Phi_0 \cos(2\pi f_e t + \varphi_0)$$

This last expression shows that the residual error beats at the same frequency as the initial one with an amplitude scaled by $2\pi f_e \Delta t$. For instance, suppose that we have a phase error of 1 000° amplitude and 0.3 Hz frequency after inertial motion compensation. Then the domain of validity of a compensation phase function characterized by a maximum residual error of 10° amplitude is less than 11 ms, that is to say, 2.2 m for a platform speed of 200 m/s. This demonstrates how critical the azimuth variability can be in strip-map mode. This prevents autofocus from exploiting information redundancy in the same way as spotlight does.

There is another difference between the two modes as far as autofocus is concerned. In spotlight mode, the measure of the error and its compensation are carried out in the same azimuth time domain. On the other hand, phase error in strip-map mode intrinsically shows up in the Doppler domain, whereas the compensation takes place in the azimuth time domain to simultaneously correct focusing and geometric defects as mentioned earlier.

In the next sections, the most well-known autofocus algorithms are described according to the open literature. The first two techniques are based on the analysis of zones in the image, whereas the last one directly deals with bright scatterers.

16.6.2 MULTILOOK REGISTRATION

The measurement of the time shift between looks (parts of the total Doppler bandwidth for strip-map mode or parts of the total illumination time for spotlight mode) and the exploitation of the relationships between the misregistration of these looks and the phase errors is probably the first autofocus technique used for SAR (Blake 1992; Carrara 1995). Figure 16.15 explains how the Doppler rate error can be estimated using two looks of the Synthetic Aperture.

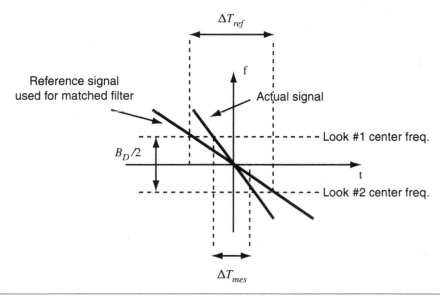

FIGURE 16.15 ESTIMATION OF DOPPLER RATE ERROR THROUGH MEASUREMENT OF THE MISREGISTRATION BETWEEN TWO LOOKS

This method can be generalized to any number of looks, N_{look}. First, the phase variation is expressed within the Synthetic Aperture (in the time domain for spotlight mode, in the Doppler domain for strip-map mode) as a Taylor expansion. For instance, for strip-map mode we may write

$$\varphi(f) = 2\pi \sum_k \alpha_k f^k ,$$

where we suppose that the Doppler centroid has been brought back to zero for simplicity. The first two terms of this model correspond respectively to a constant and to the position of the target in azimuth, and are therefore not involved in autofocus.

The center frequencies of the N_{look} adjacent looks are given by

$$f_i = \frac{B_d}{N_{look}}\left(i - \frac{N_{look}-1}{2}\right), 0 \le i < N_{look}.$$

The generation of the looks consists of selecting N_{look} regions of width B_d/N_{look} centered around f_i, bringing them back around zero and processing them. Let $\varphi_i(f)$ be the phase of look i after processing:

$$\varphi_i(f) = \varphi(f + f_i) - \varphi_{trt}(f + f_i),$$

where $\varphi_{trt}(f)$ is the phase associated with the Doppler parameters used by the processing. The focus position of the target in each look, Δ_i, is computed as the opposite of the linear term in f of each $\varphi_i(f)$:

$$\Delta_i = -\sum_{k \ge 2} \delta\alpha_k k f_i^{k-1}$$

where

$$\delta\alpha_k = \alpha_k - \alpha_{k,trt}$$

Let $\Delta_{i,j}$ be the position shift of the same target between the looks i and j:

$$\Delta_{i,j} = -\sum_{k \ge 2} \delta\alpha_k k (f_j^{k-1} - f_i^{k-1}),$$

which relates the different position shifts to the error of coefficients in the phase function.

The total number of shifts between the N_{look} looks is equal to the number of combinations of two elements among N_{look} elements without allowing for their order; that is to say

$$N_{shift} = \frac{N_{look}(N_{look}-1)}{2}.$$

Although only $(N_{look} - 1)$ of these shifts are theoretically independent, in practice the exploitation of all of them enables us to mitigate the influence of noise and target fluctuations.

From a practical point of view, the shifts between the looks are measured for each range bin through the computation of the correlation in azimuth of the amplitude of the associated images and the measurement of the correlation peak position. We shall come back to the exploitation of the correlation functions issue.

$\Delta_{i,j}$ terms can be summarized in the form of the following set of linear equations:

$$\Delta = A \cdot X,$$

where

- the components of the measurement vector, Δ, are built according to the following rules:

$$n = 0$$

$$\text{for } k = 0 \quad \text{until} \quad N_{look} - 2 \quad \text{do}$$

$$\text{for } j = k + 1 \quad \text{until} \quad N_{look} - 1 \quad \text{do}$$

$$\Delta_n = \Delta_{k,j}$$
$$n = n + 1$$

$$A = \left(a_{n,i} \right)_{\substack{0 \le n < N_{shift} \\ 0 \le i < N_{look} - 1}} \quad \text{with } a_{n,i} = -(i+2)\left(f_j^{i+1} - f_k^{i+1} \right),$$

where the indices j and k are those associated with index n as shown above

- X is the vector of the phase coefficients in the phase polynomial mode such that

$$X_i = \alpha_{i+2} - \alpha_{i+2,trt} = \delta\alpha_{i+2}$$

The least square solution of the previous system is immediately given by

$$X = \left(AA^T \right)^{-1} A^T \Delta.$$

After having described how the errors of the phase coefficients of our polynomial model are obtained, we now come back to the practical procedure that yields the components Δ_n. For the current defocused zone of the image under analysis, each range bin provides its own correlation function able to yield the shift information between the two considered looks. The classical results of estimation theory show that the linear unbiased minimum variance estimator is derived as the sum of the elementary estimates weighted by the inverse of their variance, scaled by the inverse of the sum of the weighting coefficients. By analogy to the range measurement accuracy from the ambiguity function, one can show that the variance of the elementary estimator is inversely proportional to the SNR and proportional to the square of the resolution of the correlation function (at 3 dB for instance). Figure 16.16 indicates the main features involved in the assessment of the quality of an elementary correlation function.

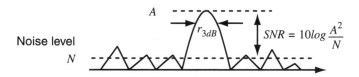

FIGURE 16.16 MAIN FEATURES OF THE CORRELATION FUNCTION BETWEEN TWO LOOKS

Practically the most reliable measures are obtained from the range bins containing bright isolated scatterers. Conversely, range bins with uniform clutter do not yield exploitable correlation peaks. That is why thresholds are generally used as far as SNR and 3 dB resolution of the correlation functions are concerned. If a particular range bin does not provide correlation features above these thresholds, it is not involved in the final estimation process.

Moreover the quality of the correlation function also depends on the initial defocusing level. The more important the error between the actual Doppler parameters and those used in the processing is, the more blurred are the looks and the less sharp are the associated correlation functions. These effects suggest an iterative application. The convergence is reached when at a given iteration the errors of the coefficients of the phase model are less than the thresholds associated with the desired image quality.

Several limits can be quoted for the use of this first algorithm. First, the expression relating the shift between two looks and the coefficients of the phase model show that all the terms of the polynomial contribute to the shift value. Therefore the degree of the polynomial, or equivalently the number of processed looks, should not be underestimated. Now, the greater the number of looks is, the less resolved is each look; hence a degradation of the estimation performance occurs. That is why a strategy to use a decreasing number of looks has been proposed (Oliver 1993) in order to start with the estimation of the high-frequency components of the phase error and to finish with the estimation of the low-frequency ones.

The correlation of two looks can only provide useful information for zones having at least several tens of samples to encompass the range of expected shifts and to obtain a good contrast for the correlation function. Consequently, the main limit of this autofocus algorithm for strip-map mode comes from its inability to efficiently track the azimuth variability of the phase errors.

16.6.3 CONTRAST MAXIMIZATION

The contrast of a zone of an image is defined as the ratio between the standard deviation and the mean of the amplitude of its pixels. The contrast maximization algorithm (Blake 1992) is based on the observation that for a bright isolated scatterer, the contrast is maximum when the Doppler rate used by the processing is identical to that of the actual signal. The underlying hypothesis of this technique is that this property remains true in case of any zone content.

To find the Doppler rate associated with the maximum contrasted image, the zone under analysis should be processed with several Doppler rate candidates. Now in order to limit the processing load, it is important not to multiply the trials. In practice we use an algorithm that approximates the contrast's maximum position by successive iterations.

At each iteration, three values of Doppler rate are considered. These values are separated by a multiple of the depth of focus at the current resolution. This multiple is a feature of the current stage. At the first iteration, the spacing between the Doppler rate values is a function of the expected uncertainty after inertial motion compensation. At the last iteration, this spacing is inferior or equal to the depth of focus. Typically the algorithm takes about three iterations to get the desired Doppler rate accuracy. The central value of the Doppler rate used at the first iteration is either the estimated value of an adjacent zone or the value obtained from geometric and inertial computations.

Once the three Doppler rate values are determined, SAR processing is carried out for each of them. If the central value has not yielded the maximum contrast, the Doppler rate that has yielded the worst contrast is replaced by a new one in the direction of highest contrast. This operation is repeated until the central Doppler rate yields the best contrasted image.

At this stage the estimate of the Doppler rate is improved through a parabolic regression from the three current Doppler rate values: $f_{d_{1-a}}$, $f_{d_{1-b}}$, and $f_{d_{1-c}}$, respectively. Let, K_a, K_b, and K_c, be the three associated contrast measures. The estimate of the optimal Doppler rate corresponds to the maximum of the parabola defined by the three couples $(f_{d_{1-x}}, K_x)$. By writing the equation of the parabola,

$$K = \alpha f_{d1}^2 + \beta f_{d1} + \gamma,$$

we immediately obtain

$$\hat{f}_{d1} = -\frac{\beta}{2\alpha}.$$

By resolving the set of three linear equations relating the measured contrasts to the current Doppler rates through the equation of the parabola, we get

$$\hat{f}_{d1} = f_{d1-b} - \frac{1}{2} \frac{(f_{d1-b} - f_{d1-a})^2 (K_b - K_c) + (f_{d1-b} - f_{d1-c})^2 (K_a - K_b)}{(f_{d1-b} - f_{d1-a})(K_b - K_c) + (f_{d1-b} - f_{d1-c})(K_a - K_b)}.$$

Once the regression is carried out and the central Doppler rate is reestimated, the following iteration is computed by updating the spacing between the three current Doppler rates. Figure 16.17 illustrates the main steps of the whole procedure.

As the multilook registration technique, the contrast maximization algorithm works with zones of a certain width. This feature makes it inefficient to precisely track the azimuth variability of the phase errors encountered in high-resolution strip-map mode.

Seeing that the contrast function presents too many local maxima when we increase the degree of the phase polynomial to estimate—and besides that, the research of extrema in a multidimensional space quickly becomes difficult—maximization contrast limits itself to the estimation of parabolic phase errors. To correct high-frequency components of the phase error, the same strategy as that mentioned for multilook registration may be used (Oliver 1993). First, the contrast for low-resolution looks is optimized and the autofocus results are used to compensate the raw data. The process is then repeated for improved resolution until the final resolution is reached. Another advantage of this strategy is to prevent the contrast maximization from being confused with an initial guess of Doppler rate that is too far from the correct value, in areas where the contrast function exhibits local maxima.

In practice we find that the most reliable contrast information stems from range bins containing bright isolated scatterers. Uniform Rayleigh distributed clutter indeed possesses no contrast at all ($K = 1$). Some zones containing accumulation of closely spaced bright scatterers may also confuse the technique by introducing interference in the position of the maximum of the contrast function. To exploit the information of contrast

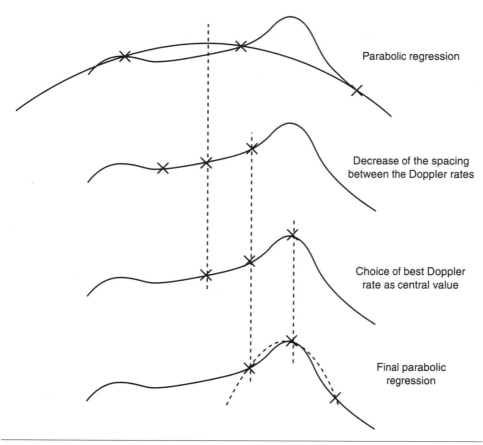

Parabolic regression

Decrease of the spacing
between the Doppler rates

Choice of best Doppler
rate as central value

Final parabolic
regression

FIGURE 16.17 PROCEDURE USED TO ESTIMATE THE DOPPLER RATE ASSOCIATED WITH
MAXIMUM CONTRAST

from the different range bins, the normalized sum of the elementary estimate is computed, weighted by the range bins' associated contrast after having cancelled those below a given contrast threshold.

16.6.4 PHASE GRADIENT

The former autofocus techniques are designed to estimate low-order polynomial-like phase errors. The phase gradient algorithm does not make any *a priori* assumption about the nature of the phase error. Theoretically this algorithm is therefore able to restore a good focusing quality to images polluted by phase errors at a much higher frequency than the inverse of the illumination time.

The main descriptions of the phase gradient algorithm (Wahl 1994; Carrara 1995) consider spotlight mode and aim at estimating in a robust way the derivative of the phase error from the defocused image by optimally exploiting the redundancy of the information in range and in azimuth.

Remaining in the traditional scope of the spotlight mode, and considering a reference target with zero Doppler shift, the temporal signal is given by

$$g(t) = ae^{j\varphi_e(t)} \, \text{rect}\left(\frac{t}{T_e}\right),$$

where $\varphi_e(t)$ designates the phase error during the illumination time.

The elementary estimator of the phase gradient for this target can be expressed as

$$\hat{\Phi}'(t) = \frac{\text{Im}(g'(t)g^*(t))}{|g(t)|^2}.$$

In reality, surrounding clutter and thermal noise are added to the useful signal,

$$n(t) = \alpha(t)e^{j\varphi_n(t)},$$

where $\alpha(t)$ is Rayleigh distributed and $\varphi_n(t)$ is uniformly distributed over 2π. In this model the real and imaginary parts of $n(t)$ are white Gaussian and independent processes. $\alpha(t)$ and $\varphi_n(t)$ are also considered independent. The signal in the temporal domain may be rewritten as

$$g(t) = ae^{j\varphi_e(t)} + a(t)e^{j\varphi_n(t)} = ae^{j\varphi_e(t)}\left\{1 + \frac{\alpha(t)}{a}e^{j(\varphi_n(t) - \varphi_e(t))}\right\}.$$

Let $\Phi(t)$ be the phase of $g(t)$:

$$\Phi(t) = \varphi_e(t) + \arg\left\{1 + \frac{\alpha(t)}{a}\cos(\varphi_n(t) - \varphi_e(t)), \frac{\alpha(t)}{a}\sin(\varphi_n(t) - \varphi_e(t))\right\}$$

By supposing a high signal-to-noise ratio, that is,

$$\frac{\alpha(t)}{a} \ll 1,$$

we get

$$\Phi(t) = \varphi_e(t) + \frac{\alpha(t)}{a}\sin(\varphi_n(t) - \varphi_e(t)) = \varphi_e(t) + \phi(t).$$

Thus the instantaneous phase of the signal in the time domain is equal to the desired phase error plus an extra term that is weaker when the signal-to-noise ratio is high. From our former assumptions, $\varphi_n(t) - \varphi_e(t)$ is uniformly distributed over 2π, and therefore $\varphi(t)$ is white Gaussian. Its power density spectrum is given by

$$S_\phi(f) = \frac{N_0}{a^2} \text{rect}\left[\frac{f}{F_e}\right].$$

The phase gradient algorithm is computing the derivative of the phase; it actually provides $\Phi'(t) = \varphi_e(t) + \phi(t)$. Moreover, by using the standard properties of Fourier transforms, we get

$$E[\phi(t)] = 0$$

$$S_{\phi'}(f) = 4\pi f^2 S_\phi(f)$$

$$E[\phi^2(t)] = Const/a^2$$

By supposing that the noise is uncorrelated from one range bin to another and by denoting $\Phi'(t)$ to be the vector of components, we get

$$\Phi_k'(t) = \frac{\text{Im}(g_k'(t)g_k^*(t))}{|g_k(t)|^2},$$

where $g_k(t) = a_k e^{i\varphi_e(t)} + n_k(t)$ corresponds to the signal at range gate, k. We can summarize the procedure of elementary estimation under the following vector form:

$$\Phi'(t) = \varphi_e t \cdot 1 + \phi'(t)$$

From the former results, the covariance matrix of $\Phi'(t)$ can be expressed as

$$R_\phi = Const \begin{pmatrix} 1/a_1^2 & 0 & 0 \\ 0 & \dots & 0 \\ 0 & 0 & 1/a_K^2 \end{pmatrix}.$$

Finally, according to classical results of estimation theory, the linear unbiased minimum variance estimator is given by

$$\hat{\varphi}_e(t) = \left(1^T R_\phi^{-1} 1\right)^{-1} 1^T R_\phi^{-1} \Phi'(t)$$

$$= \frac{\sum_k a_k^2 \dfrac{\text{Im}\left(g_k'(t)g_k^*(t)\right)}{|g_k(t)|^2}}{\sum_k a_k^2} \approx \frac{\sum_k \text{Im}\left(g_k'(t)g_k^*(t)\right)}{\sum_k |g_k(t)|^2}$$

under the hypothesis of a high signal-to-noise ratio.

Figure 16.18 presents the block diagram of the phase gradient algorithm. To reach the optimal conditions of use of this estimator and allowing for its limitations, the phase gradient algorithm consists of the following four steps:

- For each range bin, select the brightest pixel in order to increase the probability to get a high signal-to-noise ratio, then cancel the influence of the position of the candidate by using a circular shift of the samples to put it at the center of the azimuth line. (In this respect all the selected targets have zero Doppler)

- For each range bin, window around the central position to keep W useful samples for phase gradient estimation. The size of the window must be sufficiently large to capture the blur due to the defocused ambiguity function but the smallest possible in order to limit the noisy influence of its environment. It may be derived from an analysis of the imaging system used or automatically estimated. Wahl (1994) proposes to post-integrate the samples around the central position for each range bin. W is then chosen as the size when the amplitude of the pixel drops 10 dB below the maximum multiplied by 1.5

- Estimate the phase gradient according to the last expression. To obtain the temporal signal, we just take the FFT of the selected window for each range bin, after having zero-padded the array to reach a convenient size, W_{FFT}. A solution to compute the derivative of the temporal signal consists of using the properties of the Fourier transform by multiplying the samples by $j2\pi x$ (where x denotes the variable of the selected samples before taking the FFT). Once the phase error is estimated for W_{FFT} samples, it is interpolated over the N_{FFT} samples of the initial temporal signal (for instance, by taking a zero-padded FFT followed by an inverse FFT). The correction of the defocused image is accomplished by multiplying the initial temporal samples by the term $\exp(-j2\pi\hat{\varphi}_e(t))$ and taking an FFT of the corrected samples

- Iterate the preceding estimation/correction step with a smaller window size to improve the focusing quality. We note that for each iteration the maximum frequency corrected is about $WF/2N_{FFT}$. (The windowing filters out the high-frequency components of the phase error but improves the signal-to-noise ratio for the lowest spectral components as window size decreases)

According to Jakowatz (1993) and Chan (1998), the phase gradient algorithm usually takes five or six iterations to converge, that is to say, to bring the standard deviation of the phase error below a certain threshold. The need to iterate arises from the intrinsic error term, $\phi'(t)$, of the phase gradient estimator and from the interactions of the environment of the selected targets for each range bin. To improve the performance of the technique, two kinds of methods have been proposed. The first one

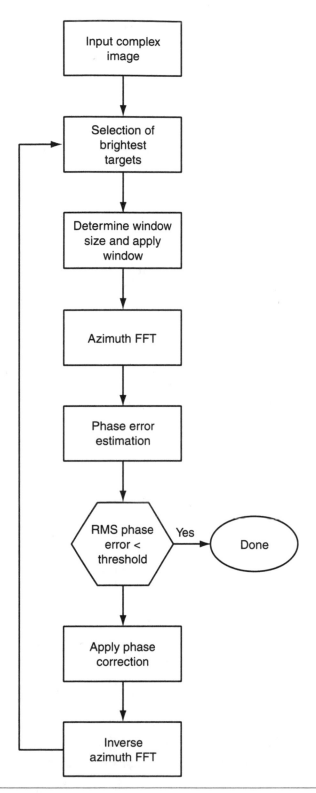

FIGURE 16.18 BLOCK DIAGRAM OF PHASE GRADIENT ALGORITHM

consists of deriving a more accurate estimator of the phase error. The second one deals with the strategy of choosing the range bins and associated targets that contribute to the global estimation of the phase error.

Jakowatz (1993) shows how the maximum likelihood principle can be invoked to solve the estimation problem where M temporal samples are used to estimate $M-1$ phases. In practice this method is more powerful than the phase gradient estimator because it does not introduce the assumption of high signal-to-noise ratio in its derivation, which is necessary to cancel the interfering term, $\phi'(t)$. The improvement of the performance brought by the maximum likelihood estimator is interpreted as a lower number of required iterations to converge.

To derive the expression of the estimator of the phase error according to the maximum likelihood principle, the same signal model as earlier is used:

$$g_{k,j} = a_k e^{j\varphi_j} + n_{k,j},$$

for the temporal sample of index, j, at the range bin of index, k. Here φ_j denotes the phase difference between the temporal tap of index j and the temporal tap of index zero. By denoting

$$g_k = (g_{k,0}, ..., g_{k,M-1})^T$$

and

$$G = (g_0^T, ..., g_{K-1}^T)^T,$$

the conditional probability density function is given as

$$p(G/\varphi) = \prod_{k=1}^{K} \frac{1}{\pi^M |R_k|} \exp(-x_k^H R_k^{-1} x_k),$$

where

$$R_k = E[x_k x_k^H] = \sigma_n^2 c I_M + \sigma_a^2 v v^H = R,$$

σ_a^2 denotes the average power of the selected targets, σ_n^2 denotes the average power of noise, and $v = (1, e^{j\varphi_1}, ..., e^{j\varphi_{M-1}})$.

The maximum likelihood estimate maximizes this conditional probability density function or, equivalently, its logarithm. Therefore,

$$\hat{\varphi} = \max_{\varphi} \left| -\sum_{k=1}^{K} \ln(\pi^M |R|) - \sum_{k=1}^{K} x_k^H R^{-1} x_k \right| .$$

To find $\hat{\varphi}$ we begin to derive the eigenvalues of R. By ascertaining that v is an eigenvector, as is also every vector orthogonal to v, we immediately obtain the eigenvalues of R as

$$\lambda_0 = \sigma_n^2 + M\sigma_a^2, \ \lambda_1 = \sigma_n^2, \ \cdots, \ \lambda_{M-1} = \sigma_n^2 .$$

From this we derive that

$$|R| = \prod_{j=0}^{M-1} \lambda_j = \sigma_n^{2M}\left(1 + M\frac{\sigma_a^2}{\sigma_n^2}\right),$$

which is independent of φ. That is why the derivation of the maximum likelihood estimator reduces to the maximization of

$$-\sum_{k=1}^{K} x_k^H R^{-1} x_k .$$

Now from the expression of the autocorrelation matrix, R, we also have

$$R^{-1} = \frac{1}{\sigma_n^2} I_M - \frac{\sigma_a^2 / \sigma_n^2}{\sigma_n^2 + M\sigma_a^2} v v^H .$$

We thus have to maximize the expression:

$$\sum_{k=1}^{K} x_k^H v v^H x_k = \sum_{k=1}^{K} v^H x_k x_k^H v = v^H\left(\sum_{k=1}^{K} x_k x_k^H\right)v$$

Finally, by denoting

$$\hat{R} = \frac{1}{K}\sum_{k-1}^{K} x_k x_k^H ,$$

the estimate of the autocorrelation matrix of the samples under analysis, one has to find the v, or equivalently φ, that maximizes $v^H R v$. If we change the constraint on v—that is to say, instead of looking for a phase-only vector, we just require that $\|v\| = M$—we can use a classical theorem of linear algebra. This theorem says that the vector that maximizes the former Hermitian form is the eigenvector of $\hat{R}$ of norm M associated with the largest eigenvalue.

To justify the change of constraint on v, first note that $\|v\| = M$ is a necessary condition to get a phase-only vector. Furthermore the desired v is the eigenvector of R associated with its largest eigenvalue. Thus, for a sufficiently large number of range bins, K, $\hat{R}$ closely approximates R, and therefore the eigenvector of norm M associated with its largest eigenvalue also approximates the optimal v.

In the case where one only wishes to use two adjacent temporal samples ($M = 2$) to estimate each value of the phase error during the Synthetic Aperture, an analytical expression of the maximum likelihood estimator can easily be found. We just have to find the current φ_j that maximizes

$$\frac{1}{K}(1\, e^{-j\varphi_j}) \begin{pmatrix} \sum\limits_{k=1}^{K} |g_{k,j}|^2 & \sum\limits_{k=1}^{K} g_{k,j}\,g_{k,j+1}^* \\[2ex] \sum\limits_{k=1}^{K} g_{k,j+1}\,g_{k,j}^* & \sum\limits_{k=1}^{K} |g_{k,j+1}|^2 \end{pmatrix} \begin{pmatrix} 1 \\ e^{j\varphi_j} \end{pmatrix},$$

or equivalently,

$$e^{j\varphi_j} \sum_{k=1}^{K} g_{k,j}\,g_{k,j+1}^* + e^{-j\varphi_j} \sum_{k=1}^{K} g_{k,j+1}\,g_{k,j}^* = 2\left| \sum_{k=1}^{K} g_{k,j}\,g_{k,j+1}^* \right| \cos\!\left(\varphi_j - \arg\!\left(\sum_{k=1}^{K} g_{k,j}\,g_{k,j+1}^* \right) \right).$$

One immediately derives

$$\hat{\varphi}_j = \arg\!\left(\sum_{k=1}^{K} g_{k,j}\,g_{k,j+1}^* \right).$$

In contrast to the phase gradient estimator, this new estimator does not require the high signal-to-noise ratio assumption in its derivation. That is why its use enables us to save some iterations.

Another aspect to investigate to improve the iterative behavior of the phase gradient algorithm consists of only retaining in the estimation the targets that have high signal-to-noise ratio and that do not suffer interference from their environment. In this way only the most reliable information participates in the estimation of the phase error function. Such a technique is described in Chan (1998).

For each range bin, instead of only retaining the brightest scatterer, one introduces the additional degree of freedom to keep the L brightest scatterers. Indeed a particular range bin may contain several interesting targets, whereas some others only encompass uniform diffuse scatterers. This first feature enables us to increase the total number of potentially interesting targets.

The second feature of the method consists of applying a quality test to the windowed temporal samples of the selected targets. At this stage it is intended to remove the synchronization points that are too corrupted by their environment to yield reliable phase information. The chosen criterion to measure the quality of a particular scatterer is the contrast of its amplitude temporal samples, defined for the range bin of index k as

$$
K_k = \frac{\dfrac{1}{T_e}\displaystyle\int_{-T_e/2}^{T_e/2}\left|g_k(t)\right|^2 dt - \left(\dfrac{1}{T_e}\displaystyle\int_{-T_e/2}^{T_e/2}\left|g_k(t)\right| dt\right)^2}{\dfrac{1}{T_e}\displaystyle\int_{-T_e/2}^{T_e/2}\left|g_k(t)\right|^2 dt} .
$$

As a matter of fact, a flat amplitude, and hence a low contrast, is representative of a bright isolated scatterer. Among all the brightest scatterers of a given range bin, only those that exhibit a contrast inferior to a certain threshold are included in the expression of the phase error estimator. Thus the bad points do not reduce the accuracy of the estimator, and the need to iterate becomes less stringent.

Seeing that the phase gradient algorithm exploits information based on well-localized reference targets, it is intrinsically suited to track the variability of the phase error in strip-map mode. However, the core of the estimation process has to be updated because it can no longer rely on the redundancy of the information in azimuth (also in range in case of large swath or short imaging range).

Jakowatz (1994) proposed to compute the second derivative of the phase error (homogeneous to the Doppler rate). These second derivatives are interpolated along the illumination time associated with each selected reference target and further averaged over the whole set of such targets. The phase error is then obtained after double integration and used to compensate the raw data as described in Section 16.6.1.

16.6.5 ASYMPTOTIC PERFORMANCE OF AUTOFOCUS

Autofocus has become a major contributor to SAR processing. Image quality of a very high-resolution SAR imaging system can only be assessed if the performance of autofocus is quantified. In this respect it is interesting to predict the PSLR as a function of the frequency of a sinusoidal phase error:

$$\varphi_e(t) = \Phi_0 \sin(2\pi f_e t + \varphi_0).$$

So that the procedure is compatible with any of the previous autofocus algorithms and with both strip-map and spotlight modes, the Doppler rate is retained as the prime measure of autofocus. The instantaneous error in Doppler rate associated with the previous phase error is given as

$$f_{dr}(t) = -2\pi f_e^2 \Phi_0 \sin(2\pi f_e t + \varphi_0).$$

Thus, let $\sigma_{f_{dr}}$ be the standard deviation of the assumed unbiased estimator. The residual phase error at f_e after autofocusing becomes

$$\delta\varphi_e(t) = \frac{1}{2\pi} \frac{\sigma_{f_{dr}}}{f_e^2} \Phi_0 \sin(2\pi f_e t + \varphi_0).$$

From previous results, we know that the PSLR associated with such residual phase error is

$$\text{PSLR} = \frac{4}{\left(\dfrac{1}{2\pi}\dfrac{\sigma_{f_{dr}}}{f_e^2}\right)^2} = \frac{16\pi^2 f_e^4}{\sigma_{f_{dr}}^2}.$$

Now the asymptotic variance of the Doppler rate estimator may be evaluated through the computation of its Cramer-Rao lower bound. The signal under analysis has the form

$$\begin{cases} g = a e^{j\varphi} s + n \\ s = (e^{j(\alpha_1 t_j + \alpha_2 t_j^2)})_{-\frac{T_a}{2} \leq t_j = \frac{j}{F_e} \leq \frac{T_a}{2}}, \end{cases}$$

where $\beta = (a, \varphi, \alpha_1, \alpha_2)^T$ is the vector of the unknown parameters, n is a complex Gaussian noise whose correlation matrix is given by $R_n = \sigma_n^2 I$, and T_a denotes the duration of analysis.

To compute the Cramer-Rao lower bound we first derive the expression for the components of the Fisher information matrix, $J_{k,n}$, such that

$$J_{k,n} = -E\left[\frac{\partial^2 Q(x/\beta)}{\partial\beta_k \partial\beta_n}\right],$$

where Q denotes the conditional likelihood given by

$$Q(x/\beta) = -(g - ae^{j\varphi}s)^H R_n^{-1}(g - ae^{j\varphi}s).$$

By differentiating this expression twice, we get

$$J_{k,n} = 2\mathrm{Re}\left[\frac{\partial(ae^{j\varphi}s)^H}{\partial\beta_k}R_n^{-1}\frac{\partial(ae^{j\varphi}s)}{\partial\beta_n}\right] = \frac{2}{\sigma_n^2}\mathrm{Re}\left[\frac{\partial(ae^{j\varphi}s)^H}{\partial\beta_k}\frac{\partial(ae^{j\varphi}s)}{\partial\beta_n}\right].$$

By computing the elementary derivatives we also have

$$\begin{cases}
\dfrac{\partial(ae^{j\varphi}s)}{\partial\beta_0} = e^{j\varphi}s \\[2mm]
\dfrac{\partial(ae^{j\varphi}s)}{\partial\beta_1} = jae^{j\varphi}s \\[2mm]
\dfrac{\partial(ae^{j\varphi}s)}{\partial\beta_2} = jae^{j\varphi}(t_j e^{j(\alpha_1 t_j + \alpha_2 t_j^2)})_j \\[2mm]
\dfrac{\partial(ae^{j\varphi}s)}{\partial\beta_3} = jae^{j\varphi}(t_j^2 e^{j(\alpha_1 t_j + \alpha_2 t_j^2)})_j
\end{cases}.$$

From this we immediately derive the expression for the Fisher information matrix as

$$J = \frac{2}{\sigma_n^2}\begin{pmatrix} F_e \int_{-T_a/2}^{T_a/2} dt & 0 & 0 & 0 \\ 0 & a^2 F_e \int_{-T_a/2}^{T_a/2} dt & F_e \int_{-T_a/2}^{T_a/2} tdt & F_e \int_{-T_a/2}^{T_a/2} t^2 dt \\ 0 & F_e \int_{-T_a/2}^{T_a/2} tdt & F_e \int_{-T_a/2}^{T_a/2} t^2 dt & F_e \int_{-T_a/2}^{T_a/2} t^3 dt \\ 0 & F_e \int_{-T_a/2}^{T_a/2} t^2 dt & F_e \int_{-T_a/2}^{T_a/2} t^3 dt & F_e \int_{-T_a/2}^{T_a/2} t^4 dt \end{pmatrix} = \frac{2}{\sigma_n^2}\begin{pmatrix} F_e T_a & 0 & 0 & 0 \\ 0 & a^2 F_e T_a & 0 & a^2 F_e T_a^3/12 \\ 0 & 0 & a^2 F_e T_a^3/12 & 0 \\ 0 & a^2 F_e T_a^3/12 & 0 & a^2 F_e T_a^5/80 \end{pmatrix}.$$

By examining this last expression we see that the position of the target (given by α_1) and its amplitude have no influence on the accuracy of the estimate of α_2. Conversely the uncertainty about the phase of the target degrades the performance of estimation. By inverting the Fisher information matrix we finally get

$$\sigma^2_{\alpha_2} = \frac{90\sigma^2_n}{a^2 T^5_a F_e} = \frac{90}{SNR\, T^4_a},$$

where SNR denotes the signal-to-noise ratio after SAR processing.

Thus the Cramer-Rao lower bound of the Doppler rate is given by

$$\sigma^2_{f_{dr}} = \frac{90}{\pi^2 SNR\, T^4_a}.$$

The signal-to-noise ratio for the estimator is dependent on the scene content. To proceed we need to fix a scenario for the target distribution. Let us assume that the autofocus algorithm can rely on targets with an average radar cross section, $\bar{\sigma}$, each one surrounded by clutter of average backscatter coefficient, $\bar{\sigma}^0$. Let us denote K as the contrast at the ultimate resolution of the image under analysis, such that

$$K = \frac{\bar{\sigma}}{(Ne\sigma^0 + \bar{\sigma}^0)s_{eq}}.$$

For the current frequency of the phase error, f_e, the associated illumination time, T_a, for which the phase error may be approximated by a parabola is about $1/4f_e$. At this current resolution, the SNR for the estimator can be expressed as

$$SNR(f_e) = \frac{k}{4f_e T_e},$$

where T_e denotes the illumination time at the ultimate resolution of the image under analysis.

The variance of the Doppler rate estimator may thus be derived as a function of the current frequency of the phase error:

$$\sigma^2_{f_{dr}} = \frac{90 T_e (4f_e)^5}{\pi^2 K}.$$

Finally, the PSLR can be written as

$$\text{PSLR} = \frac{\pi^4}{5760 f_e T_e} K.$$

If the analysis of inertial motion compensation is carried out in the same way, we get two PSLR curves as a function of f_e that cross at $f_{e-cross}$, as illustrated in Figure 16.19. Above $f_{e-cross}$, the image quality performance is provided by the inertial motion compensation. Below $f_{e-cross}$, the imaging SAR system relies on the presence of bright isolated scatterers to get the required image quality.

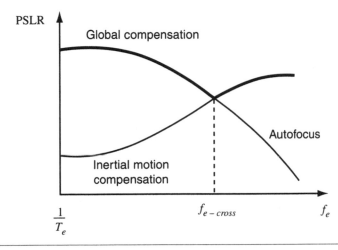

FIGURE 16.19 AUTOFOCUS VERSUS INERTIAL MOTION COMPENSATION

16.7 POWER BUDGET

The power budget of the SAR can be computed as for any radar with processing matched to the received waveform:

$$S/N = E/N_0$$

This is the ratio between received power and noise spectral density.

16.7.1 POWER BUDGET FOR POINT TARGETS

A target is point-like if it is totally contained inside an image resolution cell. In this case, it is characterized by its radar cross section, σ. In relation to the principal parameters of the SAR, the power budget can be written as

$$\frac{S_p}{N} = \frac{P_k G_t G_r \lambda^2 \sigma}{(4\pi)^3 F K T_0 R^4 L} T \, PRF \, T_e.$$

Note that because the illumination time, T_e, is proportional to target range, the signal-to-noise ratio varies in fact as a function of $1/R^3$.

G_e and G_r are the transmission antenna and the reception antenna gains in the direction of maximum radiation. In general, these antennas are physically merged, but the radiation pattern on transmission can be different than that on reception. The parameter L represents the sum of the losses. In particular, it includes the lobe modulation loss. Depending on the operating mode, a target is not always in the direction of maximum gain throughout illumination.

16.7.2 POWER BUDGET FOR DIFFUSE TARGETS

Targets are considered as diffuse, or spread, if they are spread over several resolution cells. Diffuse targets are characterized by a backscatter coefficient, σ_0. The signal-to-noise ratio can be written in relation to the ground surface of the resolution cell. It can be shown that the choice of weighting has no influence on the signal-to-noise ratio:

$$\frac{S_d}{N} = \frac{P_k G_t G_r \lambda^2 \sigma_0 s}{(4\pi)^3 FKT_0 R^4 L} T\, PRF\, T_e,$$

where

$$s = \frac{c}{2B \sin i} \cdot \frac{v_g \sin \theta_A}{B_D}$$

and

$$G_e = \frac{1}{\delta\varphi_e} \int_{-\delta\varphi_e/2}^{+\delta\varphi_e/2} G_e(\theta)d\theta \quad \text{and} \quad G_r = \frac{1}{\delta\varphi_e} \int_{-\delta\varphi_e/2}^{+\delta\varphi_e/2} G_r(\theta)d\theta.$$

$\delta\varphi_e$ is the variation of the target angle of sight during illumination. When gains are expressed in this way, the loss of lobe modulation is already accounted for.

16.7.3 MULTILOOK PROCESSING

Calculation of the power budget makes it possible to compare the image quality produced by the various types of multilook processing. Table 16.7 gives the signal-to-noise ratio for a point and a diffuse target for two cases:

- for a look generated by parallel multilook processing, with resolution degraded by a ratio n in range and a ratio m along the cross-axis
- for a full-resolution image; that is, matched processing is performed over the entire illumination time

TABLE 16.7. SIGNAL-TO-NOISE RATIO FOR ONE LOOK USING PARALLEL MULTILOOK PROCESSING AND FULL-RESOLUTION PROCESSING

	Image With Resolution Multiplied By n x m	Full Resolution Image
SNR for point target	$(S_p/N)_0/nm$	$(S_p/N)_0$
SNR for diffuse target	$(S_d/N)_0$	$(S_d/N)_0$

RADIOMETRIC RESOLUTION

The expression of radiometric resolution after multilook summation, as defined in Chapter 14, concerns only signal-to-noise ratio and the number of looks added. As there is the same signal-to-noise ratio for the images prior to summation, radiometric resolution obtained is consequently the same regardless of the process used, at the first order at least.

POINT TARGET-TO-DIFFUSE TARGET CONTRAST

The best contrast is generally obtained when the size of the resolution cell and the size of the target are of the same order, with possibly a slight over-sampling of the image.

When resolution is clearly finer than target size, it can be useful to perform multilook summation.

16.8 LOCALIZATION ACCURACY

16.8.1 LOCALIZATION MODEL

The model described in this section provides accuracy values suitable for the preliminary design phase of a SAR system. The localization of the target is estimated using

- the position of the platform
- the range measured by the radar
- the azimuth angle measured by the radar

Computation is performed in a geographical coordinates system (Figure 16.20) whose origin is the platform, and whose axes are the north (horizontal), the east (horizontal), and the vertical axis (positive downwards):

$$x_N = x_{N_P} + R\cos(\theta_0 + \theta_B)\cos\theta_D$$
$$x_E = x_{E_P} + R\sin(\theta_0 + \theta_B)\cos\theta_D \quad,$$

where (x_N, x_E, z) are the coordinates of the point observed, and (x_{Np}, x_{Ep}, z_p) are the coordinates of the platform. θ_0 is the platform course.

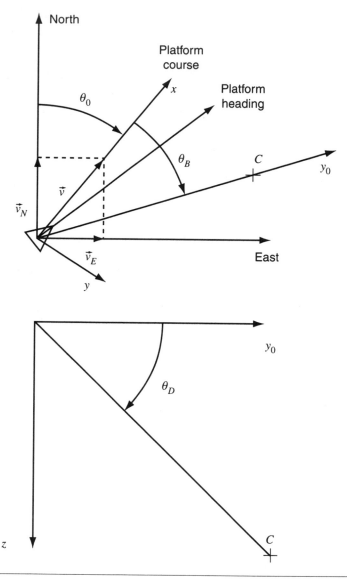

FIGURE 16.20 GEOMETRY IN THE HORIZONTAL PLANE AND IN THE VERTICAL PLANE

All parameters are known in the system. They are supplied by the navigation system or by the radar. The bearing angle, θ_B, is estimated using the Doppler frequency of the target.

16.8.2 BEARING MEASUREMENT ACCURACY

The Doppler frequency of a ground target is proportional to the projection of the velocity vector of the platform on the radar-to-target axis:

$$f_D = \frac{2}{\lambda} \vec{v} \cdot \vec{u}$$

In the geographical reference system linked to the platform and oriented to its course (Ox: horizontal forwards; Oy: horizontal, in the direction of the right wing; Oz: vertical downwards), the coordinates of these two vectors are

$$\vec{v} \begin{pmatrix} v_x \\ v_y \\ v_z \end{pmatrix} \text{ and } \vec{u} \begin{pmatrix} \cos\theta_B \cos\theta_D \\ \sin\theta_B \cos\theta_D \\ \sin\theta_D \end{pmatrix}.$$

The Doppler frequency is

$$f_D = \frac{2}{\lambda} \left(v_x \cos\theta_B \cos\theta_D + v_y \sin\theta_B \cos\theta_D + v_z \sin\theta_D \right). \tag{16.1}$$

Using differentiation, Equation (16.1) gives (16.2):

$$\cos\theta_B \cos\theta_D \, dv_x - v_x \sin\theta_B \cos\theta_D \, d\theta_B - v_x \cos\theta_B \sin\theta_D \, d\theta_D + \sin\theta_B \cos\theta_D \, dv_y$$

$$+ v_y \cos\theta_B \cos\theta_D \, d\theta_B - v_y \sin\theta_B \sin\theta_D \, d\theta_D + \sin\theta_D \, dv_z + v_z \cos\theta_D \, d\theta_D = \frac{\lambda}{2} \, df_D$$

$$\tag{16.2}$$

Depression is a function of range R and height H:

$$H = R \sin\theta_D$$
$$dH = \sin\theta_D \, dR + R \cos\theta_D \, d\theta_D \tag{16.3}$$
$$d\theta_D = \frac{dH}{R \cos\theta_D} - \operatorname{tg}\theta_D \frac{dR}{R}$$

Replacing the depression by Equation (16.3), depending on range and height (16.2) gives

$$\cos\theta_B\cos\theta_D dv_x + \sin\theta_B\cos\theta_D dv_y + \sin\theta_D dv_z \qquad (16.4)$$

$$-(v_x\sin\theta_B - v_y\cos\theta_B)\cos\theta_D d\theta_B$$

$$-((v_x\cos\theta_B + v_y\sin\theta_B)\sin\theta_D - v_z\cos\theta_D)\frac{dH}{R\cos\theta_D}$$

$$+((v_x\cos\theta_B + v_y\sin\theta_B)\sin\theta_D + -v_z\cos\theta_D)tg\theta_D\frac{dR}{R} = \frac{\lambda}{2}df_D$$

As the Doppler frequency measurement accuracy is extremely good, it is supposed to be negligible:

$$df_D \approx 0$$

If all errors are considered independent and with zero mean value, the bearing mean square value is a function of the mean square values of the other parameters:

$$f^2\sigma_{\theta_B}^2 = a^2\sigma_{v_N}^2 + b^2\sigma_{v_E}^2 + c^2\sigma_{v_z}^2 + d^2\sigma_H^2 + e^2\sigma_R^2,$$

where the coefficients a, b, c, d, e, and f are those of Equation 16.4.

To simplify the expressions, the assumption in the next section is that the platform is flying north ($v_E = 0$, $\theta_0 = 0$, $v_z = 0$):

$$\cos\theta_B\cos\theta_D dv_N + \sin\theta_B\cos\theta_D dv_E + \sin\theta_D dv_z$$

$$-v_N\sin\theta_B\cos\theta_D d\theta_B$$

$$-v_N\cos\theta_B\sin\theta_D\frac{dH}{R\cos\theta_D}$$

$$+v_N\cos\theta_B\sin\theta_D tg\theta_D\frac{dR}{R} = 0$$

16.8.3 COMPUTATION OF THE GEOGRAPHICAL LOCALIZATION ERROR

The course error can be expressed as a function of the north and east velocity errors:

$$tg\theta_0 = \frac{v_E}{v_N} \Rightarrow d\theta_0 = \frac{v_N\,dv_E - v_E\,dv_N}{v_N^2 + v_E^2} = \frac{dv_E}{v_N}$$

The first position equation is differentiated; it gives

$$dx_N = dx_{N_P} + \cos\theta_B(\cos\theta_D + \sin\theta_D tg\theta_D)dR$$

$$-R\sin\theta_B\cos\theta_D\frac{dv_E}{v_N}$$

$$-R\sin\theta_B\cos\theta_D d\theta_B$$

$$- \cos\theta_B tg\theta_D dH$$

and

$$dx_E = dx_{E_P} + \sin\theta_B(\cos\theta_D + \sin\theta_D tg\theta_D)dR \quad.$$

$$+ R\cos\theta_B\cos\theta_D\frac{dv_E}{v_N}$$

$$+ R\cos\theta_B\cos\theta_D d\theta_B$$

$$- \sin\theta_B tg\theta_D dH$$

With the same assumption as in the previous section, all parameter errors having a zero mean value and being independent, the positioning error is expressed as

$$\sigma_N^2 = \sigma_{N_P}^2 + \alpha_N^2 \ \sigma_D^2 + \beta_N^2 \ \sigma_{v_E}^2$$
$$+ \gamma_N^2 \ \sigma_{\theta_B}^2 + \delta_N^2 \ \sigma_H^2$$

$$\sigma_E^2 = \sigma_{E_P}^2 + \alpha_E^2 \ \sigma_D^2 + \beta_E^2 \ \sigma_{v_E}^2$$
$$+ \gamma_E^2 \ \sigma_{\theta_B}^2 + \delta_E^2 \ \sigma_H^2 \qquad.$$

16.8.3 EXAMPLE

An example is computed for a set of parameters suitable for a fighter radar operating at medium altitude in air-to-ground mode:

$$\sigma_{x_N} = 5 \text{ m} ; \ \sigma_{x_E} = 5 \text{ m} ; \ \sigma_z = 50 \text{ m}$$

$$\sigma_{v_N} = 0.05 \text{ m}/\text{s} ; \ \sigma_{v_E} = 0.05 \text{ m}/\text{s} ; \ \sigma_{v_z} = 0.5 \text{ m}/\text{s}$$

$$v = 300 \text{ m}/\text{s} ; \ v_z = 0$$

$$\theta_B = 45° ; \ H = 5\,000 \text{ ft} ; \ R = 40 \text{ km}$$

$$\sigma_R = 1 \text{ m}$$

With these values, the geographical localization accuracy of a ground target is 10 m on each horizontal axis. This error does not depend on any boresighting angle accuracy.

This shows that with a very accurate navigation, such as the one provided by certain GPS systems, a Doppler imaging radar can localize a target much better than any other technique at far range.

16.9 OTHER PROCESSING METHODS

16.9.1 MOVING TARGET DETECTION

Synthetic Aperture processing is suited to fixed targets. The Doppler parameters of a moving target (central Doppler frequency and Doppler slope) can be mistaken for those of a fixed target at a different location, or for parameters that do not correspond to any possible fixed target.

Assuming that the pulse-repetition frequency of the radar is sufficiently high and the target has a radial velocity such that its Doppler spectrum does not fit into the Doppler bandwidth of ground echoes, the most suitable processing for moving target detection is of the *Moving Target Indicator (MTI)* type. However, if the target moves slowly, its Doppler spectrum has a non-zero intersection with the Doppler spectrum of ground echoes, and it is no longer sufficient to suppress these in order to detect the target. The same phenomenon occurs if the pulse-repetition frequency of the radar exactly matches the Doppler bandwidth of ground echoes; whatever the velocity of the target, its spectrum intersects that of the ground echoes. Nevertheless, detection is still possible using a matched transfer function. Let us assume that the target has velocity v_t, which can be broken down into v_n along the radar line of sight and v_c along the cross-axis. The matched transfer function is

$$H(f_1, f_2) = e^{j\pi \frac{T_{ec}}{B_{Dc}} \left(\frac{f_0}{(f_1 + f_0)} f_2^2 - 2 f_{Dc} f_2 + \frac{(f_1 + f_0) f_{Dc}^2}{f_0} \right) + j \frac{\pi T}{B} f_1^2}.$$

The illumination time, the central Doppler frequency, and the Doppler bandwidth of a moving target are given by

$$T_{ec} = \frac{R_0 \, \theta_{0A}}{v \sin \theta_A + v_c}, \quad f_{Dc} = \frac{2}{\lambda} (v \cos \theta_A + v_n), \text{ et } B_D = \frac{2 (v + v_c)^2}{\lambda \, R_0} \sin^2 \theta_A \, T_{ec}.$$

Processing requires a different transfer function for each target velocity to be analyzed. It consists of several channels: one channel for fixed targets, and several channels for moving targets. Each target signal is present at each channel output, and its power depends on whether the transfer function of the corresponding processing channel is well matched to the target return signal. Theoretically, the maximum power level is achieved by the channel whose parameters are closest to those of the target.

This kind of processing, even though it enables detection, is very sensitive to false alarm. A fixed target at angle θ_A obviously creates a response in the fixed targets channel. When it is seen by the radar at angle $\theta_A + \delta\theta_A$ a moment later, its Doppler frequency is equal to that of a moving target located at angle θ_A with radial velocity $v_n = v\sin\theta_A \cdot \delta\theta_A$. If the radar cross section of the fixed target varies sufficiently in relation to the angle of sight, its response may be weaker in the fixed target channel when the azimuth is θ_A than in the moving targets channel at a velocity v_n when the azimuth is $\theta_A + \delta\theta_A$. The mismatching due to the difference between Doppler bandwidths and illumination times is not always sufficient to avoid this type of false alarm.

To overcome this kind of disadvantage, the radar must be equipped with one or several additional reception channels. The fixed targets are then cancelled using a combination of the signals received by the different channels: DPCA and STAP. Chapter 10 provides an example of this kind of processing.

16.9.2 HEIGHT MEASUREMENT USING INTERFEROMETRY

BASIC PRINCIPLE

In this application, two antennas are installed on the platform, both located in the same vertical plane (see Figure 16.21). The first antenna can transmit and receive signals, while the second can only receive them. As in the case described in the previous section, transmission could be performed alternatively by the two antennas or by a third one. The phase difference between the signals received by the two antennas is proportional to the range difference between the target and each of the two antennas:

$$\delta\varphi = \varphi_2 - \varphi_1 = -\frac{2\pi}{\lambda}\delta R$$

The height computation of a target M is based on the measurement of its angle of incidence, i_1, its range, R_1, and the platform height, h:

$$h_M = H - R_1 \cos i_1.$$

The angle of incidence, i_1, is obtained by solving the triangle (M, antenna n°1, antenna n°2):

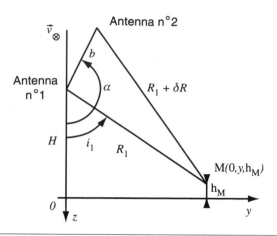

FIGURE 16.21 INTEROMETRY IN THE VERTICAL PLANE

$$\cos(\alpha - i_1) = \frac{b^2 - 2R_1\delta R - \delta R^2}{2R_1 b} = \frac{b}{2R_1} + \frac{\lambda\delta\varphi}{2\pi b} - \frac{\lambda^2\delta\varphi^2}{8\pi^2 R_1 b}$$

Note the importance of accurately measuring angle α and range R_1. Range accuracy is usually better than range resolution.

AMBIGUITY IN HEIGHT MEASUREMENT

The phase difference $\delta\varphi$ is ambiguous to within 2π, and consequently there is an ambiguity in height measurement. Let us assume that h_a and i_a are the ambiguities in height and incidence, respectively, and i_a is a small angle:

$$h_M + h_a = h - R_1\cos(i_1 + i_a) \Rightarrow h_a = R_1 i_a \sin i_1$$

If we make the calculation and neglect the small terms, we get

$$i_a \approx \frac{\lambda}{b\sin(\alpha - i_1)} \Rightarrow h_a \approx \frac{\lambda R_1 \sin i_1}{b\sin(\alpha - i_1)}.$$

ACCURACY IN HEIGHT MEASUREMENT

Height measurement depends on two types of parameters:

- the phase difference measured between the signals received by the two antennas
- the geometric parameters such as roll and platform altitude

The remainder of this section deals only with the accuracy of measurement of phase difference. The effect of errors in the geometric parameters is then easy to estimate. For example, an error in altitude corresponds directly to an error in height measurement.

The observation configuration that gives the best accuracy is the geometry in which a variation in target height corresponds to the greatest possible variation in phase (or range) difference. It is then necessary to minimize the term:

$$\frac{dh_M}{d(\delta\varphi)} = R_1 \sin i_1 \frac{di_1}{d(\delta\varphi)} = \frac{\lambda R_1 \sin i_1}{2\pi b \sin(\alpha - i_1)}$$

Two types of errors can be defined in this expression:

- A random error (noise) in the measurement of phase difference creates a random error in height measurement
- A constant error (bias) in the measurement of phase difference gives rise to an error in height proportional to range, which means locally an error in the terrain gradient

Measurement accuracy is increased by the presence of

- a short wavelength
- a short image acquisition range
- a small angle of incidence without approaching the vertical axis too closely in order to prevent serious degradation of range resolution
- a long distance between the two antenna (interferometry base)
- an angle of around 90° between the interferometry base and the line of sight

These results are obvious if we consider the interferometer as a sparse antenna composed of two radiating sources separated by distance b. In this modeling, the ambiguities are the array lobes of the sparse antenna.

MAIN DIFFICULTIES

The main difficulties encountered when using interferometric processing methods are

- the phase *calibration* of the two reception channels
- estimation of the *ambiguity step* of the phase difference measurement; the amplitude of the signals received by the two channels can be helpful in eliminating this ambiguity
- correction of the effects caused by *spurious platform motion*, especially roll in the case of aircraft
- *phase coherence* of the measurements carried out by the two channels if the measurements do not occur simultaneously (such as the case of a radar onboard a platform mapping the same site twice), or if the angles of sight are too dissimilar from each other

EXAMPLE

Consider an aircraft carrying a 3 cm wavelength radar and illuminating a
10 km distant ground patch at a 45° incidence angle. The two antennas are
spaced 1 m apart along an axis normal to the line of sight. If the phase is
measured with an accuracy of $\pi/16$ (i.e., 11.25°), then the accuracy of
target height measurement becomes

$$\Delta h_M \approx \frac{\lambda R_1 \sin i_1}{2\pi b \sin(\alpha - i_1)}\Delta\delta\varphi = 6.6\,\text{m} .$$

This accuracy can be improved by using post-detection integration of
several measurements obtained by means of multilook processing. The
error in geometrical parameters is then added.

16.9.3 POLARIMETRY

Many radar systems both transmit and receive signals in single
polarization. They therefore receive only a projection of the electrical-field
backscattered by the targets. Polarimetric radars on the other hand measure
the orientation of the electrical-field vector and thus capture more
information about target characteristics. Statistical processing methods
based on information obtained improves the image aspect, and aids
detection and recognition of natural or man-made targets.

BRIEF SUMMARY OF NOTATIONS USED

A target is characterized by its scattering matrix, which links the incident
electrical field received by the target to the reflected electrical field
measured by an antenna located at distance R_0 from the target:

$$\begin{pmatrix} E_h^r \\ E_v^r \end{pmatrix} = \frac{e^{-j\frac{2\pi R_0}{\lambda}}}{R_0}\begin{pmatrix} S_{hh} & S_{hv} \\ S_{vh} & S_{vv} \end{pmatrix}\begin{pmatrix} E_h^i \\ E_v^i \end{pmatrix}$$

The terms E_h and E_v are the coordinates of the electrical field vector in a
coordinate system contained in the plane normal to the direction of wave
propagation. E_h is the horizontal component. E_v is in the vertical plane.

The radar cross section of the target, σ_{pq}, for a polarization pair (p, q) can
be deduced from the four terms of the scattering matrix. Let us assume that
$\dot{u}_q$ is the unit vector parallel to the transmitted electrical field and $\dot{u}_p$ is the
unit vector parallel to the received electrical field. Their respective
coordinates are

$$\begin{pmatrix} u_h^q & u_v^q \end{pmatrix} \text{ and } \begin{pmatrix} u_h^p & u_v^p \end{pmatrix}.$$

With these notations, we have

$$S_{pq} = \left| S_{pq} \right| e^{j\,\varphi_{pq}} = \begin{pmatrix} u_h^p & u_v^p \end{pmatrix} \begin{pmatrix} S_{hh} & S_{hv} \\ S_{vh} & S_{vv} \end{pmatrix} \begin{pmatrix} u_h^q \\ u_v^q \end{pmatrix}, \text{and}\, \sigma_{pq} = 4\,\pi\,\left| S_{pq} \right|^2 .$$

RADAR MEASUREMENT

A radar measures the power of received signals, which makes it possible to calculate the radar cross section of targets. The phase rotation that affects the signal on its two-way path cannot be used in an absolute sense; it is not possible to discriminate between the phase shift due to the reflection onto the target and due to radar-target propagation.

However, a radar can process phase differences between two measurements made on the same target. This is the case for interferometric processing based on the use of range difference, assuming that target behavior remains identical for both measurements.

Polarimetric processing methods are based on the opposite hypothesis: the target is illuminated by waves whose polarization varies and that remain at a constant distance from the target. The measurement thus made is a complex backscatter coefficient depending on the polarization. The phase represents a difference in phase rotation during backscattering, expressed in relation to a reference phase rotation (which could be that obtained for transmission and reception in horizontal polarization).

A radar cannot illuminate a target simultaneously in two orthogonal polarizations; the two polarizations would combine and form a third one. It is necessary to proceed sequentially—to illuminate the target in one polarization, and then in the other. If the radar has only one reception channel, four successive interpulse periods are required. If the radar has two reception channels, horizontal and vertical polarizations can be received simultaneously, and two interpulse periods are sufficient to perform the measurement.

On the other hand, in the case of monostatic radar, it can be shown that polarizations hv and vh are equivalent for most natural targets. Finally, a polarimetric radar aims to extract five parameters for each target, or for each image pixel:

- $\left| S_{hh} \right|$, or the target RCS in horizontal polarization
- $\left| S_{vv} \right|$, or the target RCS in vertical polarization
- $\left| S_{hv} \right|$, or the target RCS in horizontal polarization when illumination occurs in vertical polarization
- $\Delta\varphi_D = \varphi_{vv} - \varphi_{hh}$, the phase difference between the direct polarization responses
- $\Delta\varphi_C = \varphi_{hv} - \varphi_{hh}$, the phase difference between the cross and direct polarization responses

POLARIMETRIC RADAR CALIBRATION

A polarimetric radar is in fact two radars: the first one operates in any polarization (linear horizontal, for example) and the second in the orthogonal polarization (in the example given, linear vertical). Both radar systems can have common elements such as the transmitter, the receiver, etc. Their characteristics must be identical. They must also be perfectly isolated to avoid mutual interaction. The phase centers of the two radar antennas need to be superimposed, otherwise the polarimetric measurements and the interferometry will combine.

Calibration ensures the identity of the two radars. Careful design allied with the right technological options enables isolation of more than 25 dB, which is sufficient given the level difference between the signals received in cross and direct polarization. The use of reference targets is necessary for calibration. A trihedral corner reflector and a dihedral one inclined at 45° are theoretically sufficient (Freeman 1992). Their scattering matrices are, respectively,

$$S_0 \begin{pmatrix} 1 & 0 \\ 0 & 1 \end{pmatrix}, \text{ and } S_0 \begin{pmatrix} \cos 2\alpha & \sin 2\alpha \\ \sin 2\alpha & -\cos 2\alpha \end{pmatrix} = S_0 \begin{pmatrix} 0 & 1 \\ 1 & 0 \end{pmatrix}, \text{ for } \alpha = 45°.$$

However, a 180° uncertainty remains concerning the value of the phase shift between the two transmission channels on the one hand and the two reception channels on the other. This uncertainty can be solved: the right value ensures identity of the cross polarizations after platform spurious-motion compensation.

Other calibration techniques do not require a dihedral corner reflector. They are based on the observation of homogeneous terrain (Van Zyl 1990).

16.9.4 IMAGE-ENHANCEMENT PROCESSING

Image-enhancement processing methods are specific to SAR images and aim to improve the aspect of the image. Their advantages are in the radiometric domain: reduction of speckle, reduction of the side lobe level produced by the radar pulse response.

They also benefit the geometrical domain: correction of errors due to ground configuration, geometrical linearity of the image (*georeferencing*).

The images output from this type of processing are ready for use in the interpretation stage. They can be superimposed on images from other sensors.

16.9.5 THEMATIC PROCESSING

Processing methods specifically adapted to a given mission can be used to assist image interpretation, as in automatic detection of man-made targets by statistical criteria. This enables certain data, which cannot be obtained by means of a simple visual examination of the images, to be supplied to the photo-interpreter.

17

INVERSE SYNTHETIC
APERTURE RADAR (ISAR)

17.1 OBJECTIVES AND APPLICATIONS

The Inverse Synthetic Aperture Radar technique (ISAR) provides images of objects that are in rotation with respect to the radar. It is based on analysis of the received signal as a function of time and Doppler frequency. The result is a two-dimensional image. Time analysis gives the position of bright points along the radar line of sight axis (range axis), while their position on the cross-range axis is provided by Doppler frequency analysis.

For airborne radars, the main application is the recognition of inflight aircraft or vessels at sea. Because radar and target move simultaneously, processing is not exactly SAR, but rather a kind of *generalized* SAR. It is difficult to match the receiver to the received signal; radar-target geometry cannot be completely controlled due to the uncooperative nature of the targets. Propagation characteristics are also difficult to determine.

However, this complexity may be helpful for extracting more information from the signal. Given the specific nature of the ship movements, using certain radar observation configurations, images can be produced in the three dimensions of range, bearing, and height. Processing of only three parameters is required: time, Doppler frequency, and its derivative.

17.2 PRELIMINARY DESCRIPTION OF ISAR

17.2.1 BASIC PRINCIPLES

Consider a radar with a fixed position observing an object—the vessel in Figure 17.1 for example—whose only displacement is a rotation around a fixed point G. This is a pitch motion. The term α is the rotation velocity during illumination.

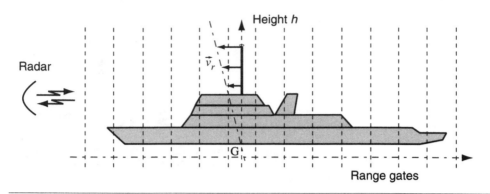

The vessel covers several range gates. If we consider the range gate that includes the mast, the relative velocity of any point on the mast whose height above the center of rotation is h is defined by

$$v_r = \omega h.$$

The Doppler frequency depends on height:

$$f_D = \frac{2}{\lambda} v_r = \frac{2}{\lambda} \omega h,$$

which gives

$$h = \frac{\lambda f_D}{2\omega}.$$

By analyzing the Doppler frequency of signals received in one range gate, it is possible to measure the height of the target reflectors relative to the center of rotation. There are two possibilities:

- The rotation speed, ω, is known and the height measurement is an absolute measurement
- The rotation speed is unknown and the measurement is relative. The scale factor along the height axis remains undetermined in the case of observation of an uncooperative target, and must be estimated from other data

When illumination time increases, the relative velocity of the bright points varies, along with their Doppler frequency. In addition, it is no longer possible to ignore the coupling effect created between range measurement and Doppler frequency, shown in the form of range migration for example.

17.2.2 RESOLUTION

If Doppler frequency analysis is performed over illumination time T_e, then the resolution along the height axis is given by

$$r_{\mathrm{h}} = \frac{\lambda\, k_r}{2\,\omega\, T_e} \approx \frac{\lambda}{2\,\omega\, T_e}.$$

With respect to variation of the target angle of view $\delta\varphi_e$,

$$r_h = \frac{\lambda k_r}{2\delta\varphi_e}.$$

This expression is identical to that established for SAR resolution along the cross-range axis in Chapter 15. It is, in fact, the same phenomenon: with SAR the moving element is the radar, whereas it is the target in the case of ISAR. Relative motion is what counts, especially variation in target viewing aspect.

It should be noted that ISAR resolution is independent of range.

17.2.3 PROJECTION PLANE

Figure 17.2 shows, using simple cases, the type of image that is obtained depending on rotation axis and position relative to the target. Figure 17.3 illustrates a more general configuration. The position of a reflecting point on the Doppler frequency axis depends on two sets of parameters (at first order):

- its coordinates x and y in the horizontal plane, which induce a Doppler frequency proportional to the yaw component of the rotation speed:

$$f_{D_l} = \frac{2}{\lambda}\omega_l(x\sin\varphi + y\cos\varphi)$$

- (φ is the ship angle of view, that is, the angle formed by the radar line of sight and the longitudinal axis of the vessel)
- its height h above the center of rotation, which induces a Doppler frequency proportional to the projection of the roll and pitch components of the rotation velocity onto the line of sight:

$$f_{D_n} = \frac{2}{\lambda}(\omega_r\sin\varphi + \omega_t\cos\varphi)h$$

For a short illumination time, the Doppler frequency of a reflecting point is

$$f_D = \frac{2}{\lambda}[(\omega_r\sin\varphi + \omega_t\cos\varphi)h + \omega_l(x\sin\varphi + y\cos\varphi)].$$

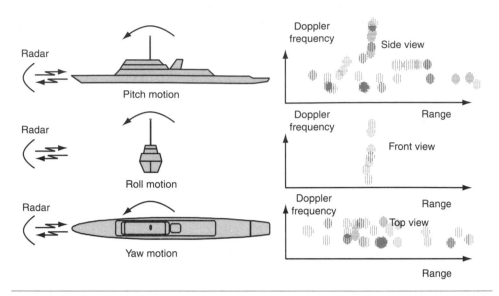

FIGURE 17.2 PROJECTION PLANE IN RELATION TO THE ROTATION AXIS OF THE TARGET
AND THE DIRECTION OF OBSERVATION

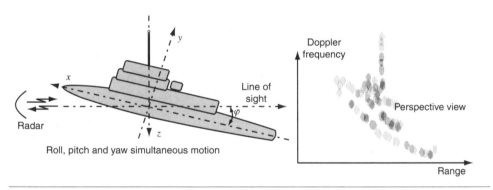

FIGURE 17.3 PROJECTION PLANE: GENERAL CASE

There is an ambiguity between the contribution of the x and y terms and that of the h term: a reflecting point located on the surface along the ship axis can have the same Doppler frequency as a point at the top of the superstructure, but on the side of the vessel. Short-duration spectral analysis (Fourier transform) provides an image in perspective, as shown in Figure 17.3.

17.3 IMAGING OF A SHIP AT SEA

17.3.1 MODELING

Figure 17.4 presents the geometry of observation. When a ship is observed by an airborne radar, the angle of view in the horizontal plane varies during the imaging procedure. This variation merges with the yaw motion. However, the ambiguity between the position of reflecting points in the horizontal plane and their height can be eliminated.

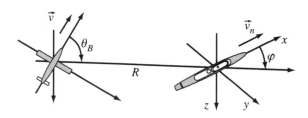

FIGURE 17.4 RADAR OBSERVATION GEOMETRY

The following hypotheses are made:

- The ship is moving in a straight line $(\omega_l = 0)$ at velocity v_n
- One of the rotating motions is dominant, roll for example
- Roll is a periodic and a sinusoidal function $(\omega_r = \omega_{r_0} \cos 2\pi f_r t)$, the time origin being the instant when roll speed is maximum
- The radar operates at a zero depression angle $(\theta_D \approx 0)$, the ship's bearing angle is θ_B, and the radar-ship range is R
- The ship is observed neither head-on nor side-on $(15° \leq \varphi \leq 75°, \text{modulo } 90°)$. If it is observed head-on, the superstructure masks a major part of the vessel. The side view also does not produce a good image; because range analysis is performed on the width of the ship, there are too few range gates to ensure a good result

The Doppler frequency of a reflecting point with $(0,y,h)$ as coordinates then becomes (when only the significant terms are retained)

$$f_D(t) = +\frac{2}{\lambda}v\cos\theta_B + v_n\cos\varphi - \frac{2}{\lambda}\frac{(v\sin\theta_B - v_n\sin\varphi)^2}{R}t + f_{D_a}(t)$$

$$-\frac{2}{\lambda}\frac{v\sin\theta_B - v_n\sin\varphi}{R}y + \frac{2}{\lambda}\omega_{r_0}h\sin\varphi\cos(2\pi f_r t)$$

The first term represents the mean Doppler frequency common to all the reflecting points. It can be compensated for, provided the speed and course of the ship are measured first (for instance, by means of a track-while-scan technique). Linear migration of the corresponding range gates must also be corrected.

The second term expresses the linear variation of the Doppler frequency during illumination, as a function of time. This represents the relative transverse displacement of the aircraft and the vessel. Like the first term, it is common to all the reflecting points and can be compensated for in the same manner.

The third term is the contribution of the spurious motion of the aircraft. It can be compensated for using data provided by the navigation system.

The most interesting terms for image production are the last two because they depend directly on the coordinates of the reflecting points. Once the first three terms have been compensated for, the Doppler frequency and its derivative can be written as

$$f_D(t) = -\frac{2}{\lambda}\frac{v\sin\theta_B - v_n\sin\varphi}{R}y + \frac{2}{\lambda}\omega_{r_0}h\sin\varphi\cos(2\pi f_r t)$$

$$= ay + bh\cos(2\pi f_r t)$$

$$\dot{f}_D(t) = -\frac{4\pi f_r \omega_{r_0}}{\lambda}h\sin\varphi\sin(2\pi f_r t)$$

$$= dh\sin(2\pi f_r t)$$

17.3.2 APPLICATION

The Doppler frequency is a sinusoidal function whose amplitude is proportional to the height of the reflecting point, and whose mean value is a function of its position in the horizontal plane. Height can, moreover, be directly deduced from the value of the Doppler frequency derivative. There are two main cases to be considered:

- At long ranges or low platform speeds, or when the radar line of sight is nearly parallel to the flight axis, the a term can be ignored. Echo scatter in the horizontal plane has very little influence on Doppler frequency, which depends solely on the height of the reflecting points. The radar is required to observe the target when rotation speed is maximum, at moment t_1 in Figure 17.5 for example
- The image is a side view of the vessel, and height resolution is

$$r_h = \frac{k_r\lambda}{2\omega_{r_0}T_e\sin\varphi}$$

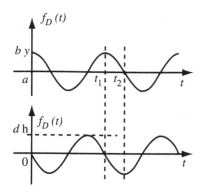

FIGURE 17.5 VARIATION OF DOPPLER FREQUENCY AND ITS DERIVATIVE DURING ILLUMINATION

- Example :

$$\omega_{r_0} = 1°/s,\ T_e = 1\,s,\ \varphi = 45°,\ \lambda = 3\,cm,\ k_r = 1.2,\ r_h = 1.5\,m$$

- The plan view cannot be obtained.
- Conversely, if the a term cannot be ignored, the height of the reflecting points is given by analysis of the Doppler frequency derivative. Radar observation must be carried out when rotation acceleration is maximum, at moment t_2 for instance. The image is a side view, and using the result obtained in Chapter 14, height resolution is

$$r_h = \frac{k_r \lambda}{\pi f_r \omega_{r_0} T_e^2 \sin \varphi}.$$

- Example: using the same parameters as in the previous section, but with a longer illumination time and a roll period of 10 s, we have

$$f_r = 0.1\,Hz,\ \omega_{r_0} = 1°/s,\ T_e = 2\,s,\ \varphi = 45°,\ \lambda = 3\,cm,\ k_r = 1.2,\ r_h = 2.3\,m.$$

Height resolution is not as good as that achieved in the previous case but it is still at a good performance level. Moreover, the plan view can be obtained by Doppler frequency analysis around the instant t_1.

18

OTHER OBSERVATION RADARS

For observation or ground mapping, radar systems based on techniques other than Synthetic Aperture can be used. A brief description of three different types of sensors is given in this chapter:

- millimeter-wave radars
- scatterometers
- altimeters

18.1 MILLIMETER-WAVE RADARS

The use of millimeter-wave radars can be considered for either airborne or spaceborne applications. In the first case, they are used for short-range air-to-ground functions, such as obstacle avoidance, landing aids (Enhanced Vision System, EVS), or high-resolution imagery. In space, millimeter-wave radars can be fit into the small available volume of a planetary exploration probe and perform very precise altimetry.

18.1.1 THE BENEFITS OF MILLIMETER WAVES

18.1.1.1 NARROW ANTENNA BEAM

For a given antenna size, the beamwidth is narrower ($k \lambda / l$) than for any other wavelengths used in radar. At short ranges, the resulting cross-range resolution is very good and often renders SAR processing unnecessary. Moreover, a very sharp antenna beam helps to improve the accuracy of the angular measurement.

Example of resolution: l = 30 cm, λ = 3 mm, R = 1 km $\Rightarrow r_c$ = 10 m.

18.1.1.2 BROAD BANDWIDTH TRANSMISSION

Bandwidth can be very broad while remaining small relative to the carrier frequency. In such cases, high-range resolution is achieved by the radar, which remains within the framework defined by the hypotheses and technological characteristics allowed by narrow bandwidth.

Example: r = 0.3 m, B = 500 MHz, f_0 = 94 GHz $\Rightarrow B / f_0$ = 0.5 %.

18.1.1.3 RCS of Air Obstacles

Millimeter wavelengths are well suited to the detection of certain types of targets that, in this case, present a maximum RCS. This is especially true for cables: electrical cables, cable car lines, etc. Centimeter-wave radars are only able to detect cables when reflection is specular, while millimeter-wave radars can detect them over a very wide angle.

18.1.2 Airborne Applications: Field of Use

In the range of millimeter wavelengths, the atmosphere offers several windows in which attenuation is not too high and where radar can operate. In two cases, around 8.5 mm (i.e., is a frequency of 35 GHz) and around 3 mm (94 GHz), respectively, the components are available and radars can be developed. The effects of atmosphere are as follows:

- *Atmospheric attenuation*, in standard atmosphere, is 0.1 dB/km at 35 GHz and 0.4 dB/km at 94 GHz
- *Smoke and dusty atmospheres* do not induce additional attenuation
- *Fog* has little influence on propagation; at 94 GHz, attenuation is 0.5 dB/km
- *Rain* has a more marked effect. When rainfall rate varies from 1 mm/h (slight rain), to 4 mm/h (moderate rain), and then to 16 mm/h (heavy rain), attenuation is, respectively, 0.8 dB/km, 3 dB/km, and 9 dB/km, at 94 GHz

Due to atmospheric attenuation, millimeter radars are limited to short-range applications: about 5 km for a 94 GHz transmission. *They are particularly useful in bad optical visibility: fog, smoke, dust.*

18.1.3 Cable RCS

The radar cross section (RCS) of a cable observed by a millimeter-wave radar has been computed and measured (Al-Khatib 1981). A cable is characterized by

- its diameter, D
- the number of interlaced strands at cable surface N
- the interlacing period, P
- the diameter of the strands, $d = P / N$

The intersection between the cable and a plane electromagnetic wave is an alignment of points; within one cable interlacing period, there is one backscattering point per strand of cable surface. For a given wavelength, the alignment of points determines an RCS pattern.

Maximum RCS is obtained when the signal reflected by all the strands are in phase:

$$d \sin \alpha = k\frac{\lambda}{2} \Rightarrow \alpha = \arcsin\!\left(\frac{k\lambda N}{2P}\right),$$

where k is an integer.

The value $\alpha = 0$ equals specular reflection onto the cable: propagation direction is perpendicular to the target plane.

The reflection lobes are spread over an angular domain that depends on the structure of the cable:

$$\alpha \in [-\alpha_{\max}, \alpha_{\max}], \text{ with } \alpha_{\max} = \operatorname{arctg}\!\left(\frac{\pi D}{P}\right)$$

Beyond this angular domain and between the reflection lobes, the cable RCS is diffuse. It is proportional to the portion of cable intersected by the antenna beam. As an example, Table 18.1 shows the RCS for a 38 mm diameter cable. A comparison is made between millimeter and centimeter waves. *With millimeter waves, the high RCS outside the reflection lobes enables cable detection over a wide angle.*

TABLE 18.1. RCS OF A CABLE OF 38 MM DIAMETER, IN VERTICAL POLARIZATION

	94 GHz Radar Frequency	18 GHz Radar Frequency
RCS inside reflection lobes	$\approx$ 10 dBm²	$\approx$ 10 dBm²
Number of reflection lobes	5 to 7, depending on the number of cable strands	1
Diffuse RCS, outside reflection lobes	−15 dBm²/m	< −30 dBm²/m

18.2 SCATTEROMETERS

A scatterometer is a radar designed for measurement of ground or sea reflectivity. A brief description of a scatterometer application is given in this section for the measurement of wind speed at sea surface level from a satellite. Sea surface wind creates undulations that modify the backscatter coefficient. The average power of the backscattered signal can be expressed as a function of

- transmission frequency and polarization
- wind speed and direction compared to line of sight
- incidence

By carrying out measurements along several boresight angles, it is possible to estimate the direction and speed of the wind in the operating area (Pike 1990).

18.2.1 ORDERS OF MAGNITUDE

A scatterometer fitted to a satellite in low orbit operates over a wide swath, typically around 500 km. Resolution is between 25 and 50 km. Wind speed can be estimated with an accuracy of 2 m/s (mean square error), and direction with 20° accuracy.

18.2.1.1 BASIC MEASUREMENT PRINCIPLES, ALGORITHM

For each sea area of the size of a resolution cell, the radar supplies the following parameters:

- the backscatter coefficient, for two or preferably three angles of view (Figure 18.1)
- the mean value of the measurement error, estimated from the signal-to-noise ratio

FIGURE 18.1 SAME PATH OF SEA OBSERVED FROM SEVERAL ANGLES

In the case of C-band (around 5.3 GHz), in vertical polarization, the backscatter coefficient and the velocity vector of the wind are linked by a model of the following type (Long 1985):

$$\sigma_0(v, \Phi, i) = A(i)\frac{1 + B(i, v)\cos\Phi + C(i, v)\cos2\Phi}{1 + B(i, v) + C(i, v)}v^{\gamma(i)} \, ,$$

where

$\qquad i \qquad = \quad$ the angle of incidence

v =   the wind speed

Φ =   the angle formed by the line of sight and the wind direction (Figure 18.2)

A, B, C, and γ are functions that vary slowly

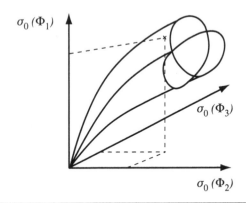

FIGURE 18.2 DETERMINATION OF WIND SPEED AND DIRECTION FROM MEASUREMENT OF THE BACKSCATTER COEFFICIENT ALONG THREE POINTING ANGLES

18.2.1.2 WAVEFORM

The necessary resolution, which is not high, requires an instantaneous bandwidth of some 10 kHz. Allowance must be made for the Doppler frequency, which is of the same order. As this frequency depends on the incidence angle, it varies significantly within the range domain. At non-lateral pointing angles, range discrimination can be performed by simple Doppler frequency analysis. In this case, the waveform transmitted can be a simple continuous wave without modulation. In side-looking mode, the Doppler frequency is zero, and it is necessary to modulate the transmitted waveform to obtain the range resolution: pulsed or linear frequency modulation, for example.

18.3 ALTIMETERS

Altimeters are used for the measurement of the altitude of aircraft above the ground. They can also be installed onboard satellites in order to measure the height of the surface of the earth or the sea.

A very high level of precision is sought: a few centimeters in the case of a spaceborne altimeter. To achieve this, the resolution must not exceed a few tens of centimeters; the range-measuring precision is a fraction of resolution. The best way to produce a radar mode with a very high resolution over a small field is to use the deramp technique, a description of which can be found in Chapter 14. The majority of altimeters are based on this method.

18.3.1 ANTENNA BEAM

There are two categories of altimeter operating modes depending on the width of the ground patch illuminated by the antenna. In Figure 18.3, the portion of ground illuminated by the left-hand altimeter and received by the range resolution cell is limited by the antenna beamwidth; this is a *beam-limited altimeter*. In the case of altimeter n°2, the limitation of the ground patch received in the resolution cell is due to the length of the compressed pulse; this is a *pulse-limited altimeter*.

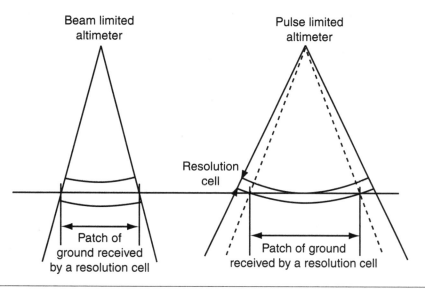

FIGURE 18.3 TYPES OF ALTIMETERS

When an altimeter is close to the ground, it operates in the beam-limited mode. This is the best configuration because it ensures that the measurement effectively corresponds to the altitude measured vertically from the platform. The higher the altitude, the larger the antenna must be. When the altimeter is high enough from the ground, the operating mode is pulse-limited. This means that altitude measurement over a sloping terrain can be subject to error. *The pulse-limited mode measures the shortest range to the ground, while the beam-limited mode measures the range to the ground at the vertical of the platform.* The result is true only above a flat horizontal terrain.

18.3.2 POWER BUDGET

18.3.2.1 BEAM-LIMITED OPERATION

If we start from the radar equation in Chapter 16 with a diffuse signal and assume that the same antenna is used for both transmission and reception, we get

$$\frac{S}{N} = \frac{P_k \, G^2 \, \lambda^2 \, \sigma_0 \, A}{(4\,\pi)^3 \, F \, KT_0 \, R^4 \, L} \, T_A \; PRF \; T_e,$$

where

- T_A is the duration of signal analysis for each interpulse period (see *Deramp*, Chapter 14)
- the product $PRF \; T_e$ indicates the gain obtained in the case of coherent integration of several successive interpulse periods
- R represents the altitude measured
- A is the ground patch illuminated:

$$A = \frac{\pi}{4}(R\theta_{0A})^2$$

This gives

$$\frac{S}{N} = \frac{P_k G^2 \lambda^2 \sigma_0 \theta_{0A}^2}{256\pi^2 F K T_0 R^2 L} T_A PRF \; T_e \, .$$

The equation we obtain is of the type $1/R^2$.

6.3.2.2 PULSE-LIMITED OPERATION

We start from the same equation:

$$\frac{S}{N} = \frac{P_k G^2 \lambda^2 \sigma_0 A}{(4\pi)^3 F K T_0 R^4 L} T_A PRF \; T_e$$

The expression for the ground surface illuminated is

$$A = \pi r^2 \text{ with } r \approx \sqrt{R\frac{kc}{B}},$$

which gives (with k being a coefficient close to 1)

$$\frac{S}{N} = \frac{k P_k G^2 \lambda^2 \sigma_0 c}{64\pi^2 F K T_0 B R^3 L} T_A PRF \; T_e \, .$$

The equation we obtain is of the type $1/R^3$.

PART IV
PRINCIPAL APPLICATIONS

MSCAN Fighter Radar (RDY Radar of the Mirage 2000)

19

RADAR APPLICATIONS AND ROLES

Chapter 2 gives an initial description of operational requirements, platforms, weapons, and systems functions. Before presenting the main specifications and technical descriptions of radar systems in Chapter 20, we list the principal aerospace applications in which radar plays a major role and discuss a number of them in this chapter.

19.1 CIVIL APPLICATIONS

19.1.1 SPACE SYSTEMS

Satellite radars are used for

- observation and surveillance of the atmosphere
- observation and surveillance of the Earth's surface (land and sea), scatterometry, and altimetry
- meteorology
- satellite docking maneuvers, etc.

19.1.2 AIR TRANSPORT APPLICATIONS

Helicopter and aircraft radars are used for

- detection of rain or hail clouds
- detection of turbulence and wind shear
- air surveillance
- in-flight collision avoidance
- landing aids
- ground collision avoidance, etc.

19.1.3 MARITIME APPLICATIONS

Aircraft or helicopter radars are used for

- maritime surveillance, maritime search-and-rescue, anti-drug traffic operations

- pollution detection
- surveillance of coastlines and areas of economic interest, etc.

19.2 MILITARY APPLICATIONS

19.2.1 SPACE SYSTEMS

Satellite imaging radars are used for

- ground observation, detection, and target reconnaissance
- cartography, creation of 3-D Digital Elevation Models (DEM) of terrain
- strategic and tactical intelligence

19.2.2 AIRBORNE APPLICATIONS

Different categories of radar fitted to aircraft, helicopters, cruise missiles, or drones are used for

- air surveillance
- air superiority, interception and combat, policing the skies
- ground observation
- battlefield surveillance
- tactical support, ground attack, exclusion
- high- and very high-altitude penetration and bombing
- low- and very low-altitude penetration and bombing
- obstacle avoidance and landing aids

19.2.3 MARITIME APPLICATIONS

Airborne radars are used for

- maritime surveillance, vessel detection, and reconnaissance
- surface vessel attack
- submarine detection and attack
- coastline surveillance, tactical support, and ground attack
- air interception and combat, etc.

19.3 EXAMPLES OF APPLICATIONS

19.3.1 GROUND OBSERVATION FROM SPACE

19.3.1.1 OPERATIONAL CAPABILITIES

One advantage of space-based radar is that it gives access to any part of the globe (depending on its orbit). It can be used to survey regions that, for various reasons, cannot be observed from an aircraft.

Its continuous presence in orbit ensures maximum coverage. The satellite is a permanent, repetitive, and systematic source of documentary intelligence. The ground segment must be sized accordingly.

Observation takes place in an incidence domain different to that of airborne radar and typically located between 15° and 80° (depending on operating modes). It minimizes the effects of projected shadows, makes it possible to observe land with very uneven relief, and, generally speaking, provides additional information.

If a small number of satellites is used, certain time performance is different to that of airborne systems. While real-time data transmission over a zone of interest is possible, the time needed to access the zone can be fairly long (several hours). Overfly time depends on orbit and cannot be changed at will. The data-refresh period is several hours.

19.3.1.2 ULTIMATE AIM OF THE MISSION

The mission purpose is the expression of the operational requirements. Some examples are

- producing maps of industrial infrastructures and ports
- plotting maps
- detecting or recognizing vehicles
- detecting or recognizing aircraft on the ground
- detecting missile systems and defenses
- detecting activity, etc.

Operational requirements determine the characteristics of the radar and are the input data used to draw up the specifications. These specifications concern two main types of performance: image quality and time performance.

Chapter 14 describes the main parameters for image quality (resolution, linearity, radiometric resolution, contrast, signal-to-noise ratio, etc.). Table 19.1 gives some examples of performance over time.

Optimal image quality for a given mission cannot usually be determined using simple analytical relations. It requires an iterative operation similar to the following:

- definition of a reference radar based on preliminary theoretical analysis (transmitted power, antenna patterns, processing algorithms, etc.)
- use of simulation to produce test patterns corresponding to previously expressed operational requirements and well-defined image quality characteristics

TABLE 19.1. TYPICAL CHARACTERISTICS OF A GROUND OBSERVATION RADAR SYSTEM OPERATING IN SPACE

Time Performance	Average image take time (time between request and image take by the satellite)	Some 10 hours*
	Average age of information (time between image take and dissemination)	Some hours, real time possible
	Average revisit time (time between two successive takes of the same target)	Some 10 hours
	Overfly time	Imposed by orbit
Observation Conditions	Range of region observed	Some 100 km
	Sighting conditions	Imposed by orbit
Quantitative Performance	Coverage (dependent on operating time per orbit and on swath)	Some 10 000 km²/h 24 h / 24 h
	Accessibility (dependent on orbit inclination)	For a polar orbit: Earth's surface (i.e., $\approx 5.10^{8}$ km² differed time)
	Swath width	5 to 500 km
Image Quality Performance	Incidence domain	15° to 80°
	Resolution	< 1 m to some 10 m
Aggression		Jammers, electromagnetic weapons
*Largely dependent on altitude and accessible incidence angle		

- psycho-visual analysis of test patterns using professional photo-analysts. This determines, for each situation, the different image quality parameters needed to meet each requirement. If necessary, the reference radar is modified to bring it into line with operational requirements.

- finally, sizing the radar to give an image quality capable of satisfying all operational requirements

19.3.2 AIRBORNE RECONNAISSANCE

With the advent of submeter resolution SAR sensors, SAR imagery has become a major source for military reconnaissance purposes. It is obvious that its all-weather/day-night capabilities make it invaluable for operational availability, especially in crisis periods when age of information is of crucial importance.

What has renewed the interest in such radars, however, is the ability of submetric modes to provide details about background and targets that were previously the unique privilege of optical sensors. In particular, at resolutions below 50 cm, the features of vehicles and planes begin to differ enough so that one is able to use SAR signature to recognize the class of a target (tank, fighter, truck, etc.) and sometimes to identify it (AMX-30 tank, Mirage F1 fighter, etc.).

Many articles on SAR automatic target recognition have been published in the past five years. The kind of performance one can expect from such technology is indicated in Novak (1997; 1999). To convince himself of the value of these SAR signatures the reader may also consult the Web site http://www.mbvlab.wpafb.af.mil, where various vehicles of different orientations at a resolution of 30 cm are freely available.

In fact, according to the various missions given to an SAR reconnaissance system, different modes have to be implemented. Table 19.2 provides some examples of different missions with the associated modes.

TABLE 19.2. EXAMPLE OF RECONNAISSANCE SAR MODES

Mission	SAR			MTI			Bandwidth (MHz)
	Res. (m)	Swath (km)	type	Res (m)	Loc. (m)	Vmin (km/h)	
Search of objectives / cartography	≈ 3	> 10	strip			NO	≈ 50
Search of objectives + activity assessment	≈ 3	10	strip	≈ 20	< 50	< 10	≈ 50
Ground- installation analysis	≈ 0.50	5	strip / spot	NO			≈ 300
Target classification	≈ 0.30	> 2	spot	NO			≈ 600
Target identification	≈ 0.15	> 1	spot	NO			≈ 1 000
Surveillance / large- area cartography	≈ 20	> 30	scan	≈ 20	< 200	< 10	≈ 7.5

To illustrate the kind of clues that SAR images can bring to fulfill reconnaissance tasks, we provide an example of an operational scenario. To remain at an unclassified level, we shall stress background content and not military-target signature analysis.

The chosen scenario is the verification of the disused state of an airfield. For this purpose we have chosen the test site of Marigny (France), which was a former *NATO* airfield. Figure 19.1 provides a global view of the zone at a crude resolution. The different zones of interest are clearly indicated within the white box.

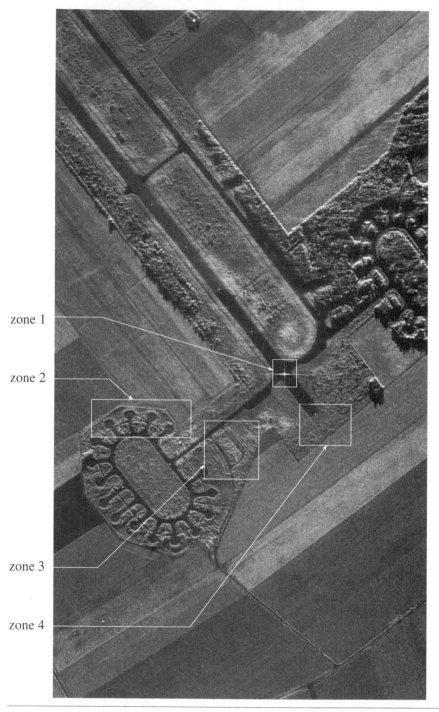

zone 1

zone 2

zone 3

zone 4

FIGURE 19.1 GLOBAL VIEW OF MARIGNY AIRFIELD AT CRUDE RESOLUTION (THALES DOCUMENT)

On the runway various trihedrals are placed for monitoring the SAR processing performance. Zone 1 corresponds to a large 5 000 m^2 trihedral. Seeing that all the images displayed are unweighted, one can see the range and azimuth side lobes associated with the 2-D sinc point-spread function.

We begin our investigation with Zone 2 displayed in Figure 19.2. This represents a part of the dispersal where the planes can park. One can clearly see that grass is growing between the patches of concrete. If we pay closer attention to the top left of the figure, we can hardly distinguish the communication paths that have begun to be occupied by the surrounding vegetation.

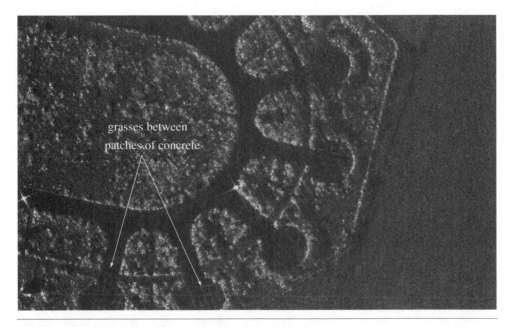

grasses between
patches of concrete

FIGURE 19.2 DISPERSAL AREA (THALES DOCUMENT)

Figure 19.3, associated with Zone 3, then shows a security area surrounded by its own fence. Now the three access gates are left open, which is another clue for us to state that this airfield is no longer in use.

Finally, Figure 19.4 shows the end of the crash runway and the airfield's external fence. One can see that shrubs are growing around the crash runway, which is incompatible with the security of the landing aircraft. The final sign that we point out is the presence of a discontinuity in the external fence.

Beyond the anecdotal aspect of this scenario, it is interesting to note the precision of the information one can get from submetric SAR imagery.

FIGURE 19.3 SECURITY AREA PROTECTED BY ITS OWN FENCE (THALES DOCUMENT)

FIGURE 19.4 END OF THE CRASH RUNWAY AND PART OF THE EXTERNAL FENCE
(THALES DOCUMENT)

19.3.3 AIR SURVEILLANCE

The purpose of air surveillance is to establish a target position over a territory or a theater of operations. It thus aims to detect, track, and identify all forms of aircraft (planes, helicopters, drones, and possibly missiles) flying over the zone observed and then to correlate tracks with information drawn from other sources (flight plans, other radar, sightings, etc.). If an aircraft is unknown or identified as an enemy aircraft, interception aircraft are alerted and sent in its direction. Air surveillance thus has a function similar to that of air traffic control. However, air traffic control is mainly concerned with navigation corridors and areas around airports, while air surveillance must cover all air zones, including targets flying at very low altitude with the aim of using land relief as a mask. This, added to the need to place the radar at high altitude to extend the radar horizon (curvature of the Earth), has led to the introduction of high-altitude airborne radar (> 10 000 m) to complete the ground-based long-range radar network.

These systems, commonly known as Airborne Early Warning systems (AEW) have the following characteristics:

- *long range*: given the typically high speed of enemy aircraft, together with the time needed to identify them, alert a control center (which can be onboard the aircraft), prepare interception aircraft for take-off (e.g., two-minute warning), and the flight time needed to reach interception range and for interception itself, the system must have a range of several hundred kilometers (> 300 km)
- *wide angular coverage*: the AEW has to cover several areas that can be spread over a wide angle in azimuth. Therefore, it requires an angular coverage of at least 150° on each side of the aircraft
- *very high tracking capabilities*: given the size of the area covered by the radar, several hundred aircraft can be present at one time
- *highly efficient ground-return rejection capabilities*: Chapter 7 showed how the need for ground return rejection is a major source of constraint for this type of application. The quality of spectral purity of the transmission-reception chain, the linear reception dynamic range, and the level of the side and far lobes of the antenna must be the highest possible (see Chapter 7).

The platform characteristics are

- *long endurance*: the need for continuous surveillance creates the problem of having a sufficient number of systems permanently operating in-flight to cover the zone in question
- *large payload*: apart from the fact that the platform must ensure an adequate power supply and be capable of carrying the required mass, installation of the radar antennas is a major problem. These must ensure the necessary coverage (> 300° and ≈ 20 000 m altitude slices) regardless of obstruction by the airframe, wings, or engines

In certain contexts, the platform must be able to be flown from an aircraft carrier.

In addition to their detection function, these radar platforms can also act as command and control centers capable of management of the military air traffic, weapon systems allocation, and, in particular, handling the interception process itself by guiding the fighters.

Finally, an AEW system is a high-value target and must therefore ensure its own protection. Its large range and coverage domains mean it can detect air targets that pose a threat to its safety and control interception aircraft assigned to protect it. Moreover, it is usually fitted with self-protection equipment, such as threat detectors, chaff-and-flares dispensers, and towed decoys.

19.3.4 MARITIME SURVEILLANCE

Maritime surveillance radar are used for the missions listed in Chapter 11.

19.3.4.1 MARITIME SURVEILLANCE MISSIONS

A maritime surveillance system usually comprises several devices, radar being one of the most important. The others are

- electronic warfare equipment—Electronic Support Equipment (ESM)
- active or passive acoustic equipment, launchable from the aircraft
- winched acoustic equipment when a helicopter is used as a platform
- forward-looking infrared (FLIR)
- cameras
- a Magnetic Anomaly Detector (MAD)
- a tactical computer
- operator workstations
- weapons (missiles, torpedoes, etc.)

The radar can be used continuously or intermittently, depending on the mission. The following examples, which do not seek to reproduce actual operations, aim to

- show different concepts of radar use
- show the complementary nature of the different items of equipment in the system

SEARCH FOR SUBMARINES IMMERSED AT PERISCOPIC DEPTH

The challenge is to detect the submarine while remaining undetected. This requires the combined action of the radar and the ESM. The plane flies at very low altitude (less than 1 000 feet) to minimize sea clutter. The radar completes one antenna revolution transmitting in panoramic mode every 5 to 10 minutes. This revolution is not repeated, as it would give enemy ESM the chance to confirm detection. Should an enemy radar be detected by the aircraft ESM, the radar is immediately switched over to sector scan mode in the direction of the signal. If target presence is confirmed, the maritime surveillance aircraft can head towards it and carry out its mission. If the detected target is a submarine periscope, the submarine will probably have dived before the plane arrives in the area. The chase thus continues using acoustic and magnetic equipment.

SURFACE SITUATION INITIALIZATION AND UPDATE

Discretion is not one of the main priorities for this type of mission. The radar operates in panoramic scan mode and tracks all detected targets using track-while-scan. If the radar is fitted with an ISAR mode, it can produce images of targets, thus building up its library. Generally speaking, the radar indicates worthwhile targets in the zone, as well as their speed and course. Maritime surveillance aircraft fly at medium altitudes (between 3 000 and 10 000 feet), compatible with the desired radar range and the constraints imposed by the other means of recognition, such as FLIR, photos, etc. If the system measures a temperature inversion while climbing (i.e., temperature increases instead of decreases after a given altitude), the aircraft should avoid flying above it in order to prevent signal extinction due to abnormal propagation of radar waves.

SURFACE VESSEL ATTACK

In this type of mission, the aircraft must come close enough to the ship to get within the firing envelope of the air-to-surface missile. Of course, this must be achieved as discreetly as possible to avoid triggering the anti-aircraft response. To do so, the radar platform stops transmitting and flies at very low altitude (less than 100 ft.) to stay below the radio horizon of the ship's radar. Once within firing distance, the aircraft climbs to lock on to the target with its radar. The aircraft can then

- either use a *fire-and-forget* type missile, in which case it returns to low altitude and pulls away from the target as quickly as possible after firing
- or fire a radar-guided missile, in which case the aircraft must remain within radar visibility of the target until impact

19.3.4.2 THE RADAR PLATFORM

A maritime surveillance radar system is relatively lightweight, between 100 and 150 kg for a modern radar. It can therefore be installed on a number of different aircraft. The three main types of platforms are

- turbo-prop aircraft (or even jet aircraft) weighing over 40 ton on take-off. These planes can be used for very long missions (up to 24 hours endurance in extreme situations) and have a wide radius of action (e.g., from France to the North Pole and back). They carry comprehensive weapon systems, with all the above mentioned equipment, and can store a lot of weaponry. They are designed to give optimum results when flying at very low altitude and very low speeds over the sea (very long wings). The system is operated by a crew of more than ten
- general aviation twin-engine aircraft weighing 10 to 15 ton on take-off. Usually designed for other missions, they can nevertheless provide a useful complement to a heavy aircraft flotilla. They can perform missions requiring less equipment or less endurance at a lower cost than heavy planes. They carry a crew of four to five
- helicopters: their advantage is their capacity to land on a vessel. They can therefore provide two types of protection:
  - against other ships, using their altitude to extend the ship's radar horizon
  - in anti-submarine combat, acting as remote ship equipment
  They also extend the vessel's weapon system by designating over the horizon targets for surface-to-surface missiles or by themselves carrying weapons. Their low speeds and limited radius of action means they are best suited to coastline or flotilla surveillance. The crew can be reduced to two or three only.

19.3.4.3 INSTALLATION

The best position for a maritime surveillance radar antenna is on an elevator located underneath the fuselage. The antenna can then be lowered during flight beneath the masks formed by the airframe. Vision is panoramic.

Installing an elevator, which means making a hole of approximately 1 m diameter in the airframe, is not always possible, particularly in pressurized planes. As a result, less advantageous locations, such as the aircraft or helicopter nose cone, must be used.

The radar units (transmitter, receiver, processing equipment, etc.) are installed in the cabin. The lack of major volume constraints in most situations has resulted in the development of standard unit formats. (Interception radars, on the other hand, are optimized to fit into the nose cone of a combat aircraft). Low power consumption (1 to 2 kW) is compatible with air-cooling systems.

19.3.4.4 SECONDARY MISSIONS

A maritime surveillance radar also enables secondary missions, although the resulting performance is not as good as that of specialized radars.

For example, the long-range surface-target detection mode can be used to detect air targets, preferably in look-up mode. The low mean power of the radar and the presence of sea clutter limit range.

This mode can also be used to detect ground targets in desert zones or for reconnaissance over ground coastal areas. The presence of SAR/ISAR imaging can also prove advantageous for this type of mission.

Finally, a weather function (cloud and rain detection) is also generally available, using a main mode waveform combined with a specific processing method.

19.3.5 BATTLEFIELD SURVEILLANCE

In peace time or crisis, strategic intelligence services seek to acquire information concerning the potential enemy, e.g., infrastructure, concentration of forces, command posts, communications, etc. This information is obtained by various sensors such as visible or infrared cameras, SAR, and MTI radar. These sensors are placed on a variety of platforms such as planes, helicopters, and drones.

During conflict, the battlefield surveillance mission is to detect, locate, identify, and monitor changes in the enemy's resources: tanks and other vehicles, helicopters, motorized troops, air defense batteries, etc. It must enable target acquisition for the artillery or for tactical aviation forces. Information gathered by the different sensors is then transmitted to the ground, analyzed, and, in certain cases, correlated, etc. Results are then "merged" to facilitate decision making by command posts.

Radar modes adapted to the battlefield surveillance mission can be divided into two categories:

- MTI mode able to detect moving targets: vehicles, helicopters, and slow UAVs. The slower the platform speed, the greater the MTI effectiveness. Indeed, the best results are obtained when the platform is stationary. However, STAP technique, combined with long antenna, can compensate for aircraft velocity and give the required performance level to MTI radar on-board planes. It should be noted that in MTI mode, the detection domain is covered by the azimuth scan of the radar antenna and not, as with SAR, by platform displacement

- high-resolution SAR mode able to detect non-moving targets. The platform used must be able to achieve certain minimum speeds (see Chapters 14 and 15)

Ground surveillance radar systems on-board planes are now fitted with both types of modes. They are interleaved to provide simultaneously the moving targets map and radar images of the ground in strip-map and spotlight modes. Planes have the advantage of flying at high speed and high altitude (between 30 000 ft. and 50 000 ft.), allowing high performance SAR imaging and coverage of wide areas. The length of the antenna limits observation to both sides of the aircraft; this means that tracking of moving targets is not continuous during the turns of the radar platform.

Helicopters are specialized in MTI, covering 360° in azimuth. An advantage is continuous observation along the whole orbit trajectory, even during the turns. Thanks to the very low speed, observation axis are chosen to minimize the masks of terrain; the helicopter can be maintained continuously in view of a given area. Finally, it does not required an equipped airfield; it is operated from any type of terrain or Army base, where armored vehicles or other helicopters could be located.

Given that radar is an active and therefore non-discrete sensor, in order to reduce the vulnerability of its platform, it must be stationed well to the rear of combat zones, e.g., at 50 to 150 km stand-off range. In addition, the platform must be at a fairly high altitude (4 000 m to 15 000 m) to reduce shadowing by the terrain.

Note that information supplied by airborne radars can be frequently renewed under any weather conditions, which is vital for providing real-time information for maneuvering corps and tactical air units. The same is not true for space-based sensors; they are subject to the temporal and spatial constraints of satellite orbits.

INSTANTANEOUS COVERAGE

Figure 19.5 shows observation conditions with respect to the operational altitude of various platforms, where

- H is the platform altitude
- θ_D is the depression angle
- θ_{0E} is the elevation angle beamwidth
- R is the radar slant range
- R_S is the swathwidth
- $R = H \cdot \operatorname{cosec} \theta_D$, $R_S = R \cdot \theta_{0E} \operatorname{cosec} \theta_D$ (on ground level)

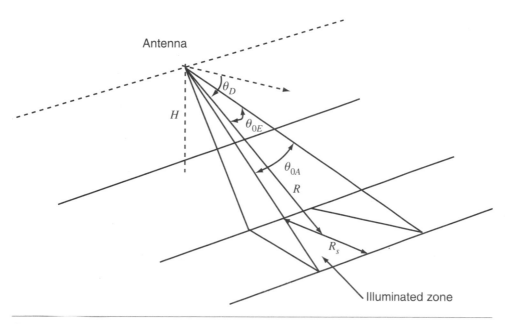

FIGURE 19.5 RADAR FOOTPRINT ON GROUND

Take two examples:

- SAR is on-board a *plane*, where $H = 10$ km, $\theta_D = 8°$ and $\theta_{0E} = 10°$. This gives $R = 72$ km and $R_S = 90$ km

  Platform safety can be ensured. There is some shadowing and the data update rate is sufficient given that detection and reconnaissance concern mainly stationary or slow-moving targets

- MTI is on-board a *helicopter*, where $H = 3$ km, $\theta_D = 3°$ and $\theta_{0E} = 3°$. This gives $R = 58$ km and $R_S = 58$ km

  Platform safety can be ensured, the terrain and the infrastructure provide numerous masks, and, if safety is ensured, detection and positioning of moving targets can be almost continuous

19.3.6 AIR SUPERIORITY, INTERCEPTION, AND COMBAT

Air superiority, interception, and combat are mainly air-to-air missions. There are others, some of which cover all or part of previously described missions. Their names and definitions vary according to the situation, and army and weapons system in question (coverage, general destruction, etc.). We shall not dwell on this subject, although the escort and airspace policing missions merit special mention.

The airspace policing mission, operational in peacetime, in all weather and after approach navigation, consists of visually identifying doubtful aircraft, both civil and military, entering or overflying territory.

The escort mission, which usually occurs at medium altitude, involves the protection of a friendly ground-attack airborne formation.

Using target designation supplied by ground-based or airborne equipment (AEW), the interception mission requires aircraft, either landed or in flight, to engage and destroy one or several planes that enter a friendly zone at any altitude. This mission requires a weapons system with highly effective components (platform: high climb rate, radar: long capability detection range, weapons: medium-range missiles with large differential height, etc.).

The air superiority mission aims to destroy all types of enemy aircraft. It is mainly performed at medium and low altitudes. It is an offensive mission that involves the location and engagement of an enemy aircraft in order to destroy it while in flight. It takes place over friendly territory or following penetration of enemy territory.

Combat, which can be the final phase of the two previously described missions, takes place at short range (< 10 NM). It may begin face-to-face (front attack), but one of its aims is to take up position behind the enemy (tail attack) to reduce relative speed and cross speed and to ensure its own safety. Weapons used are very high-speed short-range passive (IR) or active (EM) missiles, or possibly a gun, which requires the attacking aircraft to come into line with the enemy aircraft. If the enemy aircraft is destroyed, the proximity of the attacking aircraft puts its safety at risk, and it must therefore be able to "pull out" rapidly. Both the platform and the missiles for this type of mission must be highly adaptive. The radar must be able to detect and track the target in a wide angular domain, with high cross speed and rotation velocities (see Figure 19.7).

In order to carry out these air-to-air missions, the radar must be placed inside the aircraft nose cone, where it is not masked. It is "protected" by a radome that is transparent to EM waves (in the radar bandwidth), and whose shape complies with the requirements of the airframe manufacturers and radar engineers.

The role of the radar in these missions is to

- search, either autonomously or using target designation, in given angular and range domains
- detect targets in all weather (with a very low false-alarm probability), regardless of platform and target altitude
- track one or several targets simultaneously
- identify the parameters of the tracked target: range, radial velocity, acceleration, velocity vector, etc.
- transmit these parameters to the weapon system so as to
  - establish the degree of threat posed by the targets
  - identify the target to be attacked first
  - select the most suitable available weapon
  - continuously compute the firing envelope of the chosen weapon
  - draw up a navigation function and provide orders to be followed by the pilot
  - trigger firing once the target is inside the firing envelope
  - determine the disengagement order
  - select the second target to be engaged, etc.

If the selected missile is semiactive, only one continuous tracking action can be engaged, as the missile homing head requires permanent target illumination. With missiles with active seeker (radar), several targets can be simultaneously engaged (several missiles in simultaneous flight). This requires the use of a multitracking radar that supplies the missiles with the parameters of the targets for interception.

For some targets, at certain presentations, a radar in velocity search or continuous tracking mode can identify the target through spectral analysis of its RCS, this being modulated by its turbine blades (see Chapter 20).

Figure 19.6 shows an example of wide-range autonomous search (without target designation) using a radar fitted with a mechanical scanning antenna. The altitude interval covered, h, is proportional to the range and angular domain scanned in elevation.

A radar with a 50 NM range for targets of 5 m^2 RCS, fitted with a 60 cm diameter circular antenna whose beam at 3 dB is 3.5°, and scanning with four elevation lines (crossing over at 2 dB), will thus cover an angular elevation domain of 11.5° at 3 dB, or 0.2 radian. At 50 NM, this corresponds to an altitude segment, h, of 60 000 feet. This is far from covering all possible penetration altitudes, which can attain 100 000 feet. If

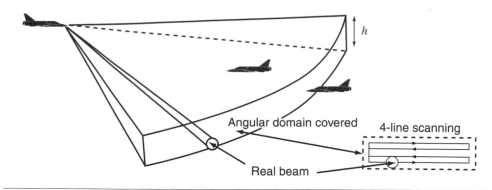

FIGURE 19.6 Autonomous Wide-range Search

the RCS of the target is less than 5m^2, the detection range and altitude segment are reduced. This simple example shows the limits of mechanical scanning, and the advantages of target designation or of patrol flights with breakdown of the altitude segments to be monitored. It also illustrates the advantages of electronic scanning. Moreover, with antenna scanning of ± 60° in bearing at 100°/s, the total exploration time can be as long as 5.6 seconds (including the time taken to reverse the antenna scanning direction, which is approximately 0.2 s).

Figure 19.7 shows an example of search domains required for rapid acquisition of a hostile aircraft (in a close combat situation). The pilot selects these domains with regard to the conditions in which he can engage the enemy. They are suited to the use of a mechanical scanning antenna. In order to reduce reaction time, the three-axis tracking lock-on feature of the radar is automatic as soon as a target is detected. In a dogfight situation, the roll axis (if it is mechanical) is blocked as it has limited angular possibilities (e.g., ± 110°).

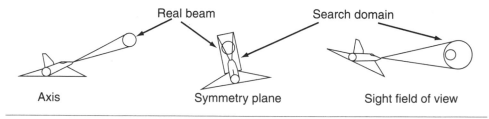

FIGURE 19.7 Search Domains in Close Combat

19.3.7 TACTICAL SUPPORT, GROUND ATTACK, AND INTERDICTION

Tactical support, ground attack, and interdiction are all air-to-ground missions. These offensive missions are generally performed by aircraft in formation at medium to low altitude (> 1 000 feet). Their aim is to destroy or neutralize ground targets such as bridges, infrastructures, airfield runways, tank formations, ground-to-air batteries, etc. These missions require

- detailed mission preparation (knowledge of the tactical situation)
- high-performance navigational resources
- air protection (escort)
- sophisticated countermeasures for the entire formation (powerful accompanying jammers)
- weaponry adapted to the targets, such as antiradar missiles, runway destruction bombs, rockets, laser-guided bombs, laser-guided air-to-ground missiles, etc.

In these missions, apart from the escort that is provided by air superiority aircraft (radar air-to-air function), the role of the radar is to

- perform ground mapping with monopulse sharpening of the real beam
- update the inertial control system with characteristic echoes if GPS is not available
- provide assistance for low-altitude navigation by means of "contour mapping" modes
- detect and lock on to contrasted fixed echoes in continuous tracking mode
- detect and track moving ground targets
- perform telemetry on an optically selected target (air-to-ground ranging)

A radar system uses only air-to-ground modes for these missions. However, in order to ensure self defense in close-combat situations using short-range missiles (IR, EM) or gun fire, the radar must have the air-to-air modes required for enemy acquisition and tracking; that is, the radar must be able to search along the axis, sight field of view, and symmetry plane and do continuous tracking. Each of the previously described air-to-air and air-to-ground modes can use low-PRF waves.

Should the tactical support aircraft be required to engage enemy helicopters, its on-board radar must be able to detect them either from the RCS of the airframe, or from the RCS of the main rotor blades. In the first case, moving ground-target detection and tracking modes are sufficient (except when hovering). In the second case, the use of a specific air-to-air mode adapted to the RCS characteristics of the blades is required. Such characteristics are

- a very brief appearance (between 50 and 400 µs)
- a wide spectrum (between 10 and 20 kHz at X-band)
- a stable repetition frequency (between 10 and 40 Hz, depending on the helicopter)

On the actual battlefield, tactical support can also be provided by helicopters flying in formation and armed with machine guns, guns, rockets, or air-to-ground missiles. However, the sensors used for fire control are optical or optronic and a radar can be used to ensure omnidirectional surveillance and target designation in air-to-air mode.

Figure 19.8 illustrates the "contour mapping" modes used at low altitude (500 to 2 000 feet). In this case, radar's role is to detect natural and man-made obstacles along the aircraft flight path and to determine their altitude so as either to enable the aircraft to overfly the obstacles with sufficient clearance to ensure its safety, or to avoid them. Although the mission is generally prepared in sufficient detail, some obstacles may not be listed. Similar, the plane may have to fly off-line outside the zone covered by mission preparation. Contour mapping provides a means of specifically displaying (e.g., in color) echoes located above and below the mapping areas. Three forms of contour mapping can be used:

- contour mapping (1) stabilized in the horizontal plane

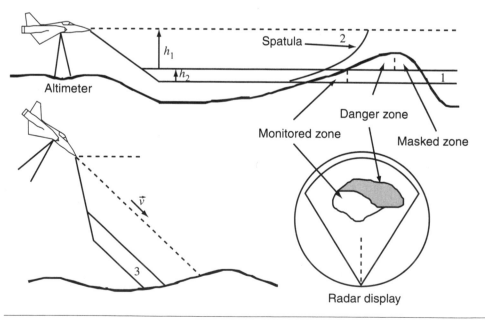

FIGURE 19.8 CONTOUR MAPPING MODES

- contour mapping with spatula (2) stabilized in the horizontal plane but whose action is reduced depending on range to avoid identifying high-altitude obstacles too early. Indeed, in enemy territory, safety is increased when the plane can fly at low altitude to take advantage of masks formed by land relief.
- contour mapping slaved to the aircraft velocity vector (3). This allows perfectly safe blind penetration, e.g., through clouds.

The pilot is free to choose the clearance altitude. He must navigate so that the detected ground is located between h_1 and h_2, thus optimizing safety with regard to land relief and enemy weapons. The pilot navigates in the vertical and horizontal planes; the radar simply supplies information. The altimeter, whose measurements arrive too late to be able to anticipate navigation, helps to increase safety over certain zones for which the radar supplies practically no information, such as a calm lake. Moreover, in situations of blind penetration, altimeter measurements are practically meaningless, unless correlated with a digital elevation model of terrain.

The contour mapping planes are horizontally stabilized with inertial navigation system data. The range domain covered must be at least 10 NM and the possible bearing domain must be at last ± 60°. In elevation, two scanning lines covering 10 mR can be used for detection within an altitude segment of 6 000 feet at 10 NM. There is no detection below 3 NM, but previously detected echoes are stored and displayed as the plane advances. The map is thus complete.

19.3.8 VERY LOW-ALTITUDE PENETRATION

Low-altitude navigation (> 500 ft.) cannot be used to penetrate far into enemy territory and still maintain required safety levels with regard to enemy fire power. However, in order to best take advantage of ground masks and to increase velocity, etc., the aircraft should fly as low as possible (between 200 and 300 ft.).

Three complementary actions can be taken:

- *terrain following*, which corresponds to navigation in the vertical plane
- *terrain avoidance*, which takes place in the horizontal and vertical planes
- *threat avoidance*, which involves navigating along a flight path in order to fly past enemy ground-to-air installations at sufficient range to ensure safety

If mission preparation were perfect—that is, if each element required for very low-altitude navigation were known (such as land relief, pylons, cables, ground-to-air batteries, including moving batteries, etc.)—and if navigation resources were also perfect ("digital file of terrain," inertial navigation unit updated using GPS, etc.), the role of the radar would be reduced to a minimum, especially as it is not discrete.

However, because mission preparation is never perfect, and because the penetrating aircraft must ensure its own protection, the specialized radar with which it is fitted plays a major role. Not only must it detect all forms of land relief, it must also, and most importantly, detect pylons, their tops in particular. The radar's contribution would be perfect if it were also able to detect cables strung between pylons. Finally, its performance must not be affected by rain clouds nor by rain itself. Unfortunately, this is far from being the case: a radar operating in X-band or Ku-band cannot detect cables, and a radar in W-band, although able to detect cables and pylons with sufficient resolution, is more sensitive to rain and less well-suited to ground detection over a large domain. A multipurpose system must therefore be fitted with two types of radar.

For terrain following, a mechanical antenna radar can be used. However, in order to perform the terrain avoidance mission under satisfactory conditions, an electronic scanning antenna is required because

- the angular domains to be explored are large, both in azimuth ($\pm\ 60°$) and in elevation ($\pm\ 15°$)
- a fine beam is required to obtain good resolution, particularly in elevation
- the detection-data refresh rate must be sufficient in each direction

Figure 19.9 gives an example of terrain avoidance with navigation in the vertical and horizontal planes.

FIGURE 19.9 TERRAIN AVOIDANCE

Threat avoidance is part of mission preparation, which determines the safest flight path prior to the mission. Other than in exceptional circumstances, the radar cannot identify enemy ground-to-air resources. However, the aircraft electronic counter-measures can detect, identify, and if necessary, jam enemy battery surveillance and tracking radar.

In terrain-following and especially terrain-avoidance missions, the crew must navigate under difficult conditions. However, from all the available data the weapon system can compute navigation functions used either for automatic piloting, or manual piloting should the pilot prefer to take control. In order to ensure maximum safety, clearance orders may be added to these navigation functions. These must be a trade-off between safety on the terrain being overflown, requesting a minimum altitude of flight, and enemy air defense efficiency, which increases with the flight altitude.

20

DESIGN OVERVIEW

20.1 BASIC EQUATIONS

In this chapter, we describe different types of radars that fit the applications presented in the previous chapters. In order to give a more precise idea of the values involved, we have illustrated this chapter with typical examples of generic radars. First, let us summarize the principal formulas discussed in the preceding chapters.

FORMULAS FOR RANGE CALCULATION

Monostatic radar equation:

$$P_r = P_t \frac{G^2 \lambda^2 \sigma}{(4\pi)^3 R^4 L} \qquad (20.1)$$

Noise spectral density at the optimum receiver output:

$$b = kT_0 F \qquad (20.2)$$

Signal-to-noise ratio on output from the coherent processing:

$$(S/N_0) = \frac{E_r}{b} = \frac{P_{mr} T_c}{b} = \frac{P_{kr} T N_{FFT}}{b} \qquad (20.3)$$

Signal-to-noise ratio (SNR) at the non-coherent output processing (including processing gain and sampling losses):

$$S/N = (S/N)_0 G_t \qquad (20.4)$$

Approximate value of non-coherent post-detection integration gain of N pulses:

$$G_t = 2\sqrt{N} - 1 \qquad (20.5)$$

Signal-to-noise required for the detection of a Rayleigh fluctuating target (Swerling I or Swerling II with no post detection integration):

$$S/N = \frac{\log(P_{fa})}{\log(P_D)} - 1 \qquad (20.6)$$

Antenna gain in the main beam axis:

$$G = \eta \frac{4\pi S}{\lambda^2} \qquad (20.7)$$

Radar range:

$$R = \left[\frac{P_m T_c G^2 \lambda^2 \sigma}{(4\pi)^3 (S/N)_0 kT_0 FL_h} \right]^{\frac{1}{4}} = \left[\frac{P_k T G^2 \lambda^2 (\sigma) N_{FFT}}{(4\pi)^3 (S/N)_0 kT_0 FL_h} \right]^{\frac{1}{4}} \qquad (20.8)$$

where

- P_r is the received power at the receiver input (P_{cr} peak value, P_{mr} mean value)
- P_t is the transmitted power at the transmitter output (P_k peak value, P_m mean value)
- σ is the radar cross section (RCS)
- R is the radar-target range
- L_h is the microwave losses (internal, radome, propagation)
- k is the Boltzmann's constant
- T_0 is the operating temperature (≈ 300 K)
- F is the receiver noise figure
- E_r is the energy received during coherent processing
- T_c is the coherent processing duration
- T is the transmitted pulse width
- N_{FFT} is the number of integrated interpulse periods during T_c (coherent burst)
- S is the antenna geometrical area
- η is the illumination efficiency

OTHER EXPRESSIONS

Beam aperture at 3dB (with optimized pattern):
Circular antenna of diameter d:

$$\theta_{0A} = \theta_{0E} = 1.25\lambda/d \qquad (20.9)$$

Rectangular antenna of dimensions h or l:

$$\theta_{0E} = 1.1\lambda/h \text{ and } \theta_{0A} = 11\lambda/l \quad (20.10)$$

Doppler frequency (radial velocity v):

$$f_D = 2v/\lambda \qquad (20.11)$$

Doppler frequency of the clutter signals received by the main lobe:

$$f_p = \frac{2v}{\lambda}\cos\theta_D\cos\theta_B \qquad (20.12)$$

Spectrum width of the ground echoes received in the 3dB main lobe aperture:

$$\Delta f = \frac{2v}{\lambda}(\cos\theta_D\sin\theta_B)\theta_{0A} \qquad (20.13)$$

Unambiguous range limit:

$$R_a = \frac{cT_R}{2} \qquad (20.14)$$

Unambiguous speed limit:

$$v_a = \frac{\lambda PRF}{2} \qquad (20.15)$$

Resolution in range (transmitted spectrum B):

$$r \approx \frac{c}{2B} \qquad (20.16)$$

Resolution in velocity (T_c coherent integration duration):

$$r_v \approx \frac{\lambda}{2T_c} \qquad (20.17)$$

Synthetic Aperture radar resolution:

$$r_C \approx \frac{\lambda R_0}{2vT_e\sin\theta_A} = \frac{v}{B_D}\sin\theta_A = \frac{\lambda}{2\delta\varphi_e}$$

$$(20.18)$$

20.2 GENERIC RADAR CONFIGURATION

Figure 20.1 shows the generic configuration of a modern radar. In addition to the coherent transmission and reception functions, processing is characterized by

- coherent processing, which consists of range processing (matched-filter or pulse-compression) and velocity processing (Doppler filtering or SAR processing). In some cases, these two processing procedures can be performed simultaneously with a 2-D FFT
- envelope detection, which eliminates the phase of the signal processed
- non-coherent post-detection processing (post-detection integration, ambiguity solving, or multiple-look summation)
- detection (decision whether an echo is present or not)
- exploitation of the processed data (tracking, identification, calculation of trajectories)

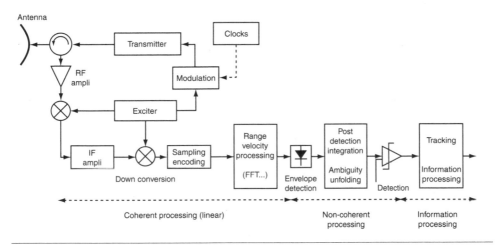

FIGURE 20.1 RADAR CONFIGURATION

20.3 SPACE OBSERVATION RADAR

Functionally, a SAR observation system can be broken down into three chains:

- a preparation and mission management chain
- an image chain
- an image exploitation chain

One of the advantages of this breakdown is that each chain can be specified easily. Each one has different but complementary performance, which enables us to talk about system performance.

On the contrary, a breakdown based on the physical location of the various elements does not necessarily produce the same benefits.

20.3.1 MISSION PREPARATION AND MANAGEMENT CHAIN

The role of the mission preparation and management chain is to

- determine the optimum observation conditions and their sequencing on the basis of the information requests coming from the customer (mission preparation)
- initiate the radar observation operation itself (management of the mission)

The first step is a ground-based operation. It is a complex task due to the potential offered by the satellite. An information request is expressed through a great number of criteria or constraints, such as geographical location (sometimes imprecise as we are working on a global scale), time constraints (duration of the radar observation period, age of the data), and varying operational requirements (detection, mapping, reconnaissance, etc.).

In addition, the satellite has its own constraints, that is, its resources and its orbitography. The combination of information requests and satellite constraints results in the creation of the satellite work plan that is itself. divided into programming messages.

This task involves a large number of computing tools, even if an entirely automatic solution is not the panacea. In fact, the framework of the problem is such that in most cases the decision is taken using fuzzy logic, on the basis of nonquantifiable criteria whose criticality evolves over time. This is why it is preferable for mission preparation to be controlled by an operator (decision maker) who has at his disposal a whole range of high-performance decision aids.

Activation of the radar observation process (second operation in this chain) is a task carried out entirely on-board and in accordance with the work plan previously defined on the ground. This task is of course totally automatic in the case of satellites.

20.3.2 IMAGE CHAIN

20.3.2.1 DESCRIPTION

The image chain is composed of five operations:

- radar transmission
- radar reception, including bandwidth matching and A/D signal conversion

- storage with data compression
- transmission of the on-board data to the ground-based station
- image formation (processing matched to the received signal)

In terms of data flow, the difficult link in this chain is the transmission of on-board data to the ground-based station.

For a satellite, whose visibility from a reception station is not always possible, radar engineers use a method combining a system of direct transmission in the flow with an on-board high-density recording system. It is possible to adapt the data rate at the coding system output to the data rate of the recorder input by means of a buffer placed between the two devices. The size of the buffer depends on the difference between their data rates and on the operational role assigned to the buffer.

The image is usually formed on the ground so as to enable a great variety of processings.

Figure 20.2 shows the structure of the image chain.

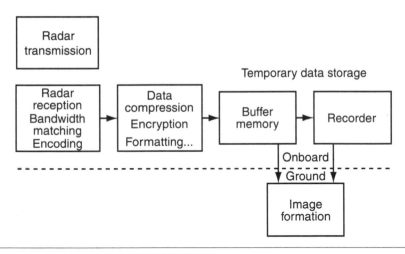

FIGURE 20.2 BLOCK DIAGRAM OF THE IMAGE OPERATING CHAIN OF AN OBSERVATION SYSTEM

20.3.2.2 EXAMPLE

The characteristics of a transmission-reception chain are given in Tables 15.2 and 15.3 in Chapter 15. A few additional characteristics are given below.

TRANSMITTED POWER

We assume that the radar is transmitting at X-band at an observation incidence angle between 20° and 60°. The interpretation of images is only possible if the power level of the mapping background is higher than that of the thermal noise. A 5 dB SNR with a diffuse mapping background having a backscatter coefficient of –15 dBm²/m² is generally sufficient to ensure that the ground is visible in the radar images.

If the radar resolution is 4.4 m, the area of the resolution cell is 20 m². The SNR is then 17 dB for a 20 m² RCS target. Under such conditions, the detection of a point target depends on the contrast between the target echo and the clutter signals present in the neighboring resolution cells. The level of thermal noise has little influence.

When a radar observes a target with a 60° angle of incidence, the illumination time is 0.6 s. With an overall loss of 7 dB (microwave, propagation, matched filter) and an antenna gain of 46 dB, the mean power to be transmitted is 700 W.

Note that in spite of the considerable radar-target range, the mean power remains of the same order as in the case of conventional airborne radars. The large physical antenna and the high processing gain avoid the increase of mean power. Taking into account transmitter efficiency and the consumption of the reception circuits, the power consumption of the sensor, during the transmission phase, is around 4 kW. Excluding transmission phases, the consumption is limited to the sensor management circuits and to possible reheating needs that correspond to some hundred Watts (300 W, for instance). With an operating time of 10 min for an orbit of 100 min, average power consumption is finally 670 W. This value allows us to define the size of the solar panel and batteries of the satellite.

DATA TRANSMISSION RATE

Ambiguous range enables us to image a 20 km swath, regardless of the angle of incidence. With a sampling width of 4 m, the number of complex signal samples to be transmitted is 5000. If pulse-compression processing is not analog processing, we must add the duration of one transmitted pulse (usually 10 to 20 µs, that is, approximately one thousand additional signal samples) to the useful reception period. If four-bit coding is used on each I and Q channel, the rate is 85 Mbits/s. Considering the formatting required for data transmission and the addition of auxiliary data, the total data rate for real-time transmission is of the order of 100 Mbit/s.

20.3.3 IMAGE EXPLOITATION CHAIN

The image exploitation chain extracts the output product of the observation system from the source image produced by the image chain. Depending on the application of the radar system, this can be information gathering, e.g., statements accompanied by plans, and also image products such as ortho-images, space-maps, or digital models of terrain.

This chain uses a series of devices available to photo-interpreters for functions such as filtering, readjustment, georeferencing, segmentation, classification, etc.

Whenever possible, the interpretation of an image is carried out using previous images of the site and the information obtained at the time. For this reason, raw data and the different products generated by the image chain and the image exploitation chain are recorded in almost every case.

The problem of data storage is therefore a direct consequence of this basic principle of intelligence gathering, which is to store the acquired data systematically. It is further complicated by the considerable amount of data generated by observation systems.

Usually, the image file of an imaging procedure is composed of

- the raw image (non-refined) and its auxiliary data
- the source image, resulting from matched processing of the radar signal
- the request and statement of information
- any products generated by the exploitation chain, such as filtered images, multiple looks, digital data maps, etc.

These hypotheses lead to two requirements: the first is to record a large volume of data, and the second is to recover useful information by means of efficient requesting procedures that enable a rapid extraction from the data bases.

20.4 AIR-SURVEILLANCE RADAR (AEW)

20.4.1 AEW SPECIFICATIONS

The overall specifications of an AEW system are

- range: this obviously depends on the application; however, considering Section 19.3.3, the minimum range would be 200 km (i.e., about 100 NM) and can exceed 500 km for a conventional target (RCS = 5 m²) with the most sophisticated systems
- velocity domain: the target velocity can reach Mach 3 (≈ 1000 ms^{-1})

- altitude segment to be covered: this extends from very low altitude (≈ 60 m) to very high altitude ($> 20\,000$ m)
- angular coverage: $300°$ to $360°$ coverage is generally required
- measurement of target altitude: the altitude segment of the target (to within some 1000 feet) must be known

20.4.2 TECHNICAL DESCRIPTION

To illustrate the main decision to be made when designing an AEW radar, we shall use the example of a generic AEW with the following overall specifications:

- range: 500 km for a target with a 5 m^2 RCS (SW 1), 1 false alarm/min, and a cumulative probability of detection $P_c = 0.85$
- data updating rate: 20 s maximum
- target velocity: 1000 m/s

The design procedure for this radar is an iterative procedure because the various parameters involved (platform, type of antenna, operating frequency, processing procedures, etc.) are interdependent. A single design solution will therefore be given without describing the intermediate stages.

20.4.2.1 CHOICE OF PLATFORM

The platform must have an in-flight endurance of several hours and the ability to fly at a sufficient altitude H such that the radio horizon remains beyond the range expected for a target flying at low altitude, that is, at least 30 000 ft. to be in sight of a target flying at 500 ft.

Also, the platform must be capable of carrying a large antenna (see Section 20.4.2.3).

The cabin must be wide enough to hold sensors equipment, operating consoles, operators, and ancillaries, and the available prime power must be adequate for all these elements.

Hence the appropriate platform to carry the system on-board is a liner jet.

20.4.2.2 CHOICE OF FREQUENCY

Atmospheric absorption is an important parameter in determining the choice of carrier frequency (see Chapter 4). Given the range domain, atmospheric losses become prohibitive at frequencies exceeding 3 GHz (S-band). As maximum angular resolution (altitude measurement and target discrimination) and a radiation pattern of excellent quality are required (clutter rejection), the sizing limitations imposed by an airborne antenna generally lead to the choice of the highest frequency, that is, the S-band, for this application.

For a range of 300 km, the C-band would be preferred, and for 150 km range, X-band (10 GHz).

20.4.2.3 CHOICE OF THE ANTENNA

The antenna is chosen on the basis of two parameters: the requirement for maximum surface area and maximum coverage. Due to the available space on the airframe, aerodynamics problems, and masking effects of the structure (wings, engine pods, tail, etc.), there are only three possibilities:

- a mechanical antenna in an aerodynamic radome (usually rotating with the antenna and called a rotodome) placed sufficiently high above the airframe to avoid masking effects and ensure 360° coverage (Figure 20.3a)
- a set of electronic conformal scanning antennas (Figure 20.3b) ensuring maximum coverage, by commutation, in spite of the masking phenomena. However, the forward and rear coverages are difficult to achieve
- an electronic scanning dorsal antenna (Figure 20.3c) ensuring coverage on both sides of the aircraft. The forward and rear coverage is achieved with a much lower antenna gain than the side coverage. The maximum radar range is in the direction perpendicular to the antenna plane. In the other directions, high range is obtained by increasing the dwell time. The dorsal antenna configuration has a reduced drag compared to the rotodome, and it is the easiest to install on-board an aircraft

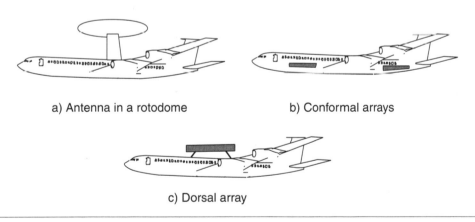

a) Antenna in a rotodome   b) Conformal arrays

c) Dorsal array

FIGURE 20.3 AEW ANTENNA INSTALLATION

If we take an electronic scanning antenna, 6 m × 0.7 m, with an aperture of

$$\theta_A \approx \frac{70 \times 0.1}{6} = 1.2°$$

in azimuth and

$$\theta_E \approx \frac{70 \times 0.1}{0.7} = 10°$$

in elevation, the gain of this antenna is

$$G \approx 10 \log 0.6 \frac{4\pi 6 \times 0.7}{(0.1)^2} = 35 \text{ dB}.$$

20.4.2.4 Choice of Waveform and Signal Processing

The choice of waveform depends essentially on rejection of clutter signals. This point has been dealt with thoroughly in Chapters 7 and 8. Either HPRF or MPRF modes are used. Considering the choice of λ, (10 cm, S-band), and the target velocities (1 000 m/s), an HPRF mode leads to a PRF higher than 20 kHz.

Processing is of the pulse-Doppler type described in Section 8.6. Chapter 7 explains the spectral purity constraints of the transmission-reception chain.

20.2.3 Performance Calculations

To determine detection performance, we shall perform the same calculations as in Section 8.6.7, the parameter to define being the power of the transmitter. The main steps of the calculation consist of determining

- the type of exploration of the search domain: antenna aperture in elevation is sufficient to enable the altitude coverage with a single elevation bar. Complete exploration of the domain is thus obtained by a 360° azimuth scan in 20 s, with an illumination time in each direction

$$T_e = \frac{20}{360} 1.2 = 0.066 \text{ s}$$

- the waveform and the type of processing: A PRF of 20 kHz (HPRF) and pulse duration T equal to 2 µs are used. Processing is of the pulse-Doppler type with FFT over $m = 128$ interpulse periods (i.e., $T_c = 128 \times 0.05 = 6.4$ ms). Given the illumination time $T_e = 66$ ms, at least eight coherent bursts can be processed, and range-ambiguity solving of the type "3/8" can therefore be used, as described in Section 20.6.6. (Eight different PRFs are used during time T_e, and a target is detected if at least three PRFs result in detection with correlation of range and velocity measurements.)
- the SNR required for detection: the specifications require a cumulative probability of detection and a false-alarm rate; these specifications should be interpreted in terms of detection probability P_{d0} and false-alarm probability P_0 on the first detection (FFT output)

The relationship between the cumulative probability, P_c, and the probability of detection, P_D, when the antenna beam illuminates the target depends on the target radial velocity and the exploration rate. If P_{D_i} is the probability at the ith illumination, the cumulative probability is then

$$P_{ci} = 1 - (1 - P_{Di})(1 - P_{Di-1})(1 - P_{Di-2})\dots,$$

with P_{D_i} increasing as the target approaches. The complete calculation is relatively laborious, but in practice, in the example in question, the operational P_D required for the confirmation of detection is $P_D \approx 0.3$, a value that will be used from now on.

From this value, we can deduce P_{d0} using the relationship from Section 8.2, that is, $P_{D_0} = 0.24$.

Calculation of the probability of overall false alarm P_{fa} is performed as in Section 2.6.7.1. The number of processed range cells during T_e, is

$$n_D = R_{max}/r = 2R_{max}/cT = 5.10^5/300 = 1666,$$

and the number of processed velocity cells is

$$n_v = v_{max}/r_v = v_{max}2T_c/\lambda = 1000 \times 2 \times 0.0064/0.1 = 128 .$$

We obtain from the above equations

$$P_{fa} = \tau_{fa}\frac{T_e}{60}\frac{1}{n_D n_v} = 1 \times \frac{0.066}{60 \times 1666 \times 128} = 5.10^{-9} .$$

The probability of elementary false alarm, P_0, is given by Equation 8.2: $P_0 \approx 4.4.10^{-4}$.

The SNR at the coherent processing output, that is, prior to envelope detection, is obtained by means of the detection curves (Figure 6.4).

For $P_{d_0} = 0.24$ and $P_{fa} = 3.10^{-4}$, the curves indicate an SNR of ≈ 7.5 dB.

The various processing losses must be added to this value:

- matched filter and range sampling losses: $L_{st} \approx 2$ dB
- weighting and velocity sampling losses: $L_{sv} \approx 2$ dB
- CFAR losses: $L_{CFAR} \approx 2$ dB
- lobe loss (the antenna gain is not maximum over the entire illumination time, T_e): $L_l \approx 1.5$ dB

The overall losses are $L \approx 7.5$ dB.

The resulting SNR required for detection is

$$S/N = 7.5 + 7.5 = 15 \text{ dB}.$$

20.4.2.5 DETERMINING THE REQUIRED TRANSMISSION POWER

This power depends on the minimal received power, P_r, required for detection; P_r must be such that S/N = 15 dB at the FFT output, assuming that all the losses occur downstream. In this case, the processing carried out upstream is entirely coherent, and the optimal receiver theory can apply.

The result is

$$S/N = \frac{E}{b} = \frac{P_{mr} T_c}{b} = \frac{mP_{kr} T}{b},$$

where E is the energy received during coherent processing, that is, the energy of the m pulses of the coherent burst on which the FFT calculation is based; P_{mr} and P_{kr} are the received mean and peak powers, respectively; and b is the noise spectral density (white noise).

This power is linked to target range by the radar equation examined in Chapter 3, and the resulting mean power is

$$P_{mr} = \frac{P_m G^2 \lambda^2 \sigma}{(4\pi)^3 R^4 L_h};$$

hence,

$$P_m = \frac{S/N k T_0 F (4\pi)^3 R^4 L_h}{T_c G^2 \lambda^2 \sigma}.$$

With $F \approx 3$ dB and $L_h \approx 4$ dB (microwave losses including the radome and atmospheric losses), we get

$$P_m = 24\ 600 \text{ W mean power},$$

that is,

$$P_k = 24600 \times 50 / 2 = 615 \text{ kW peak power}.$$

20.5 MARITIME SURVEILLANCE RADAR

The general characteristics of such a system can be found in Chapter 11.

20.5.1 SURFACE VESSEL DETECTING MODE

The signal-to-noise ratio required for such detection is calculated below with an example of radar. We are looking for a detection probability cumulated over three antenna revolutions of 0.85, or a 0.5 single-scan detection probability.

The false-alarm rate is one per minute.

A broad range domain is covered: 300 km sampled in steps of 40 m, that is, 7500 range cells.

The antenna length is 60 cm, resulting in a beam of 3.4°. The scanning rate is 6 rev/min, that is, 36°/s.

CALCULATION OF THE SNR REQUIRED FOR DETECTION
- number of decisions taken in 1 min: 7 500 x 36 / 3.4 x 60 = 4.8 x 10^6
- probability of false alarm: P_{fa} = 0.21 x 10^{-6}
- SNR for a Swerling I target: 13.3 dB
- target illumination time: T_e = θ_G / Ω = 94 ms
- number of pulses integrated in the beam (consider the post-detection integration limited to 2/3 of the illumination time): 2/3.PRF T_e = 31
- approximate post-detection integration gain: 10 dB
- SNR at the matched filter output: S/N_0 = 3.3 dB

MEAN POWER TRANSMITTED BY THE RADAR
- transmission frequency: X-band, 10 GHz
- antenna gain hypothesis: G = 32 dB
- noise factor: 3 dB
- losses taken into account in the power budget (microwave, lobe modulation, atmosphere, sampling, coding, matched filter): 10 dB
- mean power to be transmitted to detect a target with an RCS of 1000m^2 at 112 NM: 140 W

At this range the incidence angle is very small. Sea clutter signals are negligible.

Note that transmission power is low in comparison with other airborne radars.

20.5.2 Detecting Small Targets (Periscope)

In this detection mode, the antenna has a very high rotation rate (60 rev./ min), while the display rate is much slower (6 rev./min). A probability of detection of 50% is sought for an apparent antenna scan, that is, for one scan on the display. The false-alarm rate remains at one per minute.

Range domain is reduced compared to the previous case: 60 km sampled in steps of 4 m, that is, 15 000 range cells.

Calculation of the SNR Required for Detection

- criteria for detection: 5 out of 10
- number of decisions taken in 1 min:
  15 000 x 36 / 3.4 x 60 = 9.5 x 10^6
- probability of elementary detection: P_{D0} = 0.5
- probability of false alarm: P_{fa} = 105 x 10^{-9}
- probability of elementary false alarm: P_{fa0} = 13.3 x 10^{-3}
- SNR for a Swerling I target: 7.2 dB
- target illumination time: $T_e = \theta_A / \Omega$ = 9.4 ms
- number of pulses integrated in the beam (consider the post-detection integration limited to 2/3 of the illumination time) for a PRF of 2 kHz: 2/3 PRF T_e = 12
- approximate post-detection integration gain: 7.5 dB
- SNR at the matched-filter output: S/N_0 = −0.3 dB

Mean Power Transmitted by the Radar

- losses taken into account in the power budget (microwave, lobe modulation, atmosphere, sampling, coding, matched filter): 9 dB
- mean power to be transmitted to detect a target with an RCS of 1 m^2 at 20 NM: 190 W

The radar can operate in both modes provided it is equipped with a transmitter whose mean power is between 100 and 200 W.

Performance in Sea Clutter

To calculate the probability of detection actually achieved, the power of the sea clutter received in a single resolution cell must be added to the noise power. Initially we can consider that the expected signal has gain G_c over sea clutter. This gain is not linked to the total number of pulses, N, post-integrated during illumination time, but is limited to the number, N_f, of pulses of a different frequency (frequency agility transmission, with variation between frequencies greater than the instantaneous bandwidth of the transmitted pulses). Assuming that the target is located beyond the transition range of the sea clutter signals, R_{tr}, the point target signal-to-disturbance ratio, $S/(N+C)$, at the post-detection integration output is given by

$$S/(N+C) = \cfrac{1}{\cfrac{1}{S/N} + \cfrac{1}{S/C}},$$

where

$$S/C = \frac{\sigma G_C R^3}{\sigma_0 r \theta_A R_{tr}^4}.$$

As an example, consider a backscattering coefficient of $\sigma_0 = -25$ dBm^2/m^2, with a wave height of 1.25 m in the Mediterranean Sea and a flight altitude of 900 ft. The radar is transmitting on four different frequencies. The target has an RCS of 10 m^2 and 1 m height. The point signal-to-disturbance ratio at 20 NM is equal to 8.2 dB. The probability of detection is greater than 50%.

20.6 BATTLEFIELD SURVEILLANCE

20.6.1 SPECIFICATIONS

The following specifications concern the simplest case of this type of radar.

The radar is helicopter-borne and designed to detect moving ground targets. Flight is almost stationary at an altitude of about 3000 m, and the characteristics required are

- range domain coverage: 60 km, from 40 to 100 km
- angular domain coverage: sector scan up to 120°, with 360° accessibility
- radial velocity domain of the target to be detected: 2 to 20 m/s
- radar range for a target of 10 m^2 RCS: 100 km, with $P_D = 0.5$ and $P_{fa} = 10^{-7}$
- range resolution: 30 m
- angular resolution: 1°

20.6.2 TECHNICAL DESCRIPTION

Using the main technical and operational specifications, we shall attempt to optimize the technical description by means of a series of trade-offs.

The radar presented is a low-PRF radar of the MTI type, without range or velocity ambiguity.

20.6.2.1 CHOICE OF TRANSMISSION FREQUENCY

Considering the type of platform, antenna dimensions should be minimized for a high operating frequency. Also, to avoid the use of complex processing procedures, the radar will be without ambiguity. From relationships 20.14 and 20.15 and the specifications, the transmission frequency is deduced as $f_0 \leq 11.25$ GHz.

As a result, we shall use a frequency of 10 GHz, that is, $\lambda = 3$ cm.

20.6.2.2 CHOICE OF PRF

At 10 GHz, 1 m/s corresponds to a Doppler frequency of 66.7 Hz.

As the velocity domain of the targets to be detected (2 to 20 m/s) is obtained by means of a digital high-pass filter, symmetrical with respect to PRF/2 (Figure 20.4), the interpulse period frequency must be PRF $\geq$ 1 467 Hz (22 m/s).

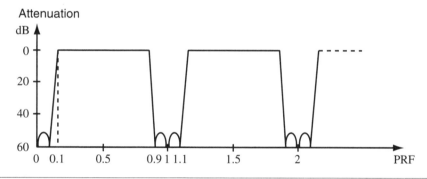

FIGURE 20.4 DOPPLER FILTER ATTENUATION

On the other hand, if we ignore the altitude of the platform, the range domain (100 km) gives a frequency of PRF $\leq$ 1 500 Hz; this is the value that will be taken.

We have now reached the limits of range and velocity ambiguities.

20.6.2.3 ANTENNA

The angular resolution requires a beam aperture in azimuth lower than 1°. With a one-line scanning, the range domain coverage gives a beam aperture of 2.3° in elevation, with the antenna elevation angle being 2.5°.

Under these conditions, $R = 70$ km and $R_{sw} = 60$ km, (Figure 19.6).

The dimensions of the antenna carried under the helicopter are determined by expression 20.10, which gives h = 80 cm and l = 200 cm; under these conditions, gain is G = 40.5 dB (relationship 20.7).

Lateral sector scanning is performed over ± 60°, at a rate of 10°/s, or an illumination time, T_e, of 100 ms. This duration cannot be significantly reduced as it is preferable for it to be long compared to the response time of the Doppler filters (recursive high-order filters), which at their lower limits process low frequencies and narrow spectra. Moreover, a long illumination time enables considerable post-detection integration gain, thereby improving range, although admittedly to the detriment of the data refresh rate (an average of 12 seconds in this case). However, this rate should remain at a sufficient level to ensure good target-plotting performance, particularly in the case of rapidly maneuvering targets.

20.6.2.4 WAVEFORM

A low-PRF waveform, unambiguous in range and in velocity, is used. As the radar operates in the standoff mode (40 km), pulse compression can be used to increase discretion by reducing transmitted peak power.

To meet the range resolution (30 m) specifications, the duration of the received compressed pulse must be less than 200 ns, which represents a transmission bandwidth of 5 MHz (see relationship 20.16).

The pulse-compression ratio can be chosen from a very large scale of values, ranging from a few units to 1320 in theory. The selection of a high value, e.g., 1000, which optimizes discretion, will depend on the compression technology and on the availability or the feasibility of a transmitter.

20.6.2.5 PROCESSING GAIN

Figure 20.5 represents processing.

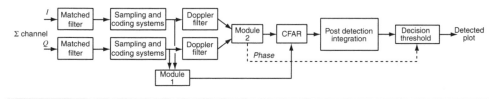

FIGURE 20.5 PROCESSING BLOCK DIAGRAM

The two components I and Q of the Σ channel provide reception signals compressed to 200 ns. These signals are filtered and then sampled at 200 ns intervals and digitally converted. The role of module n°1 is to adjust

CFAR dynamic range. The dynamic adjustment is necessary because the dynamic range of the signals received (e.g., 60 dB) exceeds the radar-visibility ratio (e.g., 40 dB) due to the presence of fixed echoes.

Coverage of the entire range domain requires 2000 cells.

Module n°2 produces the signal magnitude from the filtered I and Q components. It can also determine the phase difference of these signals, measured from one interpulse period to the next, which can be used to select or inhibit velocity intervals.

The post-detection integration constant is taken as equal to 2/3 T_e, in this case, 100 T_R.

The decision threshold is fixed by the probability of elementary false alarms (10^{-7}).

The probability of detection over one scan (50%) is obtained with S/N = 13.5 dB (relationship 20.6).

The processing gain includes

- sampling losses and pulse compression loss: 3 dB
- lobe losses: 1.5 dB
- Doppler filter gain: 0.4 dB
- CFAR losses: 4 dB
- post-detection integration gain (relationship 20.5): 12.8 dB

Under these conditions, processing gain G_p is + 4.7 dB.

20.6.2.6 TRANSMITTER POWER

As stated in Section 20.6.2.4, the transmitter can operate at a form factor close to 30%, and the pulse-compression ratio can be 1000. For the type of radar discussed in this section, we shall use more conventional values: 6% and 200.

By using the following formula or expression 20.8, it is possible to define the mean and peak powers actually supplied by the transmitter (P_m and P_k).

If the different components of this relationship are expressed in dB, we have

- $R^4 = 200$ dB (m), $(4\pi)^3 = 33$ dB, $L = 7$ dB (including 3 dB of propagation loss), $KT_0 = -204$ dB
- $F = 3$ dB, $B = 67$ dB, S/N = 13.5 dB

- G^2 = 75 dB (the range domain is obtained by the –3 dB aperture in elevation, that is, –6 dB for the two-way path)
- λ^2 = –30 dB, σ = + 10 dB, G_p = + 4.7 dB,

Thus P_k = 5.0 kW, and P_m = 300 W.

20.6.2.7 POSSIBLE DEVELOPMENTS

The operational characteristics of the generic radar described above can be greatly improved by introducing complementary processing; two short examples are given:

- expansion of the velocity domain: the radar becomes ambiguous in velocity and the average PRF is therefore reduced and varied over several values (1100, 1300, 1500 Hz, for example), in order to solve the ambiguity and attenuate the effect of the detection notch centered on PRF (due to the Doppler filter)
- achievement of velocity resolution (2 m/s, for example) by introducing an 8 point FFT which, by improving processing gain, enables either reduction of transmitted power or increased range on low-RCS targets

20.7 INTERCEPTION RADAR

20.7.1 SPECIFICATIONS

Section 19.3.6 defined the role of the interception radar: it is used in interception, air superiority, and close-combat missions. The interception mission is the most demanding in terms of range because it involves the detection and tracking of one or several targets and the engagement of one or several active "medium-range" missiles against the most threatening target. In Chapter 8, which examines air-to-air detection techniques depending on waveform, Section 8.6.7 gives an example of a medium-PRF radar design whose performances are appropriate for air superiority and close combat missions. For "gun" firing, it is preferable to use frequency agility in order to decorrelate the scanning centers of the RCS and thereby improve single-target tracking (STT) pointing precision by reducing its fluctuations.

However, the detection range of this Medium PRF radar (54 km) is insufficient for an interception mission, which requires sufficient warning to detect, track, and analyze the threat and release one or more missiles under good conditions before the enemy does so. These operations require a range of about 50 NM.

Suitable specifications for interception missions are

- range: 50 NM for a penetrating target (SW1) with an RCS of $\sigma = 5$ m^2
- $P_D = 0.5$ (for one antenna scan) with one false alarm per minute
- azimuth scan rate: 60°/s over ± 60°
- velocity search mode
- range search mode with tracking of several targets

The required performances in air superiority and close combat modes are

- range: 30 NM for a target of RCS (SW1) $\sigma = 5$ m^2
- range resolution: 150 m
- velocity resolution: 2.25 m/s
- modes: range search, TWS, and STT

20.7.2 TECHNICAL DESCRIPTION

Radars operating in the range and velocity search modes must use an high PRF without velocity ambiguity but with ambiguity in range. All other factors being equal, the best detection performances in a head-on configuration are achieved using this waveform, regardless of the altitude of the targets to be intercepted (Doppler free space). Figure 20.6 compares the performances of the various waveforms.

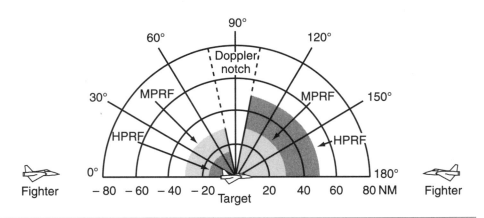

FIGURE 20.6 RANGE COMPARISON

In order to compare the ranges obtained for the different fighter-target configurations and high- or medium-PRF waveforms, target RCS is assumed to be constant.

In a fighter-target configuration of around 90°, the Doppler notch cancels main lobe clutter and thus prevents detection.

With Medium PRF, range is

- more or less independent of configuration
- more or less independent of fighter and target altitudes
- highly dependent on antenna quality (side and far lobes)

When using High PRF

- in front attacks (Doppler free space), range obeys the radar equation and is fairly independent of target and fighter altitudes
- in tail attacks, as range is intrinsically smaller than with head-on configurations, it decreases significantly as fighter altitude decreases (due to side and far lobes)

20.7.2.1 VELOCITY SEARCH

The velocity search mode, which does not enable extraction of the range of detected targets, is the mode with the maximum extended range in the "front attack" configuration (minimum processing losses). This enables rapid evaluation of the closing velocity of a possible target and the threat it may represent. Moreover, for front or tail attack configurations ($\pm 40°$), this mode enables analysis of the inherent spectrum of the target, which facilitates its identification. Except for stealth planes, the inherent spectrum of the airframe return is accompanied by the lines produced by the blades of propellers or engines (compressor blades, etc.). These lines, whose backscattered energy may be greater than the airframe return, depend on

- for frequency spacing, the number of blades viewed by the radar and their rotation speed
- the number of engines
- the aspect angle
- the masking of the blades at the front and rear of the engine
- the wave traps introduced

For a given type of aircraft, the engine rotation speed varies little (20% for example) between maximum (8500 rev/min) and cruising speed. At full speed, the blade tips approach supersonic velocity; for this reason, the greater the diameter, the slower the rotation speed. In addition, in most cases the number of blades is a prime number (such as 23) to avoid the vibrations caused by resonance phenomena.

Each aircraft has its own signature, but this signature is not unique and belongs to a family of spectra; great similarities may exist between two different aircraft. As a result, it would be impossible for a pilot to visually identify a signature on a radar display (azimuth velocity) with a high rate of confidence. To perform non-cooperative target recognition, NCTR, it is essential to use specific correlation-based shape-recognition processing, artificial intelligence, and expert systems. This is a specific radar mode.

The velocity search mode is a subproduct of the range search mode in which range ambiguity solving and range measurement are not performed, nor is there any associated tracking activity.

20.7.2.2 RANGE SEARCH

Chapter 8 describes the range search mode in detail, and Figure 8.22 shows its block diagram. In this mode, antenna scanning can be performed in various ways, depending on requirements. In the autonomous search mode (without target designation), scanning can be carried out along several different elevation lines to cover a large altitude domain (see Figure 19.6). But, when a specific target designation is given, the search domain can be significantly reduced, e.g., over two elevation lines with ± 15° bearing excursion. Under such conditions, the probability of cumulative detection increases rapidly, which increases operational range. To reduce the false alarms displayed to the radar operator, any new detection is "plotted" and is only displayed to the operator if it is confirmed after three or four antenna scans. Initial detection range is therefore slightly reduced, but operational range is maintained and even improved, making the operator's job easier.

20.7.2.3 TRACKING AND PLOTTING

Chapter 9 explains plotting, track-while-scan (TWS), and single-target-tracking (STT). Remember that in the STT mode, the antenna is continuously slaved to the target direction (e.g., in gun firing or semiactive missile-firing modes). In the TWS mode, the antenna azimuth scan is realized in accordance with a predetermined program whose average position can be slaved to the direction of the target being tracked (e.g. in the case of a reduced azimuth domain: ± 30°), whereas the elevation line is slaved to the target direction. When plotting one or more targets, the antenna is not slaved, and, in general, target(s) are off-boresight in the beam during detection. With TWS and plotting, which are more discrete than STT, several targets can be tracked simultaneously and engaged using active missiles (multitarget concept). However, as the measurement rate is slow (e.g., 0.5 /s), these tracking techniques are not appropriate for close combat, which involves a considerable number of maneuvers.

If a high-PRF waveform is used, the eclipse ratio may be high since it is linked to the duty ratio. To overcome this difficulty in tracking modes (TWS and STT), the PRF must be slaved.

20.7.2.4 Technical Characteristics

The radar uses high and medium-PRF.

The characteristics common to these waveforms are

- wavelength: λ = 3 cm
- antenna gain: G = 32 dB, beamwidth at 3 dB = 3.6°
- antenna scan velocity: $\omega(t)$ = 60°/s
- peak power transmitted: P_k = 10 kW
- noise factor: F = 3 dB
- microwave losses: L = 5 dB

Medium PRF characteristics (see Chapter 8) are

- mean power transmitted: P_m = 200 W
- pulse width: T = 1 μs
- mean: PRF = 20 kHz

High PRF characteristics are

- mean power transmitted: P_m = 1 kW
- pulse width: T = 0.5 μs
- mean PRF = 200 kHz
- 8 range gates, 512 velocity filters

Under these conditions, applying the relationships 20.4, 20.5, 20.6 and 20.8 with S/N = 15.5 dB (see Chapter 8) and taking into account processing losses and post-detection integration gain (N = 16), we have R = 92 km or approximately 50 NM. Note that processing losses represent range and velocity sampling losses, weighting losses, CFAR losses, lobe losses, eclipse losses, range extraction losses, etc. In the velocity search mode, losses are reduced by around 3 dB, enabling a range increase of nearly 20%.

20.8 Tactical Support Radar

20.8.1 Specifications

The tactical support radar, defined in Section 19.3.6, is employed in air-to-ground missions and for self-defense against airborne threats.

Assuming that the platform is flying at a speed of 300 m/s, the radar functions will be

- ground mapping over 40 NM with monopulse sharpening (σ = 1 000 m^2)
- ground mapping with DBS zoom up to 20 NM (σ = 100 m^2)
- air-to-ground ranging over 10 NM (σ = 100 m^2)

- 10 NM contour mapping for low-altitude navigation ($\sigma = 10 \text{ m}^2$)
- detection and STT over 20 NM of contrasted targets ($\sigma = 2\ 000 \text{ m}^2$)
- detection and STT over 20 NM of moving ground targets ($\sigma = 10 \text{ m}^2$)
- close combat over 10 NM ($\sigma = 5 \text{ m}^2$)

As the radar can operate in Doppler modes, it could be a coherent low-*PRF* type (MTI).

20.8.2 TECHNICAL DESCRIPTION

The radar presented in Section 20.6 is used for detection of ground moving targets and is also an MTI type, although it is fixed on a quasi-stationary platform. As a result, any ground clutter has a very narrow inherent spectrum, which is very easy to eliminate. On a platform traveling at 300 m/s, the ground clutter spectrum (given by relationship 20.13) depends on the off-boresight angle in azimuth (see Chapters 8 and 10). Using the example of the interception radar given in Section 20.7, Figure 20.7 shows the spectrum of ground clutter as a function of the off-boresight angle in azimuth (curve n°1) to which is added the error made in measurement of the ground speed (1 m/s, for example) of the platform (curve n° 2). This velocity is obtained by the inertial navigation unit.

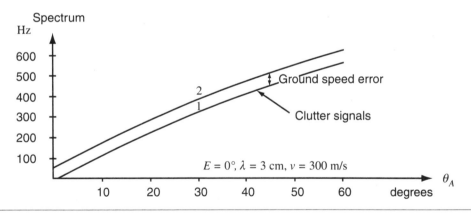

FIGURE 20.7 SPECTRUM OF FIXED GROUND ECHOES

In Figure 20.8 the level of ground clutter is given as a function of platform altitude and depression angle.

In this figure, the level of thermal noise is zero dB and σ is the RCS of a ground moving target. For a given transmitted pulse width and a given antenna beam aperture, the visibility ratio necessary for the detection of moving targets increases with platform altitude and ground σ_0. In the example in question, allowing for fluctuations in clutter, the visibility ratio

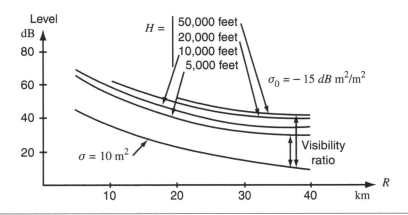

FIGURE 20.8 GROUND CLUTTER LEVEL

must be greater than 40 dB, and the improvement ratio of the Doppler filter must be in the vicinity of 50 dB (see Figure 20.4).

20.8.2.1 GROUND MAPPING

Ground mapping using the monopulse sharpening technique (see Chapter 13) is not an influencing factor in radar sizing, but the dynamic range of the ground clutter signals must be compressed (without being saturated) to match the dynamic range of the radar display (from around 50 dB to 20 dB).

20.8.2.2 GROUND MAPPING USING DOPPLER BEAM SHARPENING (DBS)

Chapter 15 examined Doppler sharpening with a rotating antenna, which provides mean cross-range resolution, suitable for the detection of ground installations. This resolution varies in relation to the azimuth angle of the antenna: it is optimum at $A = 90°$ and is worse or similar to that of a real beam with monopulse sharpening for azimuth angles below approximately 10°.

20.8.2.3 AIR-TO-GROUND RANGING

Air-to-ground ranging is a continuous range tracking loop that measures ground-radar range along the radio axis. Once a precise antenna pointing direction has been determined by the weapon, the tracking loop is slaved to the zero of the elevation angle tracking and indicates the range. The main errors come not from the range tracking loop but from angular harmonization errors arising between the radar and the weapons system, especially at low angles of incidence.

20.8.2.4 CONTOUR MAPPING

Several contour mapping modes can be used (see Section 19.3.7 and Figure 19.8), but in all cases altitude h of each ground echo detected is calculated by the radar in relation to platform altitude H, on the basis of measured range R and the elevation angle $(\theta_E \pm \Delta\theta_E)$. Altitude h is then compared with altitude h_1, h_2, etc., of the chosen mapping planes, and each echo is individualized accordingly on the radar display.

Note that the need to detect targets with an RCS of only 10 m^2 is justified by the presence of obstacles such as pylons whose tops must be located.

20.8.2.5 DETECTION AND TRACKING OF CONTRASTED TARGETS

When transmitting pulses of the order of 0.2 µs with an antenna beam of 3° or 4°, mean ground clutter produces echoes of about 200 m^2 in each range cell (Figure 20.8). As a result, an echo of 10 dB or more (i.e., 2000 m^2) will be considered contrasted. This kind of RCS is common to hangars or large metal structures. For this mode, radar search processing will consist of a CFAR associated with conventional manual designation type, continuous direction, and range tracking.

20.8.2.6 DETECTION AND TRACKING OF GROUND MOVING TARGETS

This function, which sizes the transmitted power, consists essentially of canceling ground clutter while detecting ground moving targets. Figure 20.7 shows that for a platform speed of 300 m/s, the clutter spectrum is around 650 Hz at a 60° azimuth angle (i.e., 9.7 m/s), and this spectrum is about 100 Hz (i.e., 1.5 m/s) along the flight-path axis. Given the frequency response of the Doppler filter shown in Figure 20.4, which is linked to the PRF, it is necessary to slave the radar PRF to platform speed and to the antenna azimuth angle to optimize the range of moving-target detection velocities and finally to eliminate ground clutter.

Given the components selected, an initial calculation reveals that for a platform flying at 300 m/s, the range of target-detection velocities indicated is unrealistic. However, at 200 m/s, performances are satisfactory. Table 20.1 summarizes these performances.

It should be noted that a range domain of 20 NM can be achieved provided that the azimuth angle remains below about 50°; at 300 m/s, it should remain below 30°. Moreover, the lower limit of the detected velocity range (0.15 to 0.85 PRF) rapidly becomes excessive, even at 200 m/s. In practice, this technique can only be used for slow platforms or for those fitted with an antenna with a very narrow beam aperture. However, a radar on-board a fast-moving platform, e.g., an aircraft, will perform this function satisfactorily using the DPCA or STAP techniques discussed in Section 10.3.2, Part II. This solution consists of creating displacement of the antenna phase center that compensates for forward motion of the platform.

TABLE 20.1. DETECTION OF GROUND MOVING TARGETS

θ_A	Ground Clutter Spectrum Width	0.1 PRF	PRF	Radial Velocity Range Detected
0°	0 Hz	67 Hz	2 000 Hz	4.5 to 25.4 m/s
10°	73	140	2 000 Hz	4.5 to 25.4
20°	144	211	2 200 Hz	5. to 28
30°	210	277	2 800 Hz	6.3 to 35.6
40°	270	337	3 400 Hz	7.6 to 43.3
50°	323	390	3 900 Hz	88.8 to 49.7
60°	363	430	4 300 Hz	9.7 to 54.8
$\lambda = 3$ cm, $v = 200$ m/s, $\theta_{0A} = 3.6°$, $\theta_E = 0°$				

20.8.2.7 CLOSE COMBAT

If the radar has a medium-PRF waveform, it will be used for close combat. With a low-PRF radar, range domain is limited to 10 NM; it is possible to use a PRF of up to 8 kHz without range ambiguity.

The main difficulties with this function, which includes search, automatic acquisition, and STT modes, are

- searching over a large angular domain (see Figure 19.4)
- achieving automatic target acquisition in a very short time
- tracking the target accurately despite its maneuvers and the relatively high radar-target angular velocities
- canceling ground clutter, including ground moving targets, which can be very numerous

The first and third points require the use of a highly motorized low-inertia antenna, with the roll axis blocked for dogfighting and, in particular, gun firing, which requires very high-accuracy beam-pointing. For short-range passive missiles, the requirement is simply that both range and direction tracking remain locked-on, that is, that the target remains in the antenna beam.

Canceling ground clutter and ground moving targets while retaining detection of air targets generates problems that can be overcome by a series of devices and trade-offs. If we consider the Doppler filter previously mentioned, ground clutter can be eliminated up to $\theta_A = 52°$ by using a

PRF of 6 Hz, for a platform velocity of 300 m/s. Under such conditions, air targets of radial velocity from 76 to 103 m/s are not detected. To eliminate this detection notch, a second PRF of 8 kHz is used alternatively and in bursts. Ground clutter is canceled up to 600 Hz (9 m/s), and air targets can be detected without detection notches at radial velocities between 13.5 and 340 m/s. Figure 20.9 illustrates this.

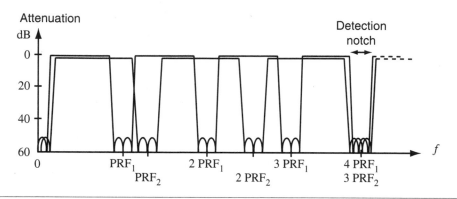

FIGURE 20.9 FILTERING WITH TWO PULSE-REPETITION FREQUENCIES

This filtering, although it eliminates ground clutter, does not enable elimination of most moving ground targets, which can reach speeds of up to 180 km/h (50 m/s). Note that the part of a wheel or a caterpillar track on a moving vehicle in contact with the ground is static while its upper part moves at twice its speed.

If the Doppler frequency, f_D, of a target is smaller than the lowest PRF used, it is not affected by the different values of PRF, and the only variation is in the phase difference measured from one interpulse period to another. To eliminate ground moving targets, the phase difference between one interpulse period and the next must be measured for each range cell and each interpulse period. If this difference, filtered for each packet of PRF_1 and PRF_2, is less than a given value, then the range cell will be inhibited. In choosing to eliminate targets whose velocity does not attain 30 m/s, we have

$$f_D = 2\,000 \text{ Hz}; \quad \Delta\varphi_1 \text{ to } PRF_1 = 120°; \quad \Delta\varphi_2 \text{ to } PRF_2 = 90°.$$

In conclusion, we would emphasize that in close combat, a non-Doppler low-PRF mode with frequency agility should be used for look-up configurations (without clutter or moving ground targets), especially when the gun-firing mode is selected or when target radial velocity is low. Under such conditions there is no velocity limitation on detection. However, close combat presents another difficulty—the dynamic range of the targets to be detected—because

- the RCS of an air target is usually between 1 and 100 m^2
- the required detection range is between 150 m and 18 km

The overall dynamic range exceeds 100 dB. However, for a radar that is unambiguous in range, this can be reduced using Sensitivity Time Control (STC) attenuators that act over about 40 dB.

20.8.2.8 TECHNICAL CHARACTERISTICS

The radar uses low-PRF waveforms only and is fitted with a three-axis (roll, relative antenna azimuth, elevation) mechanical scanning antenna.

The characteristics common to all modes are

- wavelength: $\lambda = 3$ cm
- antenna gain: $G = 32$ dB, beamwidth at 3 dB = 3.6°
- antenna scan speed: $\omega(t) = 60°/s$
- transmitted peak power: $P_k = 20$ kW
- noise figure: $F = 3$ dB
- microwave losses: $L = 5$ dB
- 500 range cells

For detection of moving ground targets, the azimuth scan speed must be slaved to the PRF used so as to maintain constant post-detection integration gain.

With $\omega(t) = 40°/s$ at 2000 Hz and 80°/s at 4000 Hz, range will be independent of antenna azimuth angle, thus optimizing transmitted power. In addition, a low pulse-compression ratio (e.g., Barker code 7) is used to avoid excessive peak power.

The calculation is performed as in Section 20.6, but with

- $T = 0.5$ µs, $B = 2$ MHz, 500 range cells, PRF from 2000 to 4000 Hz
- $G_p = 5.1$ dB, with $N = 120$ post-integrated interpulse periods
- $P_k = 20$ kW and $P_m = 280$ W at 4000 Hz

The characteristics specific to the other modes are summarized in Table 20.2.

Obviously, all these values that are given for information only would have to be optimized to accurately define a radar mission, and trade-offs would have to be found between each operational mode requirement and the corresponding technical development. However, having a single peak power on transmission is a constraint for radars with a number of operating modes. There are two solutions: pulse-compression or the use of a variable peak power transmitter.

TABLE 20.2. CHARACTERISTICS OF RADAR MODES

Radar Mode	Waveform	SNR	Processing Gain and Mean Power Transmitted	Range
Doppler Close Combat	PRF = 6000 and 8000 Hz per bursts of 48 pulses, T = 0.25 μs	13.5 dB	G_p = 6 dB and P_m = 35 W	R = 21 km
Non-Doppler Close Combat	PRF = 4000 Hz, frequency swept ± 10%, T = 0.25 μs	11 dB	G_p = 2.3 dB and P_m = 20 W	R = 20 km
Ground Mapping	PRF = 1800 Hz, frequency swept ± 10%, T = 1 μs, frequency agility on transmission	10 dB	G_p = −3.5 dB, P_m = 36 W	R = 80 km
Ground Mapping with DBS and Area Zoom	PRF = 3 600 Hz, T = 0.2 μs	10 dB	G_p = 9.7 dB (N = 128) and P_m = 14 W	R ≥ 55 km, the angular sharpening ratio is ranging from 13 at θ_A = 10° to 65 at θ_A = 60°
Ground-to-air Ranging	PRF = 3600 Hz, frequency swept ± 10%, T = 0.5 μs, frequency agility on transmission	10 dB	G_p = 10 dB and P_m = 36 W	R = 80 km
Contour Mapping	PRF = 3 600 Hz, frequency swept ± 10%, T = 0.5 μs, frequency agility on transmission	10 dB	G_p = −3.5 dB and P_m = 36 W	R = 21.5 km
Detection of Contrasted Targets	PRF = 3 600 Hz, frequency swept ± 10%, T = 0.5 μs, frequency agility on transmission	10 dB	G_p = − 3.5 dB and P_m = 36 W	R = 78 km

20.9 PENETRATION RADAR

Very low-altitude (VLA) penetration (< 500 feet) requires, as we saw in Chapter 19, a series of elements—the most important of which is the radar—for the detection of obstacles (see the previous chapter), and for self-defense.

From the simplest technical standpoint, penetration is part of the terrain-following mission, that is, navigation in the vertical plane compatible with the use of a mechanical scanning antenna. We shall now look at this configuration.

20.9.1 SPECIFICATIONS

The radar must provide

- ground mapping over 20 NM by monopulse sharpening (σ = 1 000 m^2)
- contour mapping over 10 NM for low-altitude navigation (σ = 10 m^2)
- terrain following over 10 NM for very low-altitude navigation (σ = 10 m^2)
- close combat over 10 NM (σ = 5 m^2)

20.9.2 TECHNICAL DESCRIPTION

A low-PRF radar with frequency agility will be used for all modes.

20.9.2.1 CHOICE OF ANTENNA AND WAVEFORM

In the contour-mapping and terrain-following modes, it is essential to measure the depression angle, and therefore the altitude of ground echoes (especially their peak), with great accuracy. A narrow beam aperture should be used, and a shortened λ can be helpful for an antenna of constant dimensions. In fact, if the entire vertical dimension of the ground target (i.e., pylon) is contained in the antenna beam (this is generally the case), the measurement is best performed on the barycenter, which is located in the middle or at one-third of its height. The best measurement is obtained at the lower edge of the antenna lobe. In X-band, atmospheric conditions (clouds, rain, etc.) affect detection and measurements, giving rise to spurious detection and attenuation in propagation. When λ is reduced, their effects increase (see Chapter 4). Circular polarization can be used to attenuate these effects, but at the same time it modifies the reflection characteristics of the echoes to be detected. This technique is not easy to implement, even if it is associated with switchable rectilinear polarization.

In conclusion, for the example in question, the same antenna used in the previous examples will be employed, that is, a flat-slot antenna (λ = 3 cm) with low inertia, vertical polarization, d = 60 cm, and steered by three-axis servomechanisms.

Finally, to neutralize the unwanted effects of echoes detected by the side and far lobes of the main antenna, this antenna must be fitted with an omnidirectional auxiliary antenna associated with a specific reception channel, SLS.

20.9.2.2 ANTENNA SCANNING

Although VLA navigation is theoretically carried out only in the vertical plane (terrain following, TF), the radar must produce a "radar terrain" with sufficient angular domains in either depression or bearing to enable the aircraft to navigate in the vertical plane or the horizontal plane in line with orders supplied by the weapons system (e.g., terrain file). Finally, this "radar terrain" needs to be stored, displayed, and servo-controlled to the forward motion of the aircraft and refreshed at adequate data rates. It is used to establish a navigation trajectory, and "orders" are transmitted to the automatic pilot or the pilot if he has taken over control. Figure 20.10 illustrates an example of a scanning antenna. In spite of limited angular domains and a very high scanning-rate antenna, the data-refresh rate is barely sufficient, especially for preparation of a banking maneuver; the advantage of an electronic scanning antenna in one or two planes is immediately obvious.

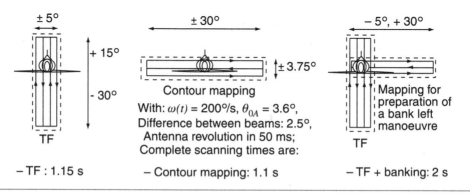

FIGURE 20.10 ANTENNA SCANNING FOR VLA PENETRATION

20.9.2.3 TRANSMISSION

Contour mapping and terrain following are the modes that size the transmitted power. Range must be obtained on the basis of raw unfiltered echoes with the best possible elevation angle tracking values, at extractor level. Smoothing of the data extracted is performed downstream. In addition the false-alarm rate should be kept very low. If we apply the data from the previous example, we obtain

$$P_k = 80 \text{ kW and } P_m = 140 \text{ W, with } T = 0.25 \text{ }\mu\text{s, and}$$
$$\text{PRF} = 7\,000 \text{ Hz} \pm 10\%.$$

With range $= 18$ km, RCS echoes $= 10$ m^2, and false-alarm probability 10^{-10}, we can cover all types of terrain, apart from large stretches of still water, for which the flight altitude is calculated from altimeter probe measurements and from the terrain data file.

21

MULTIFUNCTION RADAR

21.1 INTRODUCTION

Chapters 19 and 20 described the main missions in which radar systems are involved and the role of radar in these missions. Various specialized radars are briefly described. Naturally, it is tempting to envisage the use of a single multi-role platform capable of carrying out all these missions under ideal conditions. Unfortunately, however, certain particularly large scale missions require highly specific platforms and equipment, including the radar. Observation and surveillance missions are two such examples. Some missions, however, such as air superiority, interception, combat, tactical support, ground attack (land and sea), and low-and very low-altitude penetration, can be performed by a multi-role aircraft of the fighter-bomber type, fitted with a multifunction radar in the nose cone.

21.2 RADAR MODES AND FUNCTIONS

As seen in Chapter 2, there are four different radar functions, each of which uses several specific modes and submodes. The choice of modes and submodes depends on operational configurations, weaponry, crew preferences, etc.

21.2.1 FUNCTIONS

The four "radar" functions are

- air-to-air (A-A)
- close combat (CC)
- air-to-ground (A-G)
- air-to-surface (A-S)

21.2.1.1 THE AIR-TO-AIR FUNCTION

The main modes of the air-to-air function, which must be feasible at all altitudes and for all fighter-target configurations, are

- long-range front attack (face-to-face): velocity search, range search, and tracking (track–while-scan, multitarget tracking, single-target tracking) with high-PRF waveforms

- tail attack or medium-range front attack: range search, TWS, STT with medium-PRF waveforms
- all configurations, but with positive differential heights: long-range range search, TWS, MTT (track plotting), STT, with low-PRF waveforms (non-Doppler and thus without Doppler or velocity notch and range ambiguity)

21.2.1.2 THE CLOSE-COMBAT FUNCTION

The close combat function consists of the following modes: search within a wide angular domain and a range domain limited to 10 NM, rapid automatic acquisition, and STT. Low-PRF waveforms can be used, although medium-PRF waveforms are preferable.

21.2.1.3 THE AIR-TO-GROUND FUNCTION

The air-to-ground function, which uses low-PRF waveforms only, consists of the following modes:

- ground mapping with real beam and monopulse beam sharpening
- ground mapping with DBS zoom (medium resolution)
- high-resolution ground mapping (spotlight); antenna direction is slaved to the direction of the area of terrain to be displayed
- contour mapping
- air-to-ground ranging (AGR)
- detection and STT of contrasted fixed targets (hard target)
- ground MTI mode (GMTI) for detection and tracking of mobile terrestrial targets
- terrain following (TF), navigation in the vertical plane
- terrain avoidance (TA), navigation in the horizontal plane

21.2.1.4 THE AIR-TO-SURFACE FUNCTION

The air-to-surface function, for marine targets, consists of the following modes:

- display with the real beam and monopulse beam sharpening
- high-resolution display (analysis and identification)
- detection, TWS, track plotting, and STT
- detection of the coast boarder and ground vehicles is sometimes required

21.2.1.5 REMARKS

To the above-mentioned modes should be added modes relative to aircraft security (for example, alternated modes), electronic warfare, raid assessment, target identification, etc.

21.2.2 Sizing

The transmitter and antenna of an airborne, multifunction radar mounted in the nose cone take up approximately two-thirds of the total volume. The transmitter consumes and dissipates over half the energy supplied by the on-board power supply (e.g., 10 kVA), of which transmission microwave energy accounts for just 10% (e.g., an average of 1 kW).

21.2.3 Performance and Constraints

Chapter 25 examines the limitations of present-day radars. However, from the description already given it is possible to gain some idea of the factors that limit performance or are a source of constraint.

21.2.3.1 Antenna

The ideal antenna, combined with the radome and its pointing elements (electromechanical or electronic) must be extremely rapid, that is, with very low inertia, high scanning speed (e.g., 500°/s), and wide look angles. The side lobes of the antenna pattern must be very low. Sum and difference channel beams must be independent of the pointing direction; the same holds true for the side lobes.

An electromechanical scanning antenna cannot achieve the required velocity, which leads to operational constraints. The microwave characteristics of a passive or active electronic scanning antenna deteriorate with off-boresight angle, which in this case is a source of radar constraints.

For a given radar, Figure 21.1 compares range performance obtained for the different types of passive flat antennas used, assuming that loss and size are identical in each case, which is not obviously the real situation, especially in the case of an active ESCAN (see chapter 26).

The curves shown in Figure 21.1 take into account the fact that the maximum beam gain of a flat electronic scanning antenna is proportional to the cosine of the off-boresight angle, both in azimuth and elevation. Note that a loss of range in elevation is less problematic than in azimuth, as radar-target altitude differences are naturally limited. It should be noted, however, that at large incidence angles of the platform, the angle of elevation (or, with roll stabilization, radar depression angle) must be capable of a high value.

Compared with a passive antenna, and with all other factors being equal, the use of a two-axis planar active antenna with solid state modules, which considerably reduces microwave losses, enables an improvement of the radar power budget and thus an increase in radar range.

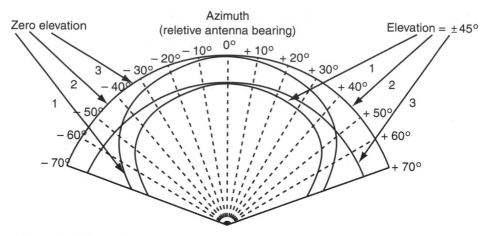

Curve 1: 2-dimensional electronic scanning
Curve 2: Mechanical scanning
Curve 3: Mechanical scanning in azimuth and electronical scanning in elevation

FIGURE 21.1 COMPARING RANGE FOR DIFFERENT TYPES OF PASSIVE ANTENNA

21.2.3.2 TRANSMITTER

The waveform should be optimized for each mode and submode of a multifunction radar, that is, for pulse-repetition frequency, form factor, and mean and peak transmitted power. As shown in previous chapters, this is no easy task. Indeed, maximum mean power is almost always determined by one mode, and maximum peak power by another. In certain cases, optimization is not directly possible. Several complementary possibilities have to be used to get around this problem, whether in terms of operational requirements—

- by fixing priority modes and functions
- by achieving acceptable trade-offs

—or in technical terms:

- by using pulse compression (high or low rate depending on the mode)
- by using a transmitter tube capable of producing variable peak power at constant mean power while maintaining spectral purity (saturated operation). Such tubes do exist and are extremely reliable. It is also possible to adjust average and peak power from one pulse to the next, with a slight deterioration in spectral purity (non-saturated operation)

21.2.3.3 OTHER CONSIDERATIONS

In the block diagram shown in Figure 22.2, the microwave and intermediate frequency receivers, frequency controller, and the signal, data, and radar map processing procedures no longer limit performance. Recent technology advances, still ongoing, have removed the main constraints, such as

- broad linear dynamic range
- identical reception channels
- performance stability
- spectral purity
- generation of a high number of frequencies
- computing power
- memory capacity
- facility of processing configuration, etc.

Chapter 26 deals with these technological aspects.

21.2.3.4 SPACE-TIME MANAGEMENT

An electromechanical scanning antenna, whatever its scanning speed, is constantly exploring space. An advantage is that the signal received is weighted by the antenna beam shape. An electronic scanning antenna operating in one or two axis can scan space in three ways: in a finely quantified and continuous manner, with beam overlapping, or in a disjointed manner. Although these last two solutions are highly tempting, they are encumbered by various constraints that considerably limit their performance, such as

- the need to initialize certain processing filters for each beam position (e.g., filters with a high time constant, such as post-detection integration filters)
- the necessity to have a dead time (1 ms) at the beginning of each burst before starting the processing in order to receive the signal from long range echoes
- the need to keep the beam pointed in a certain direction over a certain time period (e.g., 10 ms) if the Doppler effect is to be used
- rapid space exploration compared to that of electromechanical scanning; this reduces integration or post-detection integration time and thus reduces processing gain and range

In conclusion, the use of electronic scanning must be optimized for each radar mode, taking into account the waveforms and processing techniques used. The flexibility of transmitted power helping, to a certain extent, to achieve the desired range.

By way of example, consider a one-dimensional (elevation) electronic scanning antenna, mechanically roll-stabilized and used with quantified continuous electronic scanning of 2.5° (3.6° beam). For non-Doppler modes, each position is maintained for 5 ms to benefit from frequency agility. Under such conditions, scanning shown in Figure 20.11, including antenna reversal in azimuth, takes 610 ms instead of 2 seconds. It is therefore possible to double the angular domain (essential for terrain avoidance), while maintaining sufficiently rapid data refresh. For a Doppler mode similar to that shown in Figure 19.3, and with each beam position maintained for 10 ms, scanning time is reduced to 1.92 s (with bearing scanning velocity of 62.5° / s) instead of 5.6 s.

21.3 TECHNICAL SPECIFICATIONS

For simplicity's sake, let us assume that the specifications of this radar combine an interception radar, a tactical support radar, and a penetration radar, as previously described. We should then add the following:

- for air-to-air mode: long-range search at positive differential heights (non-Doppler low-PRF)
- for air-to-ground mode: spotlight (see Chapter 15), terrain avoidance
- for air-to-surface mode: long-range display with real beam: 80 NM, RCS = 1 000 m^2; the other air-to-surface modes are derived from the air-to-air and air-to-ground modes

21.4 TECHNICAL DESCRIPTION

We could not be expected to produce a detailed design project for such a highly complex radar within the contents of this chapter. Indeed, an industrial group working with adequate experienced staff and resources would need around ten years to design and develop such a radar. We shall therefore limit ourselves to the main options and a few key figures.

21.4.1 ANTENNA

Depending on performance requirements for each mode, the passive flat antenna can use either electromechanical or electronic scanning, in one or two axis, and can be combined with an auxiliary antenna.

SCANNED ANGULAR DOMAIN
- ≥ 60° conical

BEAM
- 3 dB aperture: 3.6°
- gain: 32 dB along the axis
- side lobes: see Figure 22.6

21.4.2 TRANSMITTER

The most demanding modes in terms of mean power are the high-PRF air-to-air search mode, with an average of 1 kW, and, to a lesser degree, the low-PRF air-to-air search mode and the long-range air-to-surface ground mapping mode.

The level of peak power is determined by the medium-PRF modes, and above all, by the low-PRF modes. Although, theoretically, pulse compression provides a means of solving problems associated with peak power, very high levels of pulse compression lead to long transmission times and thus make it difficult to detect very close targets whose signal is received during radar transmission.

A trade-off must therefore be made with respect to the transmission tube. The usual solutions are

- a twin peak tube: the transmitter can choose between two peak powers in saturated mode, 10 and 40 kW. It can deliver up to 1 kW on average for these two values. For long-range low-PRF modes and detection of moving ground targets, low-ratio pulse compression is used (e.g., Barker phase code at 5, 7, or 13 moments)
- a constant peak power tube: the transmitter supplies a peak power of 20 kW with a duty ratio of 5%

In the examples used for the previous radars, the following elements should be optimized in relation to the operational requirements expressed:

- the scanning velocities and angular domains explored
- the waveforms and transmitted powers used, etc.

Having more possibilities with respect to transmitted power and scanning speed makes it easier to optimize each of the numerous modes.

22

TECHNOLOGICAL ASPECTS

22.1 INTRODUCTION

For over half a century, airborne radar has benefited directly from some remarkable advances in technology. This evolution, which concerns all the components of airborne radar, has enabled the gradual adoption of new or existing technology, sometimes already applied to "ground" radar, which is subject to fewer constraints in terms of volume and mass. More recently, spaceborne radar has benefited from current technology. However, given its conditions of use, only certain specific technologies can be applied, such as components "hardened" against radiation and other technologies used to obtain very long life cycles.

Following a description of the major stages in technological innovation, we shall examine certain aspects specific to "radar components," and briefly describe how these may evolve.

22.2 THE MAJOR STAGES IN TECHNOLOGICAL INNOVATION

Using radar block diagrams as a main thread basis, the major steps in technological innovation that have marked the development of airborne radar over the last fifty years, and that have been applied on an industrial scale, are briefly summarized below. For the sake of simplicity, they have been classified as either part of the "analog age," which covers the pre-1970s, or the "digital age," which covers the 1970s onward. The major changes envisaged or already in hand for the next decade are briefly mentioned.

22.2.1 THE ANALOG AGE

Radars from the analog age, some of which are still in operation today (e.g., in many airline weather radars), did not have much processing power. They were therefore unable to exploit all the possibilities of "transmission-reception" coherence, in particular those associated with the Doppler effect. These radars, essentially unambiguous in range, used low-PRF waveforms only.

Figure 22.1, which shows the main radar components, represents a non-coherent radar with the following functions: search, target acquisition and tracking, and weapon fire at the target. The block diagram also applies to other modes, such as air-to-ground and air-to-surface. For certain radars, such as air-to-air range finders or "weather" radars, a single reception channel is sufficient.

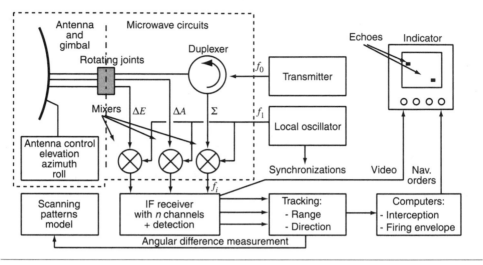

FIGURE 22.1 BLOCK DIAGRAM OF NON-COHERENT RADAR WITH ANALOG PROCESSING

The main radar components are

- the transmission-reception antenna and the gimbal
- the transmitter
- the local oscillator
- the microwave circuits
- the IF receiver
- the processing
- the display unit

The components typically used during the analog age were

- scanning or monopulse parabolic antennas with beam steering by electromechanical or hydraulic servomechanisms
- monofrequency magnetron transmitters
- klystron local oscillators frequency controlled, to within f_i, of magnetron frequency f_0. Klystron magnetron frequency deviation was compensated for by automatic frequency control (AFC) acting on the klystron reflex voltage
- microwave circuits mainly consist of crystal detector mixers which determine the global noise factor of the receiving chain

- IF receivers with one or more channels with a low instantaneous linear dynamic range (20 to 30 dB) compared to the dynamic range of detected echoes (> 60 dB), but whose gain can be controlled using voltage. Thus, for air-to-air searches, devices such as sensitivity time control (STC), and therefore range control, or instantaneous automatic gain control (IAGC)
- processing can be subdivided into four different functions:
  - servomechanisms
  - antenna scanning patterns control in search mode and for tracked target modeling
  - range and direction tracking
  - navigation interception and firing envelope computers

  This analog processing, as well as the IF amplification chains, first used miniature tubes, then subminiature tubes, and finally semiconductors. In addition to these active components, the servomechanisms, model, and scanning patterns, are also made up of electromechanical components such as motors, selsyns, resolvers, linvars, accelerometers, tachometric generators, etc.
- and finally, display units that show the radar operator information provided by the radar, such as the ground map or, in air-to-air mode, the detected target plots

22.2.2 THE DIGITAL AGE

During the 1970s, newly available "military" integrated digital circuits gradually took over from integrated analog circuits, which only offered limited processing possibilities. However, as these new components lacked density and consumed huge amounts of energy, their implementation in airborne radars was restricted by volume, energy consumption, and packaging constraints.

Increased processing possibilities, in particular for signal processing, led to the development of new radar modes using in particular the Doppler effect, T-R coherence, and a variety of waveforms.

Figure 22.2 illustrates coherent radar with digital processing. The figure shows that radar, like most of the other equipment, is connected to the weapon systems communication network. This network is composed of data buses, video links, and specific microwave links.

The radar has its own data bus to which its different subassemblies are connected. The processing equipment has its own high-speed internal bus and specific internal links.

The components typically used in the digital age are

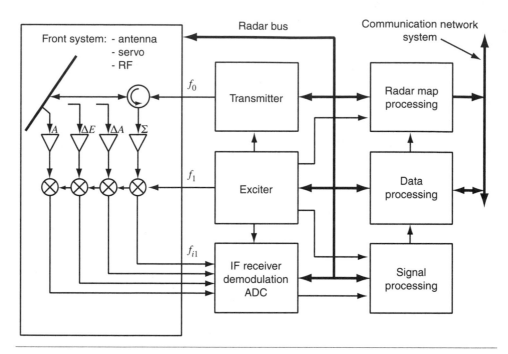

FIGURE 22.2 BLOCK DIAGRAM OF COHERENT RADAR WITH DIGITAL PROCESSING

- various types of monopulse antennas, with or without a mechanical roll axis:
  - parabolic, Cassegrain or inverted Cassegrain
  - planar slotted array
  - planar slotted array with mechanical scanning (or steering) in one plane and passive electronic scanning in the other
  - planar with passive electronic scanning
- microwave circuits located upstream from the mixers and on each reception channel and fitted with preamplifiers with a wide bandwidth and large linear dynamic range (80 dB) and a high-quality noise factor (< 3 dB)
- coherent transmitters comprising one or two power amplification stages, fitted with traveling wave tubes (TWT)
- exciters that simultaneously produce for each transmitted frequency (from over one hundred) all the signals needed for operation of the various subassemblies (coherent signals). These signals are extremely stable and of high spectral purity
- IF receivers with double frequency changes (see Chapter 12), with high instantaneous linear dynamic range (80 dB) and good separation between channels. Each reception channel is demodulated using an amplitude-phase demodulator (APD), at the IF receiver exit, which generates the signal's two filtered components, I and Q. These are then digitized using analog-digital converters (ADC) with high linear

> dynamic range. At this level, speed can exceed 200 Mbits/s per channel

- digital processing, which can be subdivided into
  - signal processing
  - data processing
  - radar map processing

Signal processing combines all the operations performed in real time on the radar signal, e.g., at the "range" quanta rate (e.g., every 0.5 μs) or at the multiplex rate of the "velocity" channels. Also covered are correction and calibration, frequency analysis (FFT), range compression, false-alarm regulation, post-detection integration, detection, extraction and ambiguity solving, calculation of angular difference measurements, etc.

Radar map processing, practically nonexistent before the introduction of digital technology, often comprises several "memory" planes. It enables conversion of radar scanning into a television standard signal. It also provides memory for the information produced by the radar and enables the use of markers, symbols, or alpha-digital characters. In addition, it carries out pixel processing for certain radar modes such as high-resolution mapping. Finally, the display unit is often separate from the radar system, integrated into a complex group of displays optimized for each platform type.

Data processing in fact carries out all the other numerous forms of processing. Digital processing requires universal standard equipment. It involves, for example, managing radar modes and exchanges, antenna scanning patterns, track plotting and tracking, elaboration of the interception navigation function and the firing envelope, management of built-in test equipment, utilities, etc.

22.2.3 THE NEW AGE

The coming generations of radars (see Chapter 27), in particular multifunction radars, will go through major changes based on

- changes in operational requirements
- facing up to stealth and discrete targets
- the use of flat active antennas, conformal or with shared apertures, and including smart skins (Baratault 1993)
- the use of spatial processing in association with active antennas
- the possibilities opened up by increasingly effective digital techniques (computing capability, memory density and capacity, connection and exchange systems, software, etc.)

At weapons-system level, the introduction of integrated modular avionics (IMA), with resource sharing between the various components (antennas, processing equipment, communication networks, etc.), will clearly alter component and system definitions as well as the respective role of system suppliers, integrators (or assembly companies), and equipment manufacturers. This concept is already being introduced in the United States, both in military (F22) and civil (Boeing 777) applications. Numerous national and international research programs are currently being developed.

Figure 22.3 shows an example of future avionics architecture. With this kind of architecture and this concept of avionics, radar, electronic countermeasures, optronic sensors, means of communication, navigation and identification, weapons system computers, etc., are no longer physical entities incorporated into the system but simply functions.

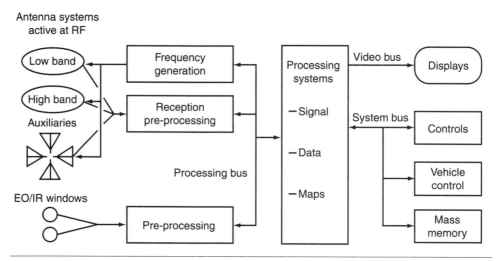

FIGURE 22.3 FUTURE AVIONICS ARCHITECTURE

22.3 ADVANCES IN RADAR COMPONENTS

The possibilities, performance, and architecture of radar have developed considerably, and these developments will continue. Most of them are linked to the evolution of the components themselves, in particular the widespread use of semiconductors (Antebi 1982).

22.3.1 ELECTRONIC CIRCUITS

From tubes to semiconductors, the density of electronic circuits has continued to increase and is set to go on increasing for the next twenty years. The densest circuits (amplification and processing) are shown Figure 22.4, with density expressed as the number of active components per liter (tubes, transistors, transistor equivalents for integrated circuits).

This density concerns industrially produced circuits and takes into consideration interconnection procedures (bundles of wires, multilayer printed circuits, thick-layer hybrid circuits, silicon connections, etc.). Similarly, "packaging" constraints are also taken into account.

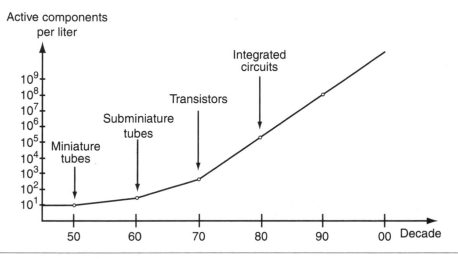

FIGURE 22.4 DENSITY OF ELECTRONIC CIRCUITS

It should be remembered that

- the first point-contact transistor dates from December 17, 1947 (Bardeen, Brattain, Shockley)
- the first junction transistor dates from 1949 (Teal)
- the first Germanium integrated circuit dates from October 1958 (Kilby)
- the first PMOS microprocessor (500 transistors) dates from 1971 (Intel)
- the first microchip microcomputer (20 000 transistors) dates from 1971 (Cochran, Boone, *Texas* Company)

Present-day submicronic integrated circuits can consist of several million CMOS transistors.

22.3.2 ELECTRONIC POWER CIRCUITS

Power analog electronics, which mainly concern the radar power supply, transmitter, and servomechanisms, is characterized by

- very high currents (several hundred amps) and high-power densities for low voltage circuits (LV)
- very high voltage in the transmitter (several dozen kV)
- production of extremely varied stabilized or regulated voltage, despite disturbance from the on-board network

- major constraints of packaging and volume, as well as the need to respect electromagnetic compatibility, etc.

Electronic circuits, whose density and power consumption have continued to increase, have required major technical and technological changes in low-voltage supplies (LV) in order to achieve high levels of efficiency (> 80%) in very limited spaces.

Figure 22.5 shows the typical development of regulated power densities produced by LV supplies.

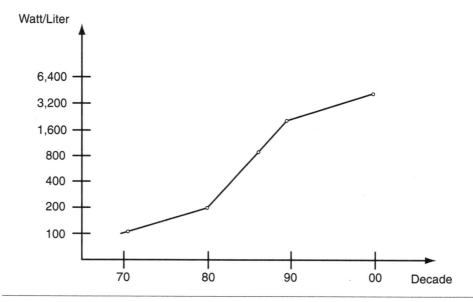

FIGURE 22.5 REGULATED POWER DENSITIES PRODUCED

22.3.3 TRANSMITTERS

Radar transmitters, combined with mechanical or electronic passive antennas, have evolved alongside microwave tubes and, more recently, semiconductors (solid-state transmitters). Active antennas, which integrate hundreds, even thousands of independent transmitters grouped in modules and combined with as many receivers, are set to challenge the traditional "transmitter-passive antenna" setup. However, both these solutions, with their advantages and disadvantages, will continue to coexist because they are complementary.

The magnetron transmitter, used for first-generation radars, acts like a triggered oscillator operating at its own frequency with a form factor of approximately $1/1000$ ($P_m/P_k = T/T_R$); when powered up, it has a high frequency drift (e.g., 10 MHz). This must be compensated for by action on the local oscillator using automatic frequency control (AFC) (see Figures 8.3 and 22.1).

This type of fairly simple transmitter is limited in terms of

- peak power, pulse length, and form factor that are practically unchangeable
- lack of phase coherence between transmitted microwave pulses
- a single, fixed transmitted frequency drifting in terms of temperature
- poor spectral purity, etc.

An amplifier tube transmitter generally consists of two cascaded amplifier stages in series with an overall gain of 70 to 80 dB. It provides a means of overcoming the limitations imposed by the magnetron transmitter and thus satisfying the demands of Doppler coherent radar. Depending on the chosen radar mode, this transmitter, which amplifies the f_0 from the master oscillator, can

- transmit over various frequencies within a bandwidth of a few percent
- facilitate selection of transmitted mean and peak powers, and therefore the wave form (HPRF, MPRF, LPRF)
- produce high mean power (> 1 kW)
- ensure coherence of the transmission-reception phase
- produce spectral purity close to that of the master oscillator
- by shaping transmitted pulses, limit the spectrum to reduce interaction between radars (see Chapter 12)

The microwave transmission tubes most frequently used (Firmain 1991) are traveling wave tubes (TWT) with coupled or helical cavities with excellent efficiency (> 40%) and life cycle (> 1 000 hours).

This type of transmitter can produce mean powers with densities of around 10 to 30 W/liter, depending on the high-voltage tube (HVT) used (50 to 20 kV).

The solid-state transmitter, which uses semiconductors (GaAs) is well suited to producing low power within a broad bandwidth (e.g., 15%). Better suited to producing mean powers than peak powers, it requires the use of high form factors (> 0.1) suited to an HPRF waveform or an LPRF waveform with pulse compression. It is used in smart munitions and in the transmission part of active antenna modules. It minimizes losses when placed close to the radiating element.

22.3.4 ANTENNAS

In most airborne applications, radar antennas are almost always combined with protective radomes whose shape is compatible with the platform aerodynamics (e.g., the nose cone of combat or civil aircraft). These radomes, in theory transparent to electromagnetic waves, disturb the wave patterns. This disturbance depends on shape and profile, materials used, beam deviation, obstructions such as the wind gauge sensors or lightning conductors, etc. Such disturbance can lead to a drop in operational performance, particularly in the air-to-air modes. Defining the antenna-radome combination should therefore be a joint decision between the radar manufacturer, the platform manufacturer, and the system manufacturer.

The antenna (including the radome and the beam steering) is a key element of the radar that has undergone great changes, and will go on changing (Baratault 1993). Designing an antenna is the work of highly specialized teams.

During the analog age, all airborne radars were fitted with a transmit-receive antenna, and beam steering was carried out mechanically using electrical or hydraulic servomechanisms.

The following types of antennas have been used (in chronological order):

- parabolic
- parabolic scanning used to track targets
- parabolic with monopulse channels
- Cassegrain, then inverted Cassegrain with monopulse channels
- planar slot antenna with monopulse channels

When monopulse angle tracking appeared in the 1960s, it was rapidly adopted because of its intrinsic qualities. Indeed, it "instantly" (i.e., during the length of the detected pulse) produces azimuth and elevation deviation of the detected target with regard to the radio axis of the antenna beam (see Chapter 9). This provides a means of sharpening the beam on reception (see Chapter 13), of precisely aiming the antenna beam at the tracked target, or of ensuring effective protection against radar jammers.

Since the dawn of the digital age, the performances required of antennas have increased, in two fields in particular:

- reduction of side and far lobes
- reduction of inertia

Side and far lobes must be reduced to obtain satisfactory performance with high-PRF and must be reduced even further for medium-PRF waveforms. This reduction also ensures adequate protection against interaction and deviated jammers.

The low inertia of a mechanical antenna is a means of reducing antenna slewing and return time. This allows dogfighting under acceptable conditions, multi-target tracking, obstacle avoidance, real-time multiplex (pilot's "perception") in air-to-air and air-to-ground modes, etc. However, only passive or active electronic scanning antennas allow unrestricted "space-time" management of beam direction.

From the 1970s onward, planar slotted arrays began to predominate in both civil and military applications. Indeed, they produced excellent side and far lobes and acceptable inertia while retaining high gain (e.g., compared to Cassegrain antennas). Figure 22.6 shows an example of the outline of the sigma channel of a planar antenna with a 20 λ diameter (60 cm at X-band), for use as a reference.

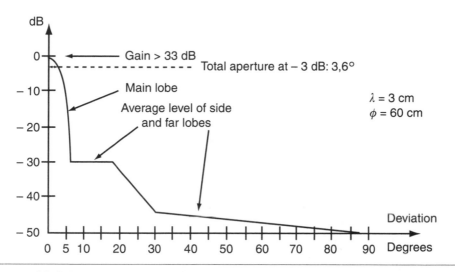

FIGURE 22.6 DIAGRAM OF THE SIGMA CHANNEL OF A FLAT SLOTTED ANTENNA

Other types of antenna have been, or are being, developed, particularly electronic scanning antennas that can be used in different configurations. Some examples are

- single-axis electronic scanning:
  - a mechanical plane (circular) carrying and feeding a passive electronic plane with a diode or ferrite phase shifter network (elevation), the whole system being mechanically roll-stabilized
- electronic scanning:
  - a passive (azimuth) lens carrying a second passive (elevation) lens with no roll axis (Chekroun 1991)
  - an active plane with modules (or mini TWT) carrying a passive plane. With modules, this can only generate low

power levels, as the active plane, which replaces the traditional transmitter, only comprises one row of modules

- two-axis electronic scanning:
  - a semi-active antenna comprising modules that are passive on transmission (phase shifters) and active on reception (amplifiers and phase shifters). This antenna, fed by a TWT transmitter, can be used with a number of waveforms
  - an antenna that is active on transmission and reception, with a high number of modules (e.g., 1000). These can be arranged on a flat or conformal structure, or can be dispersed. Each module is used to control phase and gain on transmission and on reception. This solution, currently under development in several countries, offers numerous possibilities for antenna pattern control but can only use a limited number of waveforms

22.3.5 EXCITERS

The first analog-age radars had several independent, and therefore non-coherent, frequency sources:

- a transmission source (e.g., magnetron)
- a reception local oscillator source (e.g., reflex klystron)
- sources that produced "video" signals from analog processing, e.g., triggered oscillator, blocked oscillator, relaxer, monostable or bistable multivibrator, saw tooth, sampler-blocker, etc.

All these generally unstable sources (temperature and time drift), very noisy in amplitude and time (jitter), did not offer sufficiently high performance to be used for coherent radar in the digital age.

The first change was to generate transmitter and local oscillator waves using a reference source consisting of a crystal oscillator (highly stable and with excellent spectral purity) and frequency transposition circuits combined with filters. Because each pair of transmission-reception frequencies requires specific circuits, the number of pairs was limited to a few units in the radar bandwidth.

A second development, still based on the reference source, was to use frequency synthesis to generate all the sinusoidal reference waves needed for transmission and reception as well as for the other radar components (processing, power supply converters, synchronization, tests, etc.). All elements thus function coherently and, for some, with the required spectral purity. Moreover, this technique is suitable for generating a high number of transmission-reception frequency pairs. As a result, a modern radar can easily have several dozen transmission frequencies in a bandwidth of several percent. Figure 22.7 gives an approximate profile of the spectral density of phase noise of the microwave transmission source. If this profile is not observed, the radar performance can be severely affected (see Chapter 7). The reference source must be screened, thermostatted, and suspended to protect it from outside influences.

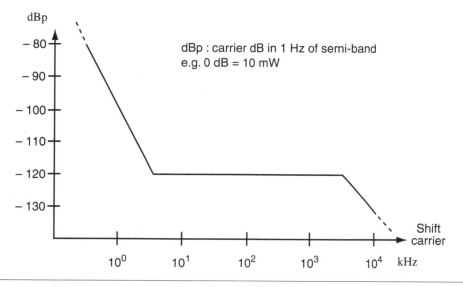

FIGURE 22.7 PROFILE OF TRANSMISSION-SOURCE PHASE NOISE

22.3.6 RECEIVERS

Microwave and intermediate frequency receivers are mainly composed of analog circuits. Because their operating frequencies are generally very high, despite advances in semiconductors, these circuits still use discrete components. Integrated circuits are used exclusively for auxiliary functions or in the second intermediate frequency amplifier operating at relatively low frequencies (e.g., 100 MHz).

Figure 22.2, showing the reception channels of a coherent radar with a passive antenna, represents a digital-age radar.

The main changes that have taken place between the analog age and the current digital age, apart from the reduction in volume and mass, concern

- a reduction in losses (from 3 dB to 1 dB)
- an improvement in the global noise factor by adding low-noise preamplifiers with broad dynamic range and wide bandwidth (from 10 dB to 3 dB)
- an increase in linear dynamic range (from 30 dB to 80 dB)
- identity of channels: gain and phase are controlled within 1 dB and 10°, respectively, in the whole transmitted bandwidth by means of calibration

In the future, if modular antennas with a high number of modules are used (e.g., 1000), the same performances will have to be maintained, after the necessary combining to constitute the channels; in other words, certain performances will have to be adapted to suit each module.

22.3.7 PROCESSING

Airborne radar processing, as described in Sections 22.2.1 and 22.2.2, has changed considerably during the development of radar systems (Marchais 1993). In the analog age, the computing capacity of a low-PRF radar able to track in range and direction was approximately 100 000 operations per second, and memory capacity was practically zero, except for the integrators and samplers. Moreover, each connection needed its own specific link, and specific electronic circuits were needed for each additional radar mode.

Since the digital age and the arrival of integrated circuits, computing power, memory capacity, and multiplex digital links have enabled major advances in processing.

Signal processing has thus resulted in

- the use of different and complex waveforms
- the optimization of performance depending on flight configuration, the nature of the ground overflown by the platform or target, the characteristics of the targets to be detected, the electromagnetic environment, etc.
- an increase in the number of radar modes without a significant increase in the "material" required
- more stable, reproducible performances, etc.

22.3.7.1 SIGNAL PROCESSING

During the 1970s and 1980s airborne radar signal processing generally used wired and microcoded logic, that is, processing carried out by a specific operator, the parameters of which may have been adjustable (e.g., FFT), and possibly with microcoded suboperators (microprocessors). This type of architecture facilitates development, improvement, and maintenance. It minimizes software but makes it difficult to carry out functional reconfiguring. It is therefore ill-suited to radars with multiple operating modes.

As a result, engineers in the 1990s who wished to produce multifunction radars developed multipurpose machines that could be reconfigured according to requirements, that is, for each radar mode. This is known as programmable signal processing (PSP).

Figure 22.8 illustrates the main difference between wired processing and programmable processing: for the former, each additional radar mode requires electronic circuits, while for the latter all that is required is software with low physical volume and production costs.

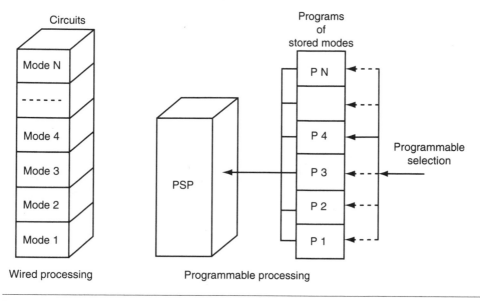

FIGURE 22.8 WIRED AND PROGRAMMABLE PROCESSING

With wired logic, signal processing consists of a series of specialized operators. The data rate decreases going from upstream to downstream. High computing power is easily obtained, but reconfiguration is difficult. The architecture of programmable processing can be of the Multiple Instructions, Multiple Data (MIMD) kind or of the Single Instruction, Multiple Data (SIMD) kind. Programmable processing can also use both types of architecture. Figure 22.9 shows the MIMD and SIMD structures. In the MIMD structure, each processor is identical, each carries out its own program, and reconfiguration is total.

In contrast, it is not always easy to achieve parallelism; programming is complex and bus management tricky as the data rate must be constant.

In the SIMD structure, all the basic processors work in parallel and execute the same program, making programming easy. However, exchanges between processors that may be required are not easy, which means that certain algorithms such as selection procedures are almost impossible.

Figure 22.10 shows the PSP architecture for a multifunction radar including, in series, one or more specialized operators, e.g., FFT, SIMD, and MIMD. The system is managed by a single CISC or RISC microprocessor. It

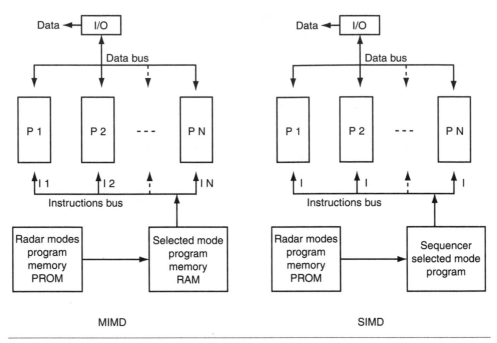

FIGURE 22.9 MIMD AND SIMD STRUCTURES

can be used to download software for the chosen radar mode. This PSP may consist of several dozen basic processors with a computing capacity greater than one gigaflop/s.

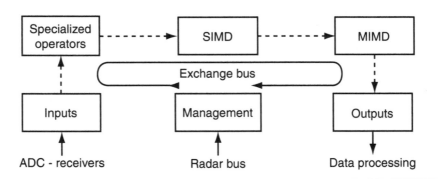

FIGURE 22.10 PSP ARCHITECTURE

Note that the processors used in SIMD structures are different than those in MIMD structures and can be created using specialized microprocessors known as digital signal processors (DSP).

Two major changes took place in the last few years:

- an increase in computing power and associated memories

- the incorporation of several resident program in the PSP in order to best exploit the possibilities of electronic scanning antennas (e.g., switching from one observation direction to another and/or switching from one radar mode to another in little more than one millisecond, bearing in mind that times of several dozen milliseconds are acceptable for mechanical scanning antennas)

22.3.7.2 DATA PROCESSING

Radar data processing can be performed by a universal CISC and RISC computer. However, for coherent radar, an RISC computer is preferable, operating autonomously and with timing and time cycles in line with the master oscillator.

This computer, which has a limited set of instructions (< 100) is pipelined and executes each instruction (or several instructions) in a single base cycle. It has several simultaneous computation rates. Computation time is independent of type of operation, programming is easy, usage rate high, and the language-related expansion rate is low. As for signal processing, performance must continue to improve in line with new requirements.

22.3.7.3 RADAR MAP PROCESSING

A distinction should be made between radar map processing and image processing used in civil and military applications whose developments have been quite remarkable, both in real and differed time.

During the analog age, radar indicators participated in signal processing if their CRTs had afterglow or memory signal function. Now that airborne displays use "television" standards, they can show images from different types of sensors, as well as the various symbols enabled either by directed scan or TV raster. The display is no longer part of the radar equipment. The radar simply has an interface used to give a standard form to information obtained for each of the different radar modes. This digital interface

- generates markers and symbols
- performs radar scanning-to-TV conversion
- filters the sharpened Doppler ground map for each pixel and over several antenna scans
- freezes and activates the map when radar transmission is shut down or during navigation update
- provides a 3-D display of the ground map at low and very low altitudes, etc.

To close this section on processing, it should be said that the needs and possibilities of these three types of processing have undergone considerable changes from one generation of radar systems to another. As an example, Figure 22.11 shows changes in computing power of multifunction radar signal and data processing. Of course, memory capacities, data-bus flows, and the number of instruction lines to be developed follow more or less the same pattern.

The density of electronic circuits and their interconnections must be increased in proportion to the above developments. Two-dimensional electronics must give way to three-dimensional electronics.

Mid- and long-term, submicronic circuits, optical connections (Vergniolle 1991), holography, and supraconductors should result in one, two, or even three orders of magnitude increases.

22.4 SPACE TECHNOLOGY

Electronic equipment fitted to space platforms, radar in particular, is subject to specific constraints, e.g.,

- life cycle
- resistance to radiation

The very concept of "electronics" must make allowance for the fact that the objective life cycle of a satellite is about ten years. Redundancy will have to be introduced combined with the use of ultra-qualified components and the elimination of design faults, etc.

22.4.1 LIFE CYCLE

Generally speaking, once the "teething problem" phase is completed, passive-component and semiconductor life cycles are adequate for normal use. However, once they are subject to radiation, certain semiconductors, and certain technological processes in particular, reveal very short life cycles that are unsuitable for space applications.

A great deal of research has been devoted to microwave power tubes, and present-day space TWTs have a life cycle of more than ten years (Firmain 1991).

22.4.2 RESISTANCE TO RADIATION

Space equipment is exposed to several forms of ionizing radiation that affects materials:

- radiation belts (protons < 300 MeV, electrons < 7 MeV)
- cosmic rays (ions from 100 to 1 000 000 MeV)
- solar wind (protons and electrons < 100 KeV)
- solar radiation (protons, α, heavy ions)
- nuclear explosions (initial and residual radiation)

These different types of radiation have varying effects. However, the inevitable radiation-material interactions considerably reduce the effectiveness of electronic components, dense integrated circuits and insulators in particular. The two main interactions take place with the peripheral electrons and lead to ionization or atomic movement as a result of collision. This interaction causes both temporary and permanent damage. Figure 22.11 summarizes the effects of radiation.

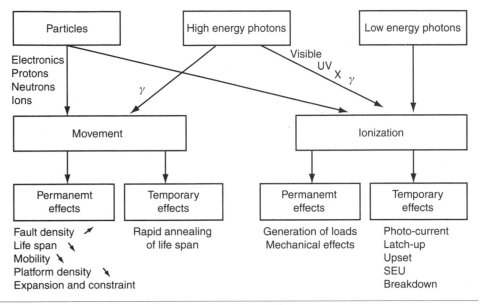

FIGURE 22.11 EFFECTS OF RADIATION

In order to eliminate or reduce the effects of radiation, space electronics must be hardened." This involves

- specific design
- the choice of passive components and materials
- the introduction of screening and shieldings capable of eliminating low energy radiation, whose flux is high
- the use of semiconductors and especially all low-density integrated circuits using hardened technologies or low-radiation sensitivity, such as SOS, CMOS/SOS, CMOS/SOI, MESFET, GaAs, etc.

PART V
RADARS OF THE FUTURE

Reconnaissance SAR POD (SLAR 2000)

23

THE CHANGING TARGET

23.1 INTRODUCTION

In the military field, the ongoing improvements to all weapon system components, particularly in aeronautics, have necessitated the introduction of new concepts and characteristics for platforms, sensors, and weapons. More specifically, in the area of radar applications for aerospace systems, potential targets, whether point targets or not, are increasingly

- equipped with highly effective self-protection methods: passive and active countermeasures, weapons
- discrete, as they have reduced electromagnetic wave transmission (suppression of leakage, space-time and frequency management of "useful" transmission)
- stealthy with regard to the various sensors deployed against them, as they have reduced signatures (RCS, IRS, LCS, sound level, visual observability, etc.)
- devoid of signature characteristics in order to avoid identification or, better still, neutralize enemy sensors

23.2 ELECTROMAGNETIC SIGNATURE

It we look at "radar" frequency bands only, we can say that the electromagnetic signatures of the great majority of modern aeronautic platforms (aircraft, helicopters, drones, missiles, etc.) are composed of three main elements.

The first is electromagnetic leakage unintentionally transmitted by a platform. These leaks can be caused by any item of on-board equipment. In principle, when equipment is taken on-board the platform, measurements are made in an anechoic chamber to detect leakage and check that the necessary precautions have been taken to eliminate it (see Chapter 12).

The second element concerns all the intentional transmissions required for the weapon system to operate. These transmissions, often at high power and with low directivity, are easy to detect and locate. They also have characteristic signatures that facilitate identification. In order to make the platform as "discrete" as possible, these transmissions should be managed by the weapon system as follows:

- strictly according to need, that is, when the passive sensors cannot meet the requirement (infrared, Electronics Support Measure receivers, radar warning receiver, etc.) or when highly directional active sensors (laser) are unsuitable (e.g., for target search within a large domain or under poor propagation conditions such as rain, fog, etc.)
- in association with other platforms, identical or not, in order to optimize global performance by carefully assigning tasks, or for a multistatic configuration

The third signature component is passive when seen by a radar. This is the radar cross section (RCS). It characterizes all potential targets, whether in the air, on land, or at sea. The RCS of any platform is thus the result of several effects and depends on a number of factors. The reduction of this RCS, which is a target in itself, involves the use of several complementary methods.

23.3 RADAR CROSS SECTION

Section 15.4 defines the RCS of a target as the energy backscattered by the target to the radar. Figure 15.6 of this same chapter shows a typical example.

23.3.1 EFFECTS THAT PRODUCE RCS

These effects that produce RCS are:

- reflection
- diffraction
- surface waves
- resonant cavities

The radar beam waves that illuminate the target totally "cover" it. The target backscatters almost all the energy received in practically all directions. Figure 23.1 illustrates some of these effects on an aircraft. Certain features of the target appear as dihedrals, corners, or discontinuities, which reflect or diffract the waves. The target, if large compared to the wavelength, behaves like a group of point targets characterized by "scattering centers" whose directions and values can fluctuate (see the discussion of angular glint in Chapter 3).

Figure 23.2 gives the RCS for an aircraft as a function of direction. The values indicated are averaged to within 10°, which eliminates the RCS "leaves" (in contrast to the RCS in Figure 3.6).

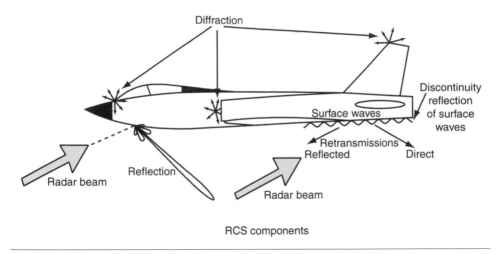

FIGURE 23.1 RCS COMPONENTS

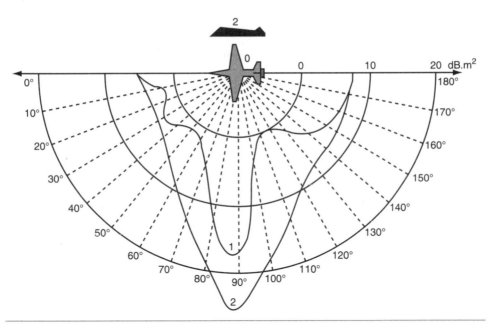

FIGURE 23.2 RCS OF A JET AIRCRAFT

Curve 1 gives the RCS, in X-band, of a jet aircraft (not fitted with a nosecone radar) seen at zero elevation in relation to the azimuth direction angle. Curve 2 gives the RCS of the same plane in relation to the elevation direction angle (seen from below).

These graphs show that in the case of this relatively small military aircraft (wingspan < 10 m), RCS can nevertheless be quite sizeable in certain directions.

However, operationally speaking, the front-view RCS values are the most important, as these determine detection ranges for ground-based or airborne fire control radars. The situation is somewhat different for surveillance radars, whose role is to detect all aircraft whatever their direction.

23.3.2 FACTORS INFLUENCING RCS

The main factors influencing RCS are

- target shape and dimensions
- materials used, how they are assembled, and their area condition
- frequency and polarization of the illumination wave
- moving, rotating, or vibrating elements (propellers, turbine blades, rotor blades, etc.)
- antennas, radomes, and the sensor structures fitted to the platform, etc.

23.3.3 SOME VALUES FOR RCS

Table 23.1 shows some typical RCS values in order to give some idea of orders of magnitude.

TABLE 23.1. SOME EXAMPLES OF RCS

Metallic Shapes	Formula	Examples where $\lambda = 3$ cm		
plane quadrilateral plate	$\sigma = 4\pi\dfrac{S^2}{\lambda^2}$	$h = 60$ cm $l = 60$ cm	$S = 0.36\text{m}^2$	$\sigma \cong 1800\text{m}^2$
plan circular plate		$d = 60$ cm	$S = 0.28\text{m}^2$	$\sigma \cong 1100\text{m}^2$
sphere	$\sigma = \dfrac{\pi d^2}{4}$ if: $\pi d/\lambda > 10$	$d = 60$ cm	$Sp = 0.28\text{m}^2$	$\sigma \cong 0.28\text{m}^2$

REMARKS

- The plate plane is at right angles to the illumination and observation directions.
- Sp is the area of the sphere projected onto a plane.
- The RCS of the plates is very nearly the same as the physical area multiplied by a factor close to the maximum gain of a plane antenna of the same size (illumination efficiency apart, see Chapter 1).
- For plane areas, RCS rapidly decreases when observation deviates.

Dihedrals, dihedral series, and trihedrals produce 0.1 to 100 m^2 RCS.

A bird, a pigeon for example, has an RCS in the X-band of approximately 30 cm^2.

Table 23.2 shows typical values of RCS at X-band for the frontal sector of different types of aircraft (Richardson 1989).

The values given show that these RCS figures are less and less dependent on target dimensions. They are above all dependent on shape, materials used, and certain techniques we shall discuss later.

TABLE 23.2. TYPICAL AIRCRAFT RCS

Type	Wingspan in m	Decade when commissioned	RCS in m^2
B 52	56	50/60	100
Blackjack	36	80	15
FB 111	11	60	7
F 4	12	60	6
Mig 21	7.2	60	4
Mig 29	16	80	3
B-1B	24	80	0.75
B 2	52.4	90	0.1
F 117 A	13	80	0.025
F 22	13.6	90	<0.1 (?)

23.3.4 RADAR RCS

Generally speaking, radar components are fitted onto the body of the platform, except for the antenna system, which is protected by a radome (or a rotodome). It is therefore the antenna system that constitutes the RCS of the radar.

Figure 23.3 illustrates the main sources of diffraction and reflection for a nosecone radar:

- reflection from the antenna, which depends on direction
- diffraction from
  - the radome itself
  - the radome-shroud junction
  - the erosion cone or wind gauge tube
  - the radar structure

Each diffraction obeys specific laws depending on

- shape and materials
- observation angle

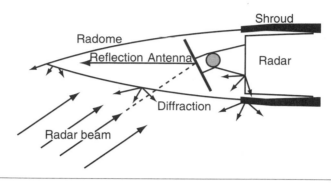

FIGURE 23.3 RADAR RCS

The frequency and polarization of the illumination wave act in different ways on the reflection and the diffraction.

23.3.4.1 DIFFRACTION

Depending on the frequency, the *radome* reflects and diffracts between 10% and 30% of incident power. In a cone of ±70° around the axis, the energy backscattered in the observation direction is mainly the energy diffracted by the erosion cone or the pole.

The *shroud* creates diffraction, mainly along its edge (where it comes into contact with the radome). It has a strong but directive effect along the axis (several m^2). In excess of a few degrees, diffraction loses coherency and is weakened by at least 10 dB.

The *radar structure*, partially masked by the shroud and the antenna system, can produce some non-coherent diffraction, which can be simply and considerably reduced by adding microwave absorbers and smooth metal covers. The resulting RCS is negligible.

23.3.4.2 REFLECTION

The reflection caused by the radar is due mainly to the antenna. In certain configurations, this reflection can produce RCSs of several hundred m^2. There are two possible situations:

- The antenna is aimed in the direction of the radar observing it.
- The antenna is aimed at least a few degrees off in relation to the direction of the radar observing it.

Results can vary depending on the type of antenna fitted to the radar, and the frequency and polarization of the observation radar. We shall limit ourselves to the examples of a planar slot antenna and an electronic scanning antenna.

The *planar slot antenna* behaves like a flat plate except for its operating frequency band and its polarization, where it absorbs most of the incident wave. If it is aimed at the observation radar, the RCS it produces depends mainly on the characteristics of the observing radar (frequency, polarization). For a 60 cm diameter antenna, this RCS can thus vary by several dozen square meters inside its band, and by several thousand square meters outside it (high band, e.g., Ku). If the antenna is aimed a few degrees off, reflection is no longer in the direction of the radar generating the illumination wave, and the RCS seen by this radar is small ($< 1 \text{ m}^2$).

An *electronic scanning antenna* is generally a narrow band structure ($< 20\%$), both in its circuits and its phase shifters. Outside its frequency band and polarization, it behaves like a flat plate that, if it is not aimed at a hostile target (or at the radar generating the illumination wave), has a small RCS. Thus a two-axis electronic scanning antenna can be designed so that it can be mechanically aimed a few degrees off from the platform axis in order to reduce RCS in the axial zone, which is the one most frequently used during operations. A single-plane electronic scanning antenna is more restrictive, as mechanical scanning of the other plane increases the high RCS spatial domain.

Inside the band and for antenna polarization, a large part of the illumination wave is absorbed. The RCS of a mechanically aimed antenna remains moderate (several dozen m^2).

23.4 REDUCING ELECTROMAGNETIC SIGNATURE

As seen in Section 23.2, the electromagnetic signature of any platform (aircraft, helicopter, missile, etc.) or any other potential target (ship, tank, hangar, surface-to-air battery, etc.) can be reduced by

- suppression of unintentional electromagnetic leakage
- time, space, and frequency management of "useful" transmissions
- reduction of transmitted power to the strict minimum
- careful selection of waveforms based on operational configurations and reduction of spectral density (widening the transmitted band)
- reduction of the RCS, infrared, and laser signatures
- elimination of the characteristics of residual signatures to avoid identification by the enemy (e.g., masking rotating elements such as engine compressor blades or helicopter blade drives)

23.4.1 ACHIEVING LOW RCS

23.4.1.1 RESOURCES

Low RCS is achieved by using a number of resources at the platform-design stage and applying them to both the platform itself and its equipment. Low RCS requires

- high-performance computer modeling using extremely powerful computers, and the optical analogies method (for D > 10 λ) or FEC (Finite Elements Calculation)
- a stock of characterized materials
- accurate measurement bases for the platform and its propulsion systems, the antennas of the integrated equipment and the weapons, etc.

23.4.1.2 SHAPES

This mainly concerns the motor drive platform, the antenna systems, and the weapon system stores, either housed in the platform or preferably in the weapons bay.

It we wish to reduce RCS to a minimum for the observing radar, we must either absorb almost all of the incident wave or reflect this wave in a direction other than that from which it arrives. This means

- avoiding vertical structures (e.g., vertical stabilizers)
- joining the wings to the frame very carefully (e.g., B2-flying wing)
- avoiding transitions between airframe materials (cavity, wedge)
- using carefully oriented plane surfaces (F117A) or surfaces whose curvature radius varies constantly
- using a reduced profile cockpit covered with a thin metal film (indium oxide and tin) that allows through 98% of light

23.4.1.3 ABSORBENT MATERIALS

In addition to shape optimization, localized use should be made of materials that can absorb electromagnetic waves depending on the radar bands used (SEE 1991; Deleuze 1992).

These materials absorb the electromagnetic waves by introducing magnetic loss (ferrites, iron-carbonyl composite) or dielectric loss (carbon, conducting polymers, metallic inclusions, etc.). They can be divided into four main categories:

- wide-band absorbers: attenuation is such that energy from the incident wave needs only to travel once and return through the thickness of the absorbent material to be almost totally absorbed

- resonant absorbers: for an incident wave close to the perpendicular and within a given frequency interval, the sum of all the waves reflected from the interfaces is practically zero
- surface-wave absorbers: surface waves develop along smooth walls, in traveling or standing groups. They behave like thread antennas. Absorbers are placed on the walls to confine these waves and prevent the creation of a group of standing waves
- shape absorbers: periodic dispersal of elements that have losses ($\lambda/2$, $\lambda/4$), makes it possible to absorb waves around λ. This type of absorber is heterogeneous and difficult to model mathematically

Heavy absorbent materials, placed at the surface, are subject to severe duty constraints, e.g., temperature and abrasion. They cause an increase in mass, and, given their limited thickness, their powers of absorption fall off with λ. Thus where $\lambda = 3m$ (100 MHz), the wavelength is the same as the dimensions of potential targets. There is a resonance effect; the RCS reaches the maximum level and the absorbers have no effect.

23.4.1.4 ELECTRONIC DEVICES

It is possible to reduce RCS or eliminate its characteristics by using either active systems (which supply energy) or controllable passive systems, e.g., by modifying RCS in terms of space or frequency. The "skin" of the platform can even be an electronic skin used by the sensors and that neutralizes the RCS.

Active cancellation: in this case, the target transmits in a given direction a signal whose amplitude is equal to that of the reflected wave but in phase opposition. To achieve this, the operator must know the characteristics of the incident wave (direction, energy, waveform, frequency, etc.) and his own RCS and transmit the same waveform in phase opposition at the correct level, with a minimum time delay. This technique, which is only feasible against monostatic radars, can apparently only be applied at relatively low frequencies, e.g., 100 MHz, and only in certain directions.

Signature modulation: this involves modulating the RCS of the entire platform. Given that it is difficult to effectively apply active cancellation, as previously described, in a large angular domain and at high frequency (> 100 MHz), the obvious reaction is to adopt the opposite solution; that is, instead of retransmitting at the same frequency, phase, energy, and direction as the incident wave, it is easier to disperse the incident wave in frequency and direction by electronically modulating the platform's skin. Thus the radar "illuminating" the platform will receive a signal backscattered by the platform's RCS. This signal is small compared to the transmitted signal and not coherent with it.

This attractive method does, however, require a lot of work before it can be applied. Its logical conclusion is the smart skin.

23.4.2 REDUCING RCS OF THE RADAR

As seen in Section 23.3.4, RCS is mainly a function of antenna size. Specific solutions, depending on the type of antenna used, can be applied to minimize this RCS.

23.4.2.1 PLANAR SLOT ANTENNA

For a planar slot antenna, RCS is maximum when illumination is perpendicular to its plane, and in this case the RCS increases with the square of the frequency (see Table 23.1), except for its operating band, where RCS is low.

When the radar is switched off, the antenna can be positioned in a chosen direction, e.g., at maximum elevation, such as $+60°$.

When operating, one way of significantly reducing RCS outside the band would be to use an active radome that absorbs all frequencies other than those used by the radar it covers (Klass 1988).

23.4.2.2 ELECTRONIC SCANNING ANTENNA

A single-plane (elevation) passive electronic antenna, or a two-times one-plane antenna, or a two-plane active antenna is mechanically fixed with respect to the radome and the platform. It behaves like a flat antenna outside its operating band. One solution is to offset the antenna plane by a few degrees (e.g., in elevation). This is easily compensated for by electronic scanning. A second solution, as with the flat slot antenna, is to use a frequency-selective radome. For an active module antenna, in addition to the previous solutions, it is also possible to add a controllable phase shifter $(0, \pi/2)$ to each module. Thus, outside the band, any incident wave is reflected and backscattered in phase opposition.

A third solution, for the future, would be to use a conformal active antenna or a network of dispersed antennas whose RCS could be controlled in the same way as the radiation patterns.

23.5 CONCLUSION

Any potential target, whether on land, sea, or air, must be discrete and stealthy if is to ensure its own safety. Discretion is achieved mainly through strict management of electromagnetic and acoustic wave transmission by making the best use of directivity, control of transmitted spectrums, frequency hopping, etc.

Apart from transmission, passive stealth, whose limits must be fixed, is achieved through coherent reduction of RCS, IRS, and LCS. Low RCS must not be achieved at the cost of IRS or LCS, for example. Similarly, in order to ensure a "determined" RCS for a weapon system as a whole, each component must have an RCS less than the required total RCS.

Creating a stealthy and discrete system thus requires well coordinated, multi-industry organization with extensive resources.

In the case of radar, we can consider that stealth has been achieved when existing enemy weapon systems can no longer fulfill their mission, even with downgraded performance (see Chapter 24). Today, stealth for an aircraft means an RCS of < 0.1 m^2. From 0.1 to 1 m^2, it is said to be low-observable.

Finally, given current resources, stealth is not possible at all frequencies, in all directions, and for all types of sensors.

24

OPERATIONAL ASPECTS

24.1 INTRODUCTION

Any potential target seeks to protect itself from enemy weapon systems. The means it uses to do so include

- detection sensors, providing the target with advance warning
- low RCS, infrared signature (IRS) and laser cross section (LCS), acoustic and visual signatures
- self-screening: passive or active jammers, decoys, decharacterization
- self-defense weapons: guns, missiles, etc.

In the case of radar, and assuming discretion is guaranteed, the reduction of aircraft RCS has an impact on

- detection systems: air-surveillance radar, interception radar, missile homing heads, surface-to-air radar, etc.
- countermeasures deployed by stealth or low observable aircraft
- missions

24.2 RCS VALUES

For a radar operating at X-band, the RCS of potential targets are of the order of

- fixed land target: 10^2 to 10^5 m^2
- mobile land target: 10 to 10^3 m^2
- sea target: 1 m^2 (periscope), 10 m^2 (snorkel), 10^2 to 10^6 m^2
- standard air target: 1 m^2 to 100 m^2
- low observable air target: 0.1 to 1 m^2
- stealth air target: < 0.1 m^2

Air targets are often situated in "clear space" in terms of distance or velocity (Doppler frequency), whereas mobile land or sea targets are only sometimes situated in "clear space" in terms of velocity and range; land and sea targets are superimposed on clutter.

24.3 DETECTION RANGE

Using the radar equation and considering clear space only, the detection range, limited by radar thermal noise, is directly related to target RCS σ:

$$P_r = P_t \frac{G^2 \lambda^2 \sigma}{(4\pi)^3 R^4 l}$$

Above the noise threshold, detection range is written as

$$R = f(\sigma)^{1/4}.$$

Figure 24.1, which illustrates the effect of σ on detection range, concerns a standard interception radar with a detection range of 100 km (54 NM) for a fluctuating target with an RCS of 5m^2. It shows how detecting a stealthy target of 0.1m^2 clear space is possible, but only at one-third the range, which considerably reduces the chances of interception. Moreover, this example represents the best possible circumstances. Chapter 25, which describes the limitations of radar, gives further details on how σ effects detection possibilities in the presence of clutter.

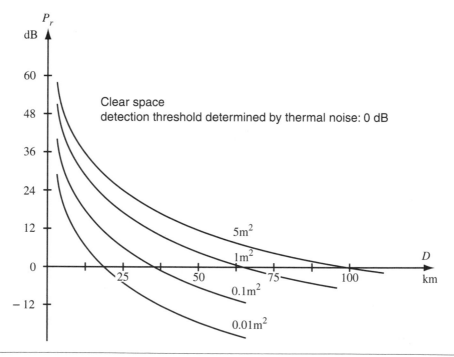

FIGURE 24.1 EFFECT OF RCS ON DETECTION RANGE

24.4 SELF-PROTECTION RANGE

The self-protection range is the range at which a platform jammer can effectively mask the vehicle RCS from a radar.

Obviously, for a given jammer, the smaller the platform's RCS, the shorter the self-protection range and the greater the level of protection. The formula below shows power received by a radar aimed at a target with its self-jammer on. We will use this formula to examine the effect of RCS on the target.

$$P_{rj}=P_{tj}\frac{G_jG\lambda^2}{(4\pi)R^2l_r},$$

where

P_{rj} =   the power received by the radar from the jammer

P_{tj} =   the power transmitted by the jammer

G_j =   the antenna gain of the jammer (e.g., 3 dB)

l_r =   the radar loss on reception (e.g., 1.5 dB)

For a radar with a detection range of 100 km for an RCS of 5m², transmitting in LPRF with frequency agility in a band 100 times its reception band, the jammer must transmit at approximately 2 kW in order to effectively protect its platform (σ = 5m²) at a range of more than 10 km.

Figure 24.2 shows changes in self-protection range with platform RCS. It shows how, for a given jammer, a self-protection range that is 10 km for a platform with an RCS of 5m² falls to 425m for a stealth platform with an RCS of 0.01m². A stealth aircraft, which is already difficult to detect, would be very easy to protect using this technique even if the power of its jammer were significantly reduced. If the jammer were fitted with a reasonably directive electronic antenna, its power could be reduced still further.

24.5 MISSIONS

The vast majority of current air-to-air and ground-to-air weapon systems using radar as their main sensor are designed for air targets with a frontal RCS of $\geq$ 5m². Reducing this RCS considerably decreases their effectiveness and can even render them totally obsolete.

Take the example of a stealth airplane with an RCS of 0.01m² penetrating at low altitude at a speed of 250 m/s. Its mission has been prepared. Its navigation aids are a terrain file, the GPS system, and an altimetric probe. This probe is the plane's only active sensor, which means that the aircraft remains discrete.

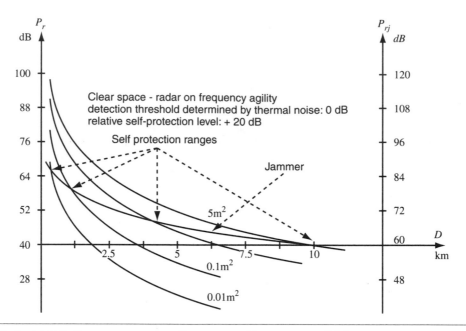

FIGURE 24.2 SELF-PROTECTION RANGE AS A FUNCTION OF RCS

If the enemy radar is designed for 5m² targets, its detection range under optimum conditions is reduced by a ratio of 4.7.

Therefore, instead of detecting up to 500 km, an airborne Early Warning System (AEW) which, for its own safety, carries out "racetrack" patterns away from the area of action, only detects within a range of 100 km. Ground-based interceptors on a two-minute alert cannot intervene, as the warning is insufficient. Only interceptors already in-flight that have received target-designation information can engage the stealth target using radar operating over a 20 km range, instead of 100 km, and using missiles with a considerably reduced homing head basket. As a result, there is very little chance of the interception mission succeeding.

Because their detection range is too limited and because of the masks created by penetrating flight at low altitude, ground-to-air systems, like air-to-air systems, receive very short advance warnings. In addition, given their detection capacity, their ground global coverage is insufficient to protect the entire area to be inhibited and they can easily be avoided (mission preparation).

In conclusion, stealth targets demand a rethinking of practically all the developed characteristics of current weapon systems, and of radar in particular. Their elimination requires the use of multisensor detection systems and radical changes in missions, tactics, weapons, etc. While acceptable levels of detection could be achieved for radar in particular, the necessary resources have yet to be developed.

25

PRINCIPAL LIMITATIONS OF PRESENT-DAY RADARS

25.1 INTRODUCTION

The factors limiting the performance of present day radars systems can be divided into two categories:

- *physical* limitations that apply to all radar in the same way
- *technological* limitations that gradually disappear as progress is made, and that may vary from one radar to another depending on how well each is designed

These two categories are dealt with separately in this chapter.

25.2 PHYSICAL LIMITATIONS

25.2.1 POWER BUDGET

One of the main physical limitations on radar range is the power budget.

Ω is the angular volume to be covered (expressed in steradians) and T_{ex} is the exploration time for this domain (determined by the data rate). If all directions are equiprobable, the observation time for each direction is

$$T_e \approx \frac{T_{ex}\theta_{Az}\theta_E}{\Omega} , \qquad (25.1)$$

where θ_{Az} and θ_E are the aperture angle of the antenna in relative azimuth (or bearing) and elevation (expressed in radians), respectively.

If S is the antenna area, we get the following relations:

$$\theta_{Az}\theta_E \approx \frac{\lambda^2}{S} , \qquad (25.2)$$

and

$$G = \frac{4\pi S}{\lambda^2}.$$

(25.3)

In Section 6.2, we established that the signal-to-noise ratio on output from the optimum receiver, in the presence of white noise, is given by

$$\text{S/N} = \text{R} = \frac{E_r}{b} = \frac{P_{mr}T_e}{kT_0F},$$

(25.4)

where P_{mr} is the mean power received from the target in question and F is the receiver noise factor. P_{mr} is linked to mean power transmitted by the radar equation (Section 3.3)

$$P_{mr} = P_{me}\frac{G^2\lambda^2\sigma}{(4\pi)^3R^4l}.$$

(25.5)

Relations 25.1 to 25.5 enable us to establish

$$R = \left[\frac{P_{me}S\,T_{ex}\,\sigma}{4\pi(\text{S/N})\Omega kT_0F\,l}\right]^{\frac{1}{4}}.$$

(25.6)

Ideally the signal-to-noise ratio required for detection is determined by the probability of detection, P_D, the false-alarm probability, P_{fa}, and target fluctuations (see Section 6.2.4). In an ideal situation (where loss l is zero and the noise factor equals one), *the physical limit on target detection range is*

$$R = k\left[\frac{P_{me}\,S\,T_{ex}}{\Omega}\sigma\right]^{\frac{1}{4}}.$$

(25.7)

It is surprising to note that, given a constant coverage volume and data rate, range depends solely on

- mean transmitted power
- antenna area

which compose the "power-aperture product."

Range does not depend on wavelength λ, except through the target's RCS, σ.

Relation 25.7 shows how a loss of spatial processing gain (drop in antenna gain) can be compensated for by an increase in the temporal processing gain (increase in integration time). This is the basis of Digital Beam Forming (DBF) processing, which uses a wide beam on transmission and several narrow beams on reception, thus enabling longer observation periods in each direction.

However, these conclusions are only valid if target coherency time T_c is greater than exploration time T_{ex}, that is, if the *target remains coherent throughout integration time*. For standard operating modes (limited resolution range and high carrier frequency), T_c is limited to several dozen milliseconds.

Similarly, Equation 25.7 does not apply to transient targets, that is, targets that are only visible or detectable for a short period, as discussed below.

25.2.2 INTERCEPTION PROBABILITY OF TRANSIENT TARGETS

Transient targets include, for example,

- a land vehicle that temporarily leaves its cover and crosses an exposed zone
- in air combat, targets crossing the angular coverage domain
- flashes from helicopter blades in hover mode (these blades have a high RCS only when they are perpendicular to the line of sight)

Target interception probability is the probability that the radar will illuminate the target during a scan. It acts as a multiplying factor for the detection probability.

In addition, if the target remains in the beam for a shorter time than the coherent integration time, the resulting loss (mismatching) for the signal-to-noise ratio must be taken into consideration.

25.2.3 LIMITS ON ACCURACY IN MEASURING TARGET PARAMETERS

The lower limit of accuracy when estimating target range, velocity, and angle (standard deviation compared to actual measurement) is determined by the Cramer-Rao limits (Le Chevalier 1989).

Target parameter measurement accuracy depends on

- signal-to-noise ratio (whatever the parameter)
- bandwidth of the transmitted signal (for range)
- coherent integration time (for velocity)
- antenna size/wavelength ratio (for angles)

25.2.4 RESOLUTION LIMITS

Angular, range, and velocity resolutions are given by Equations 2.7, 2.8, 2.14, and 2.15 in Section 24.2.

These resolutions are those obtained on output from an optimal receiver designed to maximize either probability of detection or target parameter estimation accuracy (see Chapter 6). So-called super-resolution algorithms exist, with resolution properties that are three to four times greater than those of the optimum receiver. However, their effectiveness is limited by

- the signal-to-noise ratio
- the number of targets present in the optimal receiver resolution cells

In practice, and in routine applications, resolution is considered to depend on

- the band width of the transmitted signal, for range
- the coherent integration time, for velocity
- the ratio of target wavelength to look-angle variation, for resolution, along an axis transverse to range (relative bearing or elevation, Figure 25.1)

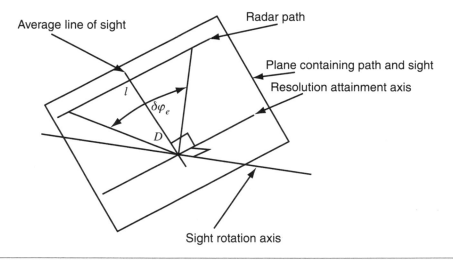

FIGURE 25.1 OBSERVATION GEOMETRY: RESOLUTION DEFINITION

The variation $\delta\varphi_e$ in the target look angle is itself proportional to antenna size (for a real or synthetic antenna):

$$r_a = \frac{\lambda}{2\delta\varphi_e} \text{ (synthetic antenna) or } r_a = \frac{\lambda}{\delta\varphi_e} \text{ (actual antenna)}$$

25.2.5 LIMITATIONS ON ANGULAR COVERAGE

Antennas designed for the most commonly used wavelengths are made up of reflecting surfaces. With a mechanical antenna, the beam is directed perpendicularly toward its surface. The angular coverage domain is limited by the mechanical constraints of the look angle. For electronic scanning antennas, the deviation angle, determined by the phase shift of each of the sources, is limited by the radiating element pattern, which is zero in the antenna plane.

The space available for the installation of antennas on an aircraft is a major constraint. Angular coverage is limited. For example, for a combat aircraft, the requirement of forward vision means fitting a front radar. Coverage is limited to a half-angle cone of approximately $60°$.

25.3 TECHNOLOGICAL LIMITATIONS

25.3.1 WAVEFORM

As seen in Chapter 6, waveform properties are characterized by the form of the ambiguity function.

The ideal waveform is one that provides an ambiguity function with a single and extremely narrow correlation peak (thumbtack type). Its advantages are

- no distance or velocity ambiguity; this facilitates rejection of ground return and increases target measurements
- total separation of targets
- target imaging (separation of scattering centres) enabling identification in the air-to-air mode
- a single waveform and a single processing for all types of radar operating modes
- a low probability of intercept waveform, which is difficult to detect and identify

As seen in Section 1.2.3, in order to reject ground returns, the waveform in question requires a very high BT_e product (> 80 dB). This waveform can, for example, be generated by illuminating a target with a 1 GHz band signal for 0.1 second.

For a given target, the optimal receiver matched to this waveform is the correlator that performs the correlation, during the observation time, between the received signal and the reference signal describing the desired target. However, this signal cannot be simplified as described in Section 3.5— $s(t) = Au(t - t_0)e^{j2\pi f_D t}e^{j\varphi}$; on the contrary, it must be complete. It must, in particular, take into account changes in velocity during integration time, and we must give up any idea of "narrow band" approximations.

Given the high resolutions obtained, a large number of reference signals are needed for the different parameters such as distance, velocity, acceleration, etc.; the technology required to process the different distance-velocity-acceleration resolution cells does not exist today.

25.3.2 SPECTRAL PURITY AND DYNAMIC RANGE

The level of phase noise, spurious lines, intermodulation lines, and of all other similar spurious signals carried by the received signal determines the limit of the lowest RCS the radar is able to detect in the presence of clutter. This limit is independent of range. The clutter residue associated with these spurious effects fixes a detection threshold linked to the clutter level. If the signal backscattered by a target is below this level, it cannot be detected even if it is above the thermal noise threshold. In such a case, increasing power transmitted by the radar or improving receiver sensitivity does not extend the detection range.

While mean transmitted power and antenna area theoretically determine detection range in clear space, range in the presence of clutter is also determined by

- the spectral purity of the radar frequency source and transmitter
- the dynamic range of the receiver
- the quality of the antenna pattern (level of side and far lobes)

25.3.3 DATA FLOW

Data flow is one of the factors that limits radar system performance. Its influence is clearly visible when the radar sends images back to the ground using a microwave link. The capacity of the link determines the rate at which radar data can be transmitted. Generally, this rate is much lower than the rate at which data becomes available at the radar receiver output. The alternative solution is to

- limit the number of images to be transmitted by rendering the radar non-operative during a certain proportion of its flight time
- compress the data using specific algorithms, at the cost of a certain fall-off in image quality

If the data is recorded, the influence of the recording equipment is similar to that of the microwave link.

25.3.4 EXPLOITATION

The capacity to exploit radar data is another limiting factor, particularly in the case of imaging radars. A radar image is extremely different than an optical image due to

- the presence of speckle
- the highlighting of objects that rarely appear on optical images, etc.

Because of backscattering, a radar image intrinsically contains information that could constitute an invaluable operational aid if it were better understood, such as fluctuation in target RCS depending on look angle and frequency, polarimetric behavior, coherence over time, etc. A number of ongoing thematic studies are aimed at establishing models that will help interpret these images, both for civil and military applications. Computerized tools will then be available to photo-interpreters to help them with their mission.

Whatever the outcome, the time needed to interpret an image will continue to limit the number of images produced by a radar observation system.

26

ELECTRONICALLY
STEERED ANTENNAS

26.1 INTRODUCTION

Improvement of radar capabilities followed evolution of technologies for the last 50 years. After continuous power tube improvements in the first age of radars, digital signal processing made the first technological breaking in the 1970s, enabling the clutter rejection and look-down capabilities. Then Programmable Signal Processors opened the way to multifunction radars. Today a new breaking emerges with Electronically Scanned Arrays (ESA) and active ESA (AESA), which brings beam agility and spatial processing (Adaptive Beam Forming).

ESA Radar offers a number of advantages compared to Mechanically Scanned Antenna systems:

- beam agility
- enhanced performance and functionality
- increased ECCM resistance
- high availability
- LPI and covertness features

These advantages are exploited differently depending on the application. Fighter radar, AEW, Air Ground surveillance, and Maritime Patrol are airborne systems where these technologies used, or will be used in the near future.

The constraints are different for each application: performance and high-quality radiating pattern are required for fighter radar, wide bandwidth and low-cost drive AGS and Maritime Patrol radar design, high power and platform installation easiness are needed for AEW.

Competing solutions are mainly the passive phase shifter antenna, among which RADANT technology is a today powerful candidate; the Reflect Array whose cost is much lower than other solutions and the Active Array which is the antenna technology of the future. The selection of the best solution for each application is a challenging task.

26.2 OPERATIONAL AND TECHNICAL BENEFITS OF ESA FOR AIRBORNE RADARS

26.2.1 FIGHTER RADAR

26.2.1.1 SEARCH-AND-TRACK DOMAIN INDEPENDENCE

As individual pointing toward all tracked targets is possible whatever the angular positions of these targets are, targets can be tracked even outside the search domain (see Figure 26.1), which enables the weapon system to keep the tactical situation specifically at short/medium range (firing range) when a hostile strike cluster starts to maneuver in a wide angular domain. Moreover, the individual pointing allows the radar management to adapt the dwell time, the waveform, and the update rate according to target characteristics (RCS, range, velocity, maneuver, etc.), which gives much better tracking accuracy and optimizes the power management in time, power, and direction. It is also possible to use specific high-gain beams toward the launched missiles to transmit a high-power Data Link to update the target designation.

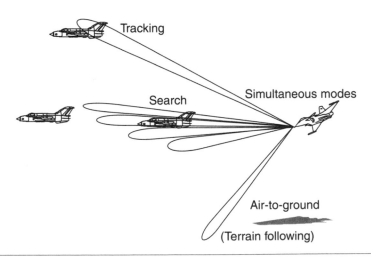

FIGURE 26.1 SEARCH-AND-TRACK INDEPENDENCE—SIMULTANEOUS MODES

26.2.1.2 NEW DETECTION STRATEGIES

Alert-then-confirm (sequential detection) increases the detection/tracking range significantly. A high-power, low-resolution waveform is used for primary detection; when a potential target is detected, a long-dwell-time, high-resolution waveform pointing is directed to it in order to confirm this detection and to initiate the track. The tracking range is then equivalent to detection range, which results in a much higher tactical situation acquisition.

Figure 26.2 shows the different probabilities of detection:

- the single-scan probability of detection, P_D (identical for both antennas)
- the cumulated probability of detection, P_C
- the probability of tracking lock-on, P_T (detection confirmed)

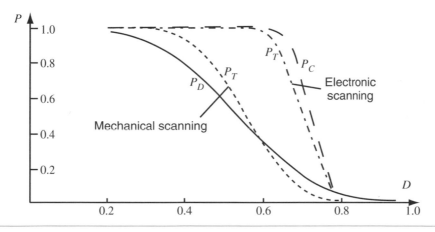

FIGURE 26.2 PROBABILITY OF DETECTION AND TRACKING

26.2.1.3 SIMULTANEOUS MODES

Beam agility enables simultaneous modes to function, such as low-level terrain following with a very high scanning rate (< 0.5 s) of a large ground area to avoid ground collision, and Air-to-Air surveillance and tracking to protect the fighter against air threats during the remaining time.

26.2.1.4 BEAM SHAPING—DIGITAL BEAM FORMING

Active ESA brings an extra feature, beam shaping, thanks to amplitude and phase control for all the Transmit/Receive Modules (TRM) connected to each radiating element of the array.

First the beam shaping provides the freedom to adapt the beam pattern to the environment, for example, creating nulls in the direction of several jammers even if they are in the antenna main lobe. This feature is obtained by partitioning the array into many subarrays to which individual receiver channels correspond.

Subarraying the antenna can also be used to implement Space Time Adaptive Processing (STAP) which enables the detection of very low-speed ground targets in MTI modes.

With multichannel architecture and subarrays, it is possible to create multiple adjacent receiving beams covering a large angular domain, provided that the transmit beam is broadened to illuminate this domain. Then this wide angular domain can be permanently covered. Transitory targets such as hostile targets crossing the search domain in combat modes, RCS flashes of stealth targets, or blade flashes of helicopters can be intercepted.

26.2.1.5 INCREASED POWER BUDGET

AESA and TRM architectures are such that the High Power Amplifier (HPA) that generates the transmitted signal, and the Low Noise Amplifier (LNA) that determines the receive sensitivity, are very close to the radiating element, avoiding a lot of losses that occur either with mechanical or passive ESA antenna. The global power budget is significantly increased (6 to 8 dB), which provides the ability either to increase the detection/ tracking range for a given RCS target or to restore the performance in the case of a stealth target.

26.2.1.6 GRACEFUL DEGRADATION

An AESA antenna keeps its performance with up to 5% of the TRM failed (i.e., 50 out of 1000). If more TRM fail, the performance degradation is progressive given the possibility to fulfill the mission; this is not the case for the single point of failure usually encountered in current systems.

26.2.2 AEW RADAR

26.2.2.1 SEARCH-AND-RACK DOMAIN INDEPENDENCE, NEW DETECTION STRATEGIES

As for Fighter Radars, beam agility is used to disconnect search-and-track domains, to adapt dwell time, to update the time to each tracked target, and to implement sequential detection.

26.2.2.2 SCANNING FLEXIBILITY/INTELLIGENT ENERGY MANAGEMENT

As detection range performance in specific areas is directly linked to the energy sent to cover this area (time-power product), the scanning flexibility provides the ability to tailor the scanning pattern according to areas of interest. Long exploration time is devoted to areas of high operational interest with enhanced detection, but much faster scanning in low-interest areas saves time (and energy).

26.2.2.3 AESA BENEFITS

Power budget and graceful degradation given by AESA technologies are, of course, of high interest for these systems.

26.2.3 AIR-TO-GROUND SURVEILLANCE

26.2.3.1 TERRAIN ACCESSIBILITY

SAR ground surveillance systems currently use Side Looking configuration (the antenna beam is maintained perpendicular to the fight path). First, this mode, known as Strip Map, has a limited cross resolution (half the antenna size); to increase the resolution, Spotlight mode is required. (In this case, the beam direction is slaved to the area to be mapped.) Moreover the side-looking configuration leads to constraints on the flight path to avoid the masking of a target of interest by obstacles (relief, for example). The ability to use Squint modes (the beam azimuth can be controlled out of the perpendicular sector) allows the expansion of the terrain accessible to mapping. The capability to implement Spotlight mode or Squint mode requires beam control in azimuth. As these SARs require large antennas either fitted in POD or in belly radome, mechanical azimuth control is not practical and an ESA is needed.

Elevation control of the antenna is needed for roll stabilization and to be able to center the swath (mapped area range) either at short range or at long range. Two-axis ESA gives the ability to have full access to the terrain without any mechanical part.

26.2.3.2 SIMULTANEOUS MODES

AGS radars use two very different modes, which need to be simultaneous operationally:

- SAR mode for fixed-echoes mapping, which requires very long illumination (several tens of seconds)
- MTI mode for detection and tracking of moving vehicles, which requires a high update rate (seconds)

These simultaneous modes need to interleave the waveforms and the beam pointing on a PRI basis. Only the beam agility given by ESA provides the ability to implement these simultaneous modes.

26.2.3.3 BEAM SHAPING

In some operational conditions (short range or large area at medium resolution), the current narrow pencil beams do not cover the entire area of interest and several scans are needed. In order to save time, it would be better to widen the beam on transmit-and-receive, which is made possible with ESA. Anti-jamming processing also requires adaptive beam forming.

26.2.4 MARITIME PATROL RADAR

26.2.4.1 SCAN-TO-SCAN INTEGRATION

Detection of a low-RCS (submarine) target requires scan-to-scan integration with a very high rotation rate of the antenna (up to 120 RPM). Roll-and-pitch stabilization of such a quickly rotating and large antenna is quite difficult, and elevation ESA makes it much easier if low-cost ESA is available.

26.2.4.2 NEW MODES

Detection/confirmation strategies are of interest for this kind of radar. The 360° azimuth coverage implies a continuous rotation, and azimuth ESA makes it possible to come back to the detection direction without stopping the antenna rotation. It also makes it possible to slow down the rotation speed (with counter rotation) in some directions to have longer dwell time needed for identification purposes (ISAR mode), for example.

26.3 COMPETING ESA SOLUTIONS

ESA principles have been known for a long time, and a number of solutions based on them have been developed in the last few decades. It is worth noting that although many naval or ground-based radars are fitted with an ESA, only a very few airborne radars are.

The reason for that situation certainly lies in the fact that, for the radars operating in X-band, the right goal in terms of efficiency, cost, and performance was not easily achievable with an ESA. Ferrite phase-shifter ESAs have, however, been realized for the B1B bomber and for the MIG 31. For the time being there is no application (with this type of technology) to a lighter combat aircraft.

For AEW radars, the lower bands (P-, L-, or S-) allow us to reach a better balance between performance, weight, and mechanical complexity with an ESA: all AEW radars are fitted with different kinds of passive or active ESAs.

The situation has been evolving for the last ten years. We can now identify three candidates for airborne ESA, with some specificities in their field of application:

- the reflectarray with an innovative design
- the RADANT antenna
- the active ESA (AESA)

26.3.1 REFLECTARRAY

Reflectarray principle is very well known and thoroughly described in the literature.

Let us just recall that it is basically an array of elementary antennas connected to reflective phase shifters. This array, spatially fed by the spherical wave coming from a feeding horn, focuses the spherical wave into a plane wave and tilts it in the desired direction (Figure 26.3) by means of the controlled electronic phase shifters.

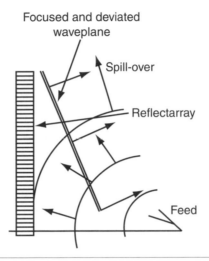

FIGURE 26.3 REFLECTARRAY PRINCIPLE

It is also known that this type of antenna has to be designed to minimize the effects of spillover (that is, the losses and the spurious lobes due to the radiation of the feed apart from the reflecting array), the effect of feed blockage on the wave radiated by the array (small losses and diffraction lobes), the effects of mismatch at the elementary antenna level (small losses and image lobe of the feed pattern). The right balance between the radiation efficiency, linked to the amplitude taper imposed by the feed pattern, and the spillover losses has to be found.

In addition, we also find the losses and the diffused lobes due to the phase quantization and the errors of the phase shifters, as in any ESA.

The balance sheet of performance of that type of antenna could be expected to be quite bad, and it generally was!

An innovative design of the reflectarray has been developed by THALES (ex-THOMSON-CSF). This new design leads designers to reconsider positively the use of reflectarray. It is characterized by

- the integration of the antenna element and the phase shifter with an excellent match over a wide angular range
- the integration of the control circuits on the rear face of a multilayer printed circuit whose front face is the active phase-shifter layer
- a total thickness of the reflectarray less than 2 cm and a very low weight per square meter
- a price lower than that of the mechanical antenna

That means that this type of reflectarray can offer the two-axis electronic beam agility and a cost reduction, compared to the conventional mechanical antenna.

The preferred field of application for this new type of ESA mainly includes rather large antennas (more than half a square meter in X-band) and air-to-ground radars (SAR, ground and sea surveillance).

26.3.2 RADANT ESA

The RADANT principle consists in adding electronic prisms to a fixed-beam antenna. The fixed-beam antenna may be an array antenna or an optical-type antenna (feed alone, or feed plus reflector). In front of this antenna, one or two electronic prisms introduce the controlled phase shift. Each prism is made of parallel plate waveguides into which are inserted parallel printed circuits loaded with controlled PIN diodes. The design of the circuits is such that each one introduces a certain amount of phase shift (perturbation phase shifter) and the same change of bias applied to the diodes of a group of two, three, or more of these circuits produces a phase shift between the two states of bias, without changing the match of the group.

So, each prism is able to add a one-dimensional insertion phase variation on the wave front coming from the antenna side. One prism leads to single-axis electronic beam scanning; two crossed prisms lead to two-axis electronic scanning.

This type of ESA is characterized by

- low-loss, high-accuracy (6 bits) phase shift
- temperature-insensitive phase shift
- very low switching time independent from power consumption
- light and cheap realization of the electronic prisms
- use of cheap PIN diodes

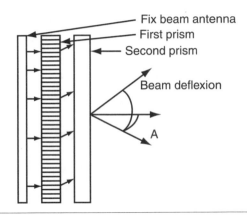

FIGURE 26.4 RADANT PRINCIPLE

- preservation of the quality and characteristics of the fixed-beam antenna pattern

This concept has been applied equally as well to large ground-based radar antennas as to much smaller airborne radar antenna, and validated from S-band to Ka-band.

It is perfectly matched to combat aircraft nose radar (two-axis ESA) and to single- or two-axis Air to Ground radar.

26.3.3 ACTIVE ESA (AESA)

Active ESA concept starts with the general architecture of ESA with phase shifters, that is, an array of elementary antennas fed, by a distribution network, through phase shifters inserted between the outputs of the distribution and the antennas. Instead of simple phase shifters, however, an AESA uses multifunction modules including phase shift, amplitude control, a transmit/receive switch, and amplification on transmit and receive (Figure 26.5).

Power microwave amplification is implemented in the modules, and the classic power amplifier, using high-voltage microwave tubes, is no longer needed.

As there are generally more than one thousand modules, each module needs only to produce about 10 w of microwave power, which is made possible with the present MMIC. Compared to passive ESA, AESA has a number of advantages, the main one being high efficiency, that is, the best use of the consumed power. AESA also offers

- high reliability of all solid-state realization

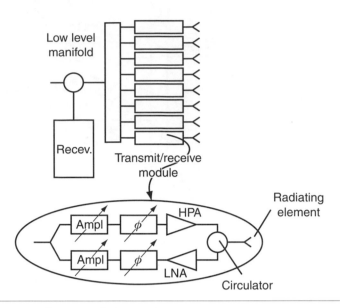

FIGURE 26.5 DIFFERENT KINDS OF ANTENNAS

- high level of availability (graceful degradation) and low maintenance cost
- low probability of interception on transmit due to power management and flexible beam shaping
- jammer resistance due to adaptive beam shaping/nulling
- high resolution due to its wide bandwidth

Microwave technology (components, assembly, packaging) is now mature enough that we can plan the realization of cost-effective AESA in the coming years, for all types of X-band airborne radars, as is already the case for L- or S-band AEW.

The growth potential of AESA must be emphasized:

- It opens the way to multifunction (Radar/EW Communication) wide-bandwidth antennas
- It is the basic concept for smart conformal antennas

26.4 CONCLUSION: ESA SOLUTIONS FOR AIRBORNE RADARS

Fighter radars have to detect targets in heavy clutter and dense jammer conditions, which means that side lobes and far lobes are very low. Power efficiency is also of great importance due to strong limitation in prime power and cooling capability of the aircraft, and to the limited size of the nose antenna area.

Short-term, a passive antenna is well suited, and RADANT technology, currently used on Rafale RBE2 radar, has proven its efficiency. All beam agility features described above are operational in this radar, which is in operational use.

As soon as active ESA becomes affordable (well before 2010), it will be the preferred solution for fighter radar. It will open the way to Adaptive Beam Forming, graceful degradation, and later to shared aperture (common antennas for radar EW and communications) and smart skin.

AEW's main problem is installation of a very large array (needed to get long range) on the airframe. Mechanical solutions require a large rotodome (like AWACS) to shelter this antenna. ESA offers a much easier way to integrate a flat dorsal antenna on any kind of platform (like Erieye). Its beam-steering flexibility allows the surveillance areas to be tailored to operational needs and to avoid wasting valuable time in areas of low operational interest. In the frequency used for these applications (L- to S-band), Active ESA is today the best choice.

New AGS and RECCO systems require large antennas with very wide bandwidth (>1 GHz in X-band) and beam steering. Antenna pattern quality is not as constraining as it is for the previous application, but cost and bandwidth are the drivers. For a medium-range, low-cost system mounted in POD, a passive solution is well suited, and the reflect array technology is a very good candidate. For longer-range stand-off systems or for spaceborne radar, Active ESA is the most promising technology.

Maritime Patrol systems currently use low-cost technologies (reflector antennas or slotted flat plane antennas). Pattern requirements are not very constraining, but bandwidth (for ISAR modes) and particularly cost are of great importance; whatever the operational benefits are, ESA will be chosen only if the global cost (antenna plus gimbals) is lower than a mechanical solution. Once again, new Reflect Array technology (as developed within THALES) offers the advantage of ESA operational benefits at the cost of the current mechanical solution.

Compared to ground-based radars, ESA technologies came late in airborne radars, but the next century will see a quick growth of ESA applications both with passive ESAs like RADANT or Reflect Array and with Active ESA, which is the technology of the future for radars and for EW systems.

27

AIRBORNE AND SPACEBORNE RADAR ENHANCEMENT

27.1 INTRODUCTION

The development of airborne radar will be influenced by

- changing operational requirements in response to the developing threat
- the possibilities offered by the evolution of existing technology and the emergence of new technologies

Foreseeable changes in requirements will include, in particular,

- a response to the reduction in target RCS
- the fight against the increasing electromagnetic threat (improved electromagnetic counter-measures (ECM) and electronic support measures (ESM), antiradiation missiles)
- the answer to multiple and evolving targets and the increase in angular coverage
- non-cooperative target classification (identification)

In this chapter, we outline possible responses to these needs.

27.2 RESPONSE TO TARGET RCS REDUCTION

27.2.1 POWER BUDGET INCREASE

The power budget can be increased (see Chapter 25), in order to compensate for the reduction in RCS, by acting on

- the mean transmitted power by either
    - increasing the power supplied by the platform to the radar
    - improving the overall efficiency of the transmitter (tubes, solid state)
    - reducing the losses: active antennas, in which power is directly generated by radiating elements and whose receiver circuits are located close to the radiating elements (Figure

26.1), minimize losses and provide a means of increasing the power budget by 6 to 10 dB

- the antenna surface. However, the limited space available for installation on an aircraft leaves little room for any increase
- the coverage volume and the refresh rate. Reducing these parameters means a change in the operational concepts. Although it is possible to increase the power budget by increasing the dwell time, this means a reduction in the area under surveillance, which then requires the predesignation of this area by another detection system with sufficient antistealth capability
- the target RCS. A number of techniques, outlined in Chapter 24, can be used to reduce target RCS. However, they can be countered by certain techniques such as
  - the use of frequency bands in which these techniques are inefficient (metric bands)
  - multistatic mode

These two techniques are described below

27.2.2 USING LOW-FREQUENCY BANDS

RCS reduction techniques are only efficient or applicable within limited frequency ranges. Figure 27.1 shows the techniques used depending on the frequency. The frequency band covering V-UHF waves (100 to 500 MHz) is on the borderline between passive (absorbent) and active (cancellation) techniques. Moreover, these metric wavelengths cause the aircraft structure to resonate, leading to a considerable increase in RCS.

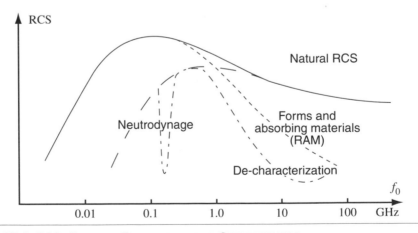

FIGURE 27.1 RCS, CARRIER FREQUENCY, AND STEALTHINESS

These frequency bands thus provide a useful means of countering stealth targets and their applications continue to increase for ground-based radar. In the case of airborne radars, the use of V/UHF frequencies raises the problem of fitting large antennas to the platforms. The relationship between antenna size and wavelength, and the interaction with the aircraft structure, result in poor pattern quality, making it difficult to eliminate ground returns (see Part VII).

Given the apparent impossibility of controlling the pattern, self-calibration techniques must be used, of which space-time processing, as previously described, is an example.

27.2.3 MULTISTATIC RADAR

Stealth targets are often designed with the aim of deceiving traditional monostatic radars, which explore space sequentially. RCS is reduced mainly in the front sector and certain observation directions are sacrificed if they only produce transient detections, which do not allow the radar to lock on for tracking. A multistatic system, comprising a transmitter and receivers situated in different places, uses different geometries for which RCS reduction is not optimized and can provide permanent target tracking by correlating the different sensor detections.

A multistatic detection system can use airborne, ground-based, or spaceborne resources (transmitters and receivers). In some cases it is even possible to use non-cooperative resources such as enemy transmissions or, for example, TV-broadcasting signals. Figure 27.2 illustrates the various possible configurations.

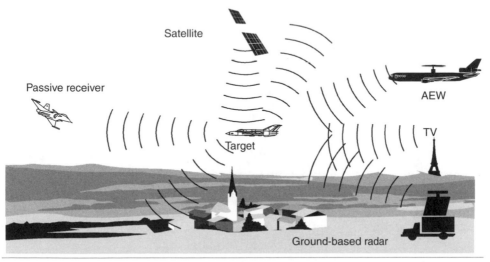

FIGURE 27.2 AIR-TO-GROUND-TO-SPACE MULTISTATIC MODE

The main difficulty associated with multistatic systems is obtaining the transmitted signal reference at receiver level. This requires

- an absolute time reference
- knowledge of the relative position of the transmitter and the receiver
- the possibility, at receiver level, of regenerating a replica of the transmitted signal to enable suitable processing (see Chapter 6). One way of solving this problem is to use extremely stable frequency references, imposed in particular by the requirement for ground return rejection (see Chapter 7), and to receive the transmitted signal by the direct transmit-receive path

27.3 Countering Electromagnetic Threats

Radar must face an increasingly hostile electromagnetic environment. Threats may be

- passive: electromagnetic radiation detection systems (electromagnetic support measures, or ESM). These are increasingly sensitive and "smart," enabling detection and identification of the radar, and threatening platform safety. Countering this danger means increasing radar discretion either through the choice of waveform or through multistatic mode or active/passive smart sensor management
- active: enemy jamming (electronic counter measures, or ECM), which aims to reduce radar effectiveness both when detecting and measuring. Countering these jammers involves either time (frequential) or space (antenna) processing. Anti-jammer processing (electronic counter measures, or ECCM) has been widely developed in the time domain. These techniques form part of industrial know-how and thus are covered by the industrial secret. The introduction of multisource/multi-subarray antennas (notably active antennas) opens the way to space processing, that is, the possibility of adapting and optimizing the receiver pattern to suit its environment

27.3.1 Waveforms

Chapter 7 described the constraints imposed on waveform selection by the need for ground clutter rejection, and the limits imposed by current computing capability (Chapter 25). The rapid evolution of signal processing technology will make it possible to use new waveforms, the choice of which will be influenced by the need

- to increase bandwidth for the transmitted signal to improve the range resolution needed to recognize targets and dilute transmitted energy by reducing spectral density. This will increase radar discretion while it reduces the effectiveness of electronic counter measures.
- to reduce peak power, which is an important parameter in radar detectability, by the use of electronic countermeasures and

electromagnetic support measures. The ultimate limit to peak power reduction is a continuous waveform. However, this waveform, which requires simultaneous transmission and reception, is difficult to use on aircraft, as the possible decoupling between the transmission and reception channels is not sufficient to avoid saturation of the receiver. Even when antennas are fitted at different points on the structure, coupling via nearby obstacles (aircraft structure, rain clutter, etc.) is sufficient to saturate the receiver
- to produce a more complex transmitted signal, making it more difficult to identify the radar

The waveform must always be chosen on the basis of the properties of its ambiguity function.

27.3.2 BEAM MATCHING (DIGITAL BEAMFORMING)

Antennas composed of a number of sensors with independent access (such as active antennas) make it possible to modify the radiation pattern on reception and/or transmission through processing, giving the radar designer greater freedom.

The first application of antenna matching is the possibility of creating nulls in the radiation pattern in the direction of the jammers (see Chapter 6). Such techniques are already used with ground-based radar. However, they create specific problems when applied to airborne radar due to

- platform movement, which means the radar-jammer direction is not stationary
- the need to control the radiation pattern of the side and far lobes, in order to reject ground returns present throughout a large part of the range-velocity domain. Generally, this constraint is not included in traditional algorithms and is made worse by the need to combine transmit-receive modules in subarrays to reduce the number of processing channels needed

  the problem a multipath created by the jammer signals reflected by the ground clutter and received by the main beam of the radar antenna (hot clutter), whose power could be equivalent to direct path signals (jammer to radar) received by the antenna side lobes

Controlling the radiation pattern is also a means of implementing new modes such as (Figure 27.3)

- illumination by a wide beam followed by formation of very thin directive beams on reception (digital beamforming, DBF)
- multistatic operation using DBF on reception so that the zone illuminated by the transmission beam coincides with reception coverage

- simultaneous modes enabling, for example, low-altitude penetration
  with terrain-following while ensuring air-to-air surveillance, for self-
  protection against air threats
- multiple beams enabling exploration of space using several
  simultaneous beams

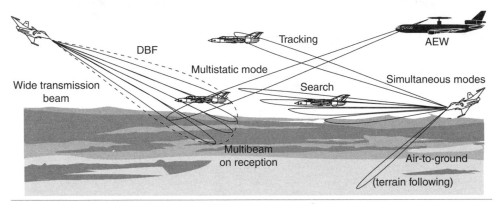

FIGURE 27.3 BEAM MATCHING AND SIMULTANEOUS MODES

27.4 MULTIPLE AND EVOLVING TARGETS; ANGULAR COVERAGE

Radar is faced with ever more complex targets: multiple targets either
grouped together or spread out in altitude, and highly evolving targets.
Whereas mechanical scanning only provides slow and systematic
exploration of space, electronic scanning, thanks to its agile beam, enables
independent and optimal adaptation of target search scanning and the
refresh rate at which target tracking is renewed. Overall effectiveness is
thus increased.

One of the major limitations of present-day radar is the angular domain
accessible to a fighter radar located in the nose cone. The need for
balconies (parts of the aircraft where an antenna can be fitted, giving
increased visibility) has led to the design of conformal antennas that
closely follow the structure of the aircraft, and to the spreading of antennas
over the structure.

27.4.1 ELECTRONIC SCANNING: DETECTION AND SCANNING STRATEGIES

As seen in the previous chapter, one of the many operational advantages of
passive or active electronic scanning is the ability to process the search and
tracking functions independently. It is possible to track targets outside the
search domain with a high *tracking refresh rate*, which ensures excellent
tracking continuity and precision, even for highly evolving targets. It is also

possible to use detection strategies of the detection-confirmation type: if an alarm is detected, confirmation pointing (possibly using a modified waveform) takes place.

In the case of electronic scanning, the tracking range is very close to the first detection range, giving a far better operational range and situation awareness than with mechanical scanning.

27.4.2 CONFORMAL ANTENNAS AND DISPERSED ANTENNAS

The angular domain can be widened using multiple plane antennas. Generally speaking, the limited amount of space available in the aircraft for antenna installation (particularly in the nose of a fighter) limits the radar performance. One solution (see Figure 27.4) is to spread the radiating elements over the whole of the usable surface (nose cone, leading edge of the wings, engine inlets) by closely following its contours. (These antennas are said to be conformal.) This solution presents the following advantages:

- maximum apparent radiating surface in all directions
- no radome or concomitant parasitic effects (reflections, aberrations, losses, etc.)
- reduction of radar RCS, where the RCS of a complex aerodynamic form is less than that of a flat surface

FIGURE 27.4 DISPERSED AND CONFORMAL ANTENNAS

This concept, which can be extended to include smart skins, requires major technological development (highly efficient, compact transmit-receive modules, polarization control, thermal packaging, broadcasting of microwave frequency signals and control data using optic links, etc.). Control of the radiation pattern also requires close attention.

Given their asymmetrical form, the radiating pattern of these antennas, which can be flat but are preferably conformal, is not satisfactory in all directions. This means pairing off the transmission and reception patterns by combining, for example, a beam that is directive in elevation on transmission (antenna situated on the tail) with a beam that is directive in bearing on reception (antenna located on the leading edge of the wings).

27.5 SPACE IMAGING RADAR

27.5.1 SHORT- AND MEDIUM-TERM DEVELOPMENT

No major changes are foreseen short- and medium-term for this type of radar. Certain parameters will, however, be improved:

- increase in resolution
- simultaneous transmission over several frequencies
- capacity for polarimetric analysis, etc.

These trends will increase effectiveness in terms of target detection and classification. They make use of existing techniques and technologies.

27.5.2 LONG-TERM DEVELOPMENT

A major change could occur if it is proven that the "non-ambiguous" waveform described in Section 7.2.3, characterized by a BT product of over 80 dB, can be applied to spaceborne radar. Ambiguity would disappear, and with it the heavy constraints it entails. Chapter 15 explains how such ambiguity determines the minimum surface area of the radar antenna. Non-ambiguous waveforms would pave the way for radars fitted with small antennas:

- lightweight radar for satellites (from several dozen to several hundred kg)
- very low-frequency radar, for which the antenna area will measure several hundred m² based on existing principles

27.5.3 AIR-SPACE COOPERATION

A satellite in orbit, whether moving or geostationary, could be a useful transmitter for a bistatic or multistatic system. It could illuminate a large area while allowing airborne receivers to remain highly discrete. Passive-mode low-altitude penetration missions could be achieved in this way.

28

CONCLUSIONS

In this final chapter we attempt to draw a number of conclusions concerning the nature of airborne and spaceborne radars.

The first concerns the prior development of platforms, radars, and weapon systems. A Vautour (or an F-94C Starfire) platform may look quite similar to a Rafale (or an F22) platform, but a radar operator knows that apart from the basic principles, the two different-generation radars fitted to these aircraft are extremely different in terms of complexity, circuit density, waveforms, processing, performance, etc. He also knows that weapon systems evolve considerably and that the overall system, including the platform and the weapon system, is an increasingly important factor in radar design. The development of new specialized platforms such as satellites, helicopters, and drones (UAV and UCAV), designed for specific missions, has led to the creation of new families of radars. These too are highly specific, although they make use of existing techniques and technologies.

The second conclusion concerns future trends, which were briefly discussed in Chapter 27. Four main factors will undoubtedly call into question the design and use of airborne radar:

- electronic warfare
- new sensors
- the integration of avionics
- stealth targets and platforms

ELECTRONIC WARFARE

The various elements used in electronic warfare, radar detectors and jammers in particular, are becoming more and more sophisticated. This, along with the voluntary or involuntary interaction between systems, means that radar must become more discrete (with the control of transmission time in a given direction and the control of transmitted power). Radar must also use waveforms with a personalized signature.

NEW SENSORS

New optronic sensors, whose performance continues to improve (in range, sensitivity, spectral domain, angular search domain, etc.) now rival radar.

These sensors are passive and therefore discrete, except for laser, which has the advantage of being highly directive. However, their performance is considerably degraded in poor atmospheric conditions such as clouds, rain, fog, airborne sand, etc. Rather than replacing radar, they are complementary elements, like passive listening systems. Combining information provided by these sensors provides a means of optimizing the effectiveness of the overall system.

INTEGRATED AVIONICS

The various specialized equipment that makes up platform avionics (radar, optronics, communications, self-screening, etc.) now uses autonomous items such as antennas, receivers, processing equipment, etc. In the future, radar, jammers, communications, etc. will use the same elements in order to reduce development, acquisition, and maintenance costs, and to make optimal use of resources. The notion of radar function will gradually replace the notion of radar equipment.

STEALTH TARGETS AND PLATFORMS

Issues raised by stealth targets and platforms can be divided into three main categories:

- true stealth platform, e.g., a penetrator. This platform must be extremely discrete. It can only make limited use of active equipment such as radar at critical moments during the mission, e.g., navigation update, target recognition and acquisition.
- platform that is not a stealth platform but that must detect stealth targets, e.g., an interceptor. Its equipment must include a radar optimized for this mission (use of low-frequency bands, multistatic mode in association with other systems).
- an interceptor that must detect both conventional and stealth targets while remaining discrete. This is the most complex situation, and many technical difficulties have to be overcome.

We hope this book has provided an interesting overview of airborne and spaceborne radars, presented not as separate entities but as a part of a weapon system with a specific role to fulfill. The authors are fully aware that this book does not give an exhaustive account of all the subjects and techniques concerning radar, and that they could be further developed. However, a limit had to be found and industrial and military secrets respected.

LIST OF ACRONYMS

A-A	Air-to-Air
AC	Alternative Current
ADC	Analog-to-Digital Converter
AEW	Advanced Early Warning
AFC	Automatic Frequency Control
A-G	Air-to-Ground
AGR	Air-to-Ground Ranging
AIMS	Avionics Integrated Modular System
A-S	Air-to-Surface, Air-to-Sea
ASAAC	Allied Standard Avionics Architecture Council
BFPQ	Block Floating Point Quantizer
CC	Close Combat
C^3I	Communication Command Control and Information System
CEPA	Common European Priority Areas
CFAR	Constant False Alarm Rate
CISC	Complex Instruction Set Computer
CMOS	Complementary Metal Oxide Semiconductor
CNI	Communication Navigation Identification
CW	Continuous Wave
DBF	Digital Beam Forming
DBS	Doppler Beam Sharpening
DC	Direct Current
DDS	Direct Digital Synthesis
DEM	Digital Elevation Model of terrain
DFT	Digital Fourier Transform
DGA	Delegation Générale pour l'Armement
DPCA	Displaced Phase Center Antenna
DSP	Digital Signal Processor
ECCM	Electronic Counter Counter Measures
ECM	Electronic Counter Measures
EFC	Electronic Flight Control

EO	Electro-Optical
ESM	Electronic Support Measure
EUCLID	European Cooperation for the Long term In Defense
EVS	Enhanced Vision System
FFT	Fast Fourier Transform
FMCW	Frequency Modulated Continuous Wave
FMICW	Frequency Modulated Interrupted Continuous Wave
GaAs	Gallium Arsenide
GPS	Global Positioning System
HA	High Altitude
HALE	High-Altitude Long Endurance (UAV)
HDD	Head-Down Display
HPA	High Power Amplifier
HPRF	High PRF
HUD	Head-Up Display
ICNIA	Integrated Communication Navigation and Identification Avionics
IF	Intermediate Frequency
IFOV	Instantaneous Field Of View
ILS	Instrument Landing System
I/O	Input/Ouput
I/Q Demodulator	In-phase and in-Quadrature Demodulator
IR	InfraRed
IRS	InfraRed Signature
ISAR	Inverse Synthetic Aperture Radar
ISLR	Integrated Side-Lobe Ratio
JIAWG	Joint Integrated Avionics Working Group
JTIDS	Joint Tactical Information Distribution System
LA	Low Altitude
LCS	Laser Cross Section
LFM	Linear Frequency Modulation
LNA	Low-Noise Amplifier
LO	Local Oscillator
LPI	Low Probability of Intercept
LPRF	Low PRF
LRU	Line Replaceable Unit

MA	Medium Altitude
MALE	Medium-Altitude Long Endurance (UAV)
LV	Low Voltage
MESFET	MEtal Semiconductor Field Effect Transistor
MIMD	Multiple Instruction Multiple Data
MPRF	Medium PRF
MTI	Moving Target Indicator
MTT	Multi-Target Tracking
NCTR	Non-Cooperative Target Recognition
Neσ_0	Noise Equivalent Sigma Zero
NM	Nautical Mile
NMOS	Negative Metal Oxide Semiconductor
PMOS	Positive Metal Oxide Semiconductor
PPI	Panoramic Plan Indicator
PRF	Pulse Repetition Frequency
PRI	Pulse Repetition Interval
PROM	Programmable Read-Only Memory
PSLR	Peak-to-Side-Lobe Ratio
PSP	Programmable Signal Processor
RAM	Random Access Memory
RCS	Radar Cross Section
RISC	Reduced Instruction Set Computer
RF	Radio Frequency
RMS	Root Mean Square
SAR	Synthetic Aperture Radar
SAW	Surface Acoustic Wave
SEAD	Suppression of Enemy Air Defense
SLAR	Side-Looking Airborne Radar
SLB	Side Lobe Blanking
SLC	Side Lobe Cancellation
SEU	Single Event Upset
SIMD	Single Instruction Multiple Data
SNR	Signal-to-Noise Ratio
SOI	Silicon On Insulator
SOS	Silicon On Saphire

SSLR	Secondary Side-Lobe Ratio
STAP	Space-Time Adaptive Processing
STC	Sensitivity Time Control
STT	Single-Target Tracking
SW1... SW4	Swerling I to Swerling IV types of target
TA	Terrain Avoidance
TF	Terrain Following
TR	Transmit Receive
TV	Television
TWS	Track-While-Scan
TWT	Travelling Wave Tube
UAV	Unmanned Aerial Vehicle
UHF	Ultra-High Frequency
UV	Ultraviolet
VCO	Voltage Controlled Oscillator
VHA	Very High Altitude
VHF	Very High Frequency
VLA	Very Low Altitude
VSWR	Voltage Standing Wave Ratio
2-D	Two-Dimensional
3-D	Three-Dimensional

LIST OF SYMBOLS

B	Bandwidth
B_D	Target Doppler bandwidth
B_{D_a}	Instantaneous Doppler bandwidth in the main antenna lobe
B_e	Equivalent bandwidth
b	White noise power spectral density
c	Velocity of the light : $\approx 3.10^8$ m / s
$c(\Delta t, \Delta f)$	Matched filter output
d	Antenna diameter
E	Energy
$E(.)$	Mathematical expectation
E_h	Horizontal component of the electric field
E_v	Vertical component of the electric field
F	Noise factor
F_s	Sampling frequency
f	Frequency
f_0	Carrier frequency
f_D	Doppler frequency
f_{D_0}	Central Doppler frequency
$\dot{f}_D$	Derivative of the Doppler frequency
f_m	Spurious motion frequency
G	Antenna gain
G_t	Transmission antenna gain
G_r	Reception antenna gain
g_0	Gravity acceleration
h	Antenna size in height
h	Height of the target above the ground (altitude)
H	Height of the radar above the ground (altitude)
$h(t)$	Impulse response of the matched filter
i	Incidence angle
k	Boltzmann constant, $1,38 \times 10^{-23}$ J K^{-1}
L	Losses

l	Antenna size in length
$L(f)$	Phase noise spectrum
N	Noise power
N_{FFT}	Number of points in the Discrete Fourier Transform
P_k	Peak power
P_D	Probability of detection
P_{fa}	Probability of false alarm
P_m	Mean power
P_r	Received power
PRF	Pulse Repetition Frequency
P_t	Transmitted power
R	Energy Ratio
R	Target range
R_a	Ambiguity range
R_s	Swathwidth, on ground
R_{sw}	Swathwidth, radar line of sight
R_0	Initial radar range; minimum range between radar trajectory and target (SAR)
$\text{Rect}_T(.)$	Rectangle function
r	Range resolution, radar line of sight
r_c	Cross-range resolution
r_{f_D}	Doppler frequency resolution
r_A	Azimuth angle resolution
r_g	Ground range resolution
S	Signal power
SNR, S / N	Signal to noise ratio
SCR	Signal to clutter ratio
$S(f)$	Power spectral density, Fourier transform of $\Re_{ss}(t)$
s	Resolution cell surface
$s(t)$	Received signal modulation
$s_r(t)$	Received signal, in RF
$\text{sinc}(x)$	Cardinal sine: $\sin x / x$
T	Transmitted pulse duration
T_e, T_{dwell}	Illumination time, dwell time
T_i	Image acquisition time
T_R	Interpulse period

T_0	Noise temperature
t	Time
t_0	2 way propagation time of RF wave between radar and target
$u(t)$	Modulation of the transmitted signal
$u_e(t)$	Transmitted signal, in RF
$y(\Delta t, \Delta f)$	Signal module output of matched filter
v or V	Velocity
v_r	Radial velocity
v_g	Velocity of the satellite or the aircraft, projected on the ground
v_E	Velocity of a point on the ground, due to the rotation of the Earth
v_c	Cross-range velocity
$x(t)$	Modulation of the received signal, including the noise
γ	Acceleration
$\delta(.)$	Dirac function
$\delta\varphi_e$	Difference of aspect angle of the target
η	Efficiency, yield
θ_{0A}	3 dB azimuth beamwidth (aperture)
θ_{0E}	3 dB elevation beamwidth (aperture)
θ_A	Azimuth angle
θ_B	Bearing angle
θ_D	Depression angle
θ_E	Elevation angle
θ_G	Grazing angle
λ	Wavelength
$\rho(.)$	Weighting function (windowing)
υ	frequency (in the ambiguity function)
σ	Radar cross section
σ	Root mean square
σ_0	Reflectivity
τ	Time
τ	Duration of the compressed pulse, output of matched filter
τ_{3dB}	3 dB width of the correlation peak
φ	Phase
$\|\chi(\tau,\upsilon)\|^2$	Ambiguity function
$\omega(t)$	Rotation rate

BIBLIOGRAPHY

Al-Khatib, H.H. "Laser and Millimiter-Wave Backscatter of Transmission Cables." In *SPIE Proceedings, Physics and Technology of Coherent Infrared Radar,* vol. 300 (1981): 219–229.

Antebi, E. *La grande épopée de l'électronique.* Paris: Editions Hologramme, 1982.

Baratault, P., F. Gautier, and G. Albarel. "Evolution des antennes pour radars aéroportés." *Revue technique Thomson-CSF,* vol. 25, no. 3 (September 1993), Paris: Gauthier-Villars-Elsevier, 749–793.

Barton, D.K. *Radar System Book.* Norwood, MA: Artech House, 1985.

Bentejac, R. *Technique du radar classique.* Paris: Masson-Dunod, 1992.

Berkowitz, R.S. *Modern Radar Analysis, Evaluation and System Design.* New York: John Wiley & Sons, 1965.

Blacknell, D., et al. "Geometric Accuracy in Airborne SAR Images." In *IEEE Transactions on Aerospace and Electronic Systems,* vol. 25, no. 2 (March 1989): 241–256.

Blake, A.P. "Autofocus techniques: multilook registration and contrast optimization—a comparison." In *RSRE Memorandum,* no. 4626 (1992).

Boudenot, J.C., and G. Labaune. *La compatibilité électromagnétique et nucléaire.* Paris: Ellipses, 1998.

Brigham, E.O. *The Fast Fourier Transform.* Upper Saddle River, NJ: Prentice Hall, 1974.

Carpentier, M. *Radars—Bases modernes.* Paris: Masson-Dunod, 1977.

Carrara, W.G., R.S. Goodman, and R.M. Majewski. *Spotlight Synthetic Aperture Radar—Signal Processing Algorithms.* Norwood, MA: Artech House, 1995.

Chan, H.L., et al. "Noniterative Quality Phase-Gradient Autofocus (QPGA) Algorithm for Spotlight SAR Imagery." In *IEEE Transactions. GE,* vol. 36, no. 5 (September 1998): 1531–1539.

Chang, C.Y., M.Y. Jin, and J.C. Curlander. "SAR Processing Based on the Exact Two-Dimensional Transfer Function." In *Proceedings of IGARSS'92, Houston* (May 1992): 355–359.

Clarke, J. *Advances in Radar Techniques.* IEE Electromagnetic Waves Series 20, 1985.

Cumming, I., F. Wong, and R.K. Raney. "A SAR Processing Algorithm with no Interpolation." In *Proceedings of IGARSS'92, Houston* (May 1992): 376–379.

Curlander, J.C., and R.N. McDonough. *Synthetic Aperture Radar, System and Signal Processing.* New York: John Wiley & Sons, 1991.

Darricau, J. *Physique et théorie du radar.* Paris: Sodipe, 1993.

Day, J.K. "Digital adaptive beamforming." In *Proceedings of National Radar Conference IEEE 93* (22 April 1993): tutorial.

Deleuze, C., A. Mathiet, P. Zamora, and G. Zerah. "Matériaux pour la furtivité." In *Revue scientifique et technique de la Direction des applications militaires,* no.6 (December 1992): 15–29.

Di Franco, J.V., and W.L. Rubin. *Radar detection.* Norwood, MA: Artech House, 1980.

Drabowitch, S., and C. Ancona. *Antennes—Applications.* Paris: Masson-Dunod, 1986.

Eichel, P.H., et al. "Phase Gradient Algorithm as an Optimal Estimator of the Phase Derivative." In *Optics Letters,* vol. 14, no. 28 (October 1989): 1101–1109.

Firmain, G. "Durée de vie des tubes spatiaux." *Revue technique Thomson-CSF,* vol. 23, no. 4 (September 1991), Paris: Gauthier-Villars-Elsevier, 1063–1086.

Firmain, G. "Les tubes hyperfréquences, état de l'art, évolution, perspectives." *Revue technique Thomson-CSF,* vol. 23, no. 4 (September 1991), Paris: Gauthier-Villars-Elsevier, 731–763.

Freeman, A., J.J. Van Zyl, J.D. Klein, H.A. Zebker, and Y.S. Shen. "Calibration of Stokes and Scattering Matrix Format Polarimetric SAR Data." In *IEEE Transactions on Geoscience and Remote Sensing,* vol. 30, no. 1 (May 1992): 531–538.

Gray, A.L., K.E. Mattar, and P.J. Farris-Manning. "Airborne SAR Interferometry for Terrain Elevation." In *Proceedings of IGARSS'92, Houston* (May 1992): 1589–1591.

Gray, G.A., et al. "Quantization and saturation noise due to analogue-to-digital conversion." In *Transactions on Aerospace and Electronic Systems,* vol. 7 (January 1971): 222–223.

Hounam, D. "Motion Errors and Compensation Possibilities." In *AGARD Lecture Series 182, Fundamental and Special Problems of Synthetic Aperture* (August 1992): 3.1–3.12.

Jakowatz, C.V., et al. "Eigenvector method for maximum-likerlihood estimation of phase errors in synthetic-aperture radar imagery." In *Journal of the Optical Society of America A,* vol. 10, no. 12 (December 1993): 2539–2546.

___. "New Approach to Strip-Map SAR Autofocus." In *Proceedings of the IEEE 6th Digital Signal Processing Workshop* (1994): 53–56.

Joo, T.H., et al. "An adaptive quantization method for bust mode synthetic aperture radar." In *Proceedings of the International Radar Conference* (1985): 385–390.

Katzin, M. "On the Mechanism of Radar Sea Clutter." In *Proceedings of the IRE*, vol. 45 (January 1957): 44–54.

Klass, P. J. "Stealth Experts Trade Design Strategies in Public Forum." *Aviation Week and Space Technology* (26 September 1988): 75.

Klemm, R. "Antenna design for airborne MTI." In *Proceedings of IEE Radar Conference 1992, Brighton U.K.* (12–13 October 1992): 296–299.

___. "Effets des repliements en distance du fouillis sur un radar Doppler." *Colloque international sur le radar, Paris* (May 1994): 121–126.

___. "Quelques propriétés des matrices de covariance espace-temps." *Colloque international sur le radar, Paris* (May 1994): 357–361.

___. *Space-time Adaptive Processing—Principles and Applications.* London: IEE Publishers 1998.

Kretschner, F. F., and B.L. Lewis. "Doppler properties of polyphase coded pulse compression waveforms." In *IEEE Aspects of Radar Signal Processing* (1983): 78–88.

Lacomme, P. "Modélisation du fouillis de sol." *Colloque international sur le radar, Paris* (April 1989): 158–163.

Le Chevalier, F. *Principes de traitement des signaux radar et sonar.* Paris: Masson-Dunod, 1989.

Levanon, N. *Radar Principles.* New York: John Wiley & Sons, 1988.

Long, A. "Toward a C-Band Radar Sea Echo Model for the ERS-1 Scatterometer." In *Proceedings of First International Conference on Spectral Signatures of Objects in Remote Sensing, Les Arcs* (16–20 December 1985): European Space Agency, 29–34.

Madsen, S.N., H.A. Zebker, and J. Martin. "Topographic Mapping Using Radar Interferometry: Processing Techniques." In *IEEE Transactions on Geoscience and Remote Sensing*, vol. 31, no. 1 (January 1993): 246–255.

Marchais, J.C. "Evolution des traitements des radars aéroportés en France." In *Revue technique Thomson-CSF*, vol. 25, no. 3 (September 1993): Paris: Gauthier-Villars-Elsevier, 813–834.

Max, J. "Quantization for Minimum Distortion." *IRE Transactions on Information Theory*, vol. IT-6 (1960): 7–12.

Meyer, D.P., and H.A. Mayer. *Radar Target Detection—Handbook of Theory and Practice*, Electrical Science Series, edited by Henry G. Booker and Nicholas De Claris. Burlington, MA: Academic Press, 1973.

Mitchell, R.L. *Radar Signal Simulation.* Norwood, MA: Artech House, 1976.

Novak, L.M., et al. "Effects of Polarization and Resolution on SAR ATR." In *IEEE Transactions on Aerospace and Electronic Systems*, vol. 33, no. 1 (January 1997): 102–115.

___. "Automatic Target Recognition Using Enhanced Resolution SAR Data." In *IEEE Transactions on Aerospace and Electronic Systems*, vol. 35, no. 1 (January 1999): 157–175.

Oliver, C.J. "High frequency limits on SAR autofocus and phase corresction." In *International Journal of Remote Sensing* vol. 14, no. 3 (1993): 495–519.

Oliver, C.J., and S. Quegan. "Understanding Synthetic Aperture Radar Images." Norwood, MA: Artech House, 1998.

Pike, T.K., J.M. Hermer, J.L. Perrot, A. Cavanie, and D. Hounam. "Polar Platform Wind Scatterometer, Final Report." *ESTEC Contract n° 8208/89/NL/JS* (April 1990). Noordwijk: European Space Agency ESTEC.

Queen, B., et al. "Advanced Targeting Improvements for Joint STARS." In *AGARD MSP Symposium* (October 1995): 4.1–4.19.

Raney, K. "Special SAR Techniques and Applications." In *AGARD Lecture Series 182, Fundamental and Special Problems of Synthetic Aperture* (August 1992): 10.1–10.15.

Richardson, D. *Stealth Warplanes.* Baltimore: Salamander Books, 1989.

Ridenour, L.N. *Radar System Engineering.* Vol. 1 of *MIT Radiation Laboratory Series,* 21. New York: McGraw Hill, 1947.

Rihaczek, A.W. *Principles of High Resolution Radar.* Norwood, MA: Artech House, 1996.

Rihaczek, A.W., and S.J. Hershkowitz. "Man-Made Target Backscattering Behavior: Applicability of Conventional Radar Resolution Theory." In *IEEE Transactions on Aerospace and Electronic Systems,* vol. AES-32, no. 2 (April 1996): 809–823.

Roubine, E., and J.C. Bolomey. *Antennes—Introduction générale.* Paris: Masson-Dunod, 1986.

Sappl, E. "Optimale quantisierer fuer complexe kreisnormal zufallsgrossieren mit unbekannter varianz." In *AUEe,* vol. 40, no. 4 (1986): 208–212.

SEE "Caractérisation microonde des matériaux absorbants." *Journées d'études des 7 et 8 février 1991, Limoges.*

Skolnik, M. *Radar Handbook.* New York: McGraw Hill, 1990.

Stevens, D.R., I.G. Cumming, M.R. Ito, and A.L. Gray. "Airborne Interferometric SAR: Terrain Induced Phase Errors." In *Proceedings of IGARSS'93, Tokyo* (August 1993): 977–979.

Stimson, G.W. *Introduction to Airborne Radar.* Park Ridge, NJ: Scitech Publishing, 1998.

Sweetmann, R. *Stealth Aircraft.* Osceola, WI: Motorbooks International, 1986.

Thomson-CSF. "Traitement d'images Tomes 1 et 2." In *Revue technique Thomson-CSF,* vol. 24 and 25 (December 1992, March 1993), Paris: Gauthier-Villars-Elsevier.

Urkowitz, H. *Signal theory and random processes.* Norwood, MA: Artech House, 1983.

Van Zyl, J.J. "Calibration of Polarimetric Radar Images Using Only Image parameters and Trihedral Corner Reflector Responses." In *IEEE Transactions on Geoscience and Remote Sensing,* vol. 28 (May 1990): 337–348.

Vergniolle, C. "Architecture et connectique de proceseur programmable de signaux radar." *Revue scientifique et technique de la Défense,* 3e trimestre 1991, Paris: Gauthier-Villars-Elsevier, 21–25.

Wahl, D.E., et al. "Phase Gradient Autofocus—A Robust Tool for High Resolution SAR Phase Correction." In *IEEE Transactions on Aerospace and Electronic Systems,* vol. 30, no. 3 (July 1994): 827–835.

Ward, K.D. "A radar sea clutter model and its application to performance assessment." In *IEE Conference Publication 216,* Radar 82 (1982): 204–207.

Ward, K.D., C.J. Baker, and S. Watts. "Maritime surveillance radar Part 1: Radar scattering from the ocean surface." In *IEE Proceedings,* vol. 137, Pt. F, no. 2 (April 1990): 51–62.

Wehner, D.R. *High Resolution Radar.* Norwood, MA: Artech House, 1995.

White, R. G. "Change detection in SAR Imagery." In *International Journal of Remote Sensing,* vol. 12, no. 2 (1991): 339–360.

Wiley, C.A. "A Paradigm for Technology Evolution." In *IEEE Transactions on Aerospace and Electronic Systems,* vol. AES-21, no. 3 (May 1985): 440–443.

Woodward, P.M. *Probability and Information Theory with Application to Radar.* New York: Mc Graw Hill, 1955.

ABOUT THE AUTHORS

PHILIPPE LACOMME

Professor Philippe Lacomme is a Senior Radar Designer with *THALES* Company. He is the Technical Director of the *Radar Unit*, which is in charge of developing and producing airborne radar systems such as *RBE2* for *Rafale* aircraft, *RDY* for *Mirage 2000*, maritime patrol *Ocean Master* radar, airborne SAR (*Raphaël TH*), terrain following radar (*Antilope*), airport surveillance radar (*Rapsodie*) or advanced radar for the next generation of fighters (*AMSAR* French-British-German active array radar).

His 30 years of experience encompasses design and development of missile seekers and fighters radar deployed on *Mirage F1*, *Mirage 2000*, *Mirage 2000-5*, *Rafale* aircraft which are in service in many countries. He has been involved in radar architecture design, signal processing and flight trials, in particular for LPRF, HPRF and MPRF Doppler modes. He has been for two years the French Administrator-Gerant of the *trinational GTDAR EEIG (GEC-THOMSON-DASA Airborne Radar)* which is developing the *AMSAR* radar.

Professor Lacomme has taught Radar theory within *Thomson-CSF* and in many Universities and Engineering Schools (*ESME, ENST, ENSTA, ESE, ENSAE*). He has lectured radar systems in many international conferences. He is co-author of a book on Air and Spaceborne Radar Systems which is published in French and which is here published in English.

JEAN-CLAUDE MARCHAIS

Jean-Claude Marchais was Technical Director of *Thomson-CSF Radars & Contre-Mesures*. During is long career, he was involved in the development of all radar systems for the *Mirage* aircraft family. Before being retired, he was Deputy Technical Director of *Thomson-CSF Aeronautics Equipment Business Group*. He was lecturing radar systems at *ESME-Sudria* engineer school.

Jean-Claude Marchais has a wide experience in publishing technical works. "L'amplificateur opérationnel et ses applications" (The Operational Amplifier and its Applications), Masson 1971, was published in French and Spanish, with four revisions and eight editions between 1971 and 1986. It has been sold by ten thousands. He also is the author of "Structures élémentaires des filtres actifs" (Elementary Structure of Active Filters), Masson 1979, and co-author of the French version of this book.

JEAN-PHILIPPE HARDANGE

Jean-Philippe Hardange studied aeronautical and space engineering at Ecole Nationale Supérieure de l'Aéronautique et de l'Espace in Toulouse. His first practical contact with radar was in military service as a Radar Officer in the French Navy on-board maritime patrol aircraft *Alizé* and *Atlantic* over the Mediterranean. Jean-Philippe then joined Thomson-CSF in 1982, and has been working there as a radar engineer on all types of airborne radar, both in the design and in the validation phases, ever since. He was involved in the development of Synthetic Aperture radar (*Arcana* for *Mirage IV* strategic bomber, *Raphaël-TH* on-board *Mirage F1CR*), maritime patrol radar (*Ocean Master*), terrain-following radar (*Antilope* for *Mirage 2000 N*), and fire-control radar (*RDY* for *Mirage 2000-5*, and *RBE2* for *Rafale*). In 1996 he was head of the Airborne Radar Engineering Department. Jean-Philippe then managed the Airborne Radar Surveillance Systems Programs Department (AEW systems and ground surveillance systems such as *Horizon*) and launched the *SOSTAR* project of ground surveillance for NATO, with an international team. He is now Projects Manager for the naval and ground-based radar systems in THALES.

Jean-Philippe Hardange has given radar lectures in several engineer schools. He is coauthor of the French version of this book.

ERIC NORMANT

Eric Normant (MIEE, CEng), received its engineering degree in 1990 from *Ecole Nationale Supérieure des Télécommunications de Bretagne*. He simultaneously got a diploma in spacecraft technology and satellite communication from *University College of London*. In 1991 he joined *Thomson-CSF Radars & Contre-Mesures*, now *THALES*, as research scientist. Since then he has worked on numerous aspects of SAR processing and system engineering. He is now head of airborne reconnaissance radar team. In 1996 he was appointed member of the Scientific & Technical Council of *Thomson-CSF* and received a Best Inventions Award in 1998 for a patent about strip/spot hybrid SAR mode. He holds a dozen of patents in the field of SAR and has been teaching general radar theory and SAR at *ENSTBr* and *ENSTA*.